Teaching and Learning of Fluid Mechanics, Volume II

Teaching and Learning of Fluid Mechanics, Volume II

Editor

Ashwin Vaidya

MDPI • Basel • Beijing • Wuhan • Barcelona • Belgrade • Manchester • Tokyo • Cluj • Tianjin

Editor
Ashwin Vaidya
Montclair State University
USA

Editorial Office
MDPI
St. Alban-Anlage 66
4052 Basel, Switzerland

This is a reprint of articles from the Special Issue published online in the open access journal *Fluids* (ISSN 2311-5521) (available at: http://www.mdpi.com).

For citation purposes, cite each article independently as indicated on the article page online and as indicated below:

LastName, A.A.; LastName, B.B.; LastName, C.C. Article Title. *Journal Name* **Year**, *Volume Number*, Page Range.

ISBN 978-3-0365-1998-2 (Hbk)
ISBN 978-3-0365-1999-9 (PDF)

Contents

About the Editor

Dr. Ashwin Vaidya is a faculty member in the Department of Mathematics at Montclair State University in New Jersey, USA. He also serves as the director of the Complex Fluids Laboratory and holds a joint appointment in the Department of Physics and Astronomy at MSU. His research interests lie in the general areas of complex systems which includes fluid mechanics, fluid structure interaction, biological fluid mechanics and non-equilibrium thermodynamics, among others. He is also interested in issues surrounding math and science education with a focus on making 'creativity' a central theme of our educational practice.

Editorial

Contributions to the Teaching and Learning of Fluid Mechanics

Ashwin Vaidya

Department of Mathematics and Complex Fluids Laboratory, Montclair State University, Montclair, NJ 07043, USA; vaidyaa@mail.montclair.edu

1. Introduction

This issue showcases a compilation of papers on fluid mechanics (FM) education, covering different sub topics of the subject. The success of the first volume [1] prompted us to consider another follow-up special issue on the topic, which has also been very successful in garnering an impressive variety of submissions.

As a classical branch of science, the beauty and complexity of fluid dynamics cannot be overemphasized. This is an extremely well-studied subject which has now become a significant component of several major scientific disciplines ranging from aerospace engineering, astrophysics, atmospheric science (including climate modeling), biological and biomedical science and engineering, energy harvesting, oceanography, geophysical and environmental science and engineering, etc. While each of these disciplines has its own nuances and specific constraints, the fundamental physics behind the kinds of 'flow' phenomena discussed remains the same. In this volume, we bring together articles from authors with diverse expertise ranging from mathematics, physics, mechanical engineering, aerospace engineering, environmental engineering, and chemical engineering to discuss topics in fluid mechanics, many of which are of multidisciplinary interest.

The focus of all articles in this issue remains on the presentation of fundamental and advanced ideas on fluid mechanics which are suitable for presentation in an undergraduate or graduate course in fluid mechanics. Overall, I would divide the collection into the following four categories: (a) Pedagogy of fluid mechanics; (b) experimental or lab-based perspectives; (c) computational approaches; and (d) mathematical fluid mechanics. The following pages provide a brief summary of each of the contributions.

2. Pedagogical Issues

Student-centered practices such as problem-based and project-based learning (PBL) are more commonly practiced in the arts. PBL related instructional methods promote a more inductive approach to learning whereby generalizations and abstractions follow from first understanding specific cases. This approach is in contrast to the deductive strategy taken in the sciences which is a more top-down approach and a possible cause of alienation towards math and science in several students. The concept of problem-based learning began more than 30 years ago in the context of medical education and has been defined as the "posing of a complex problem to students to initiate the learning process" [2,3] and as "experiential learning organized around the investigation and resolution of messy, real-world problems." [4]. PBL can be implemented at various scales in a course with a focus from a "teacher to student-centered education with process-oriented methods of learning." [5]. The recent popularity of the project-based learning approach in physics and engineering education is based on research indicating the effectives of PBL in enhancing student engagement [5–7]. This volume presents a selection of papers that speak to the efficacy of PBL-based experiences in fluid mechanics courses.

The first article in the collection [8] by authors Garrard, Bangert, and Beck discusses an innovative pedagogical approach by planning and delivering large scale, multi-disciplinary labs with as many as 80 students in a single cohort and nearly 1000 students over a year.

Citation: Vaidya, A. Contributions to the Teaching and Learning of Fluid Mechanics. *Fluids* **2021**, *6*, 269. https://doi.org/10.3390/fluids6080269

Received: 23 July 2021
Accepted: 28 July 2021
Published: 30 July 2021

Publisher's Note: MDPI stays neutral with regard to jurisdictional claims in published maps and institutional affiliations.

Topics discussed in the lab are those which would be common to students from various branches of engineering at their institution including basic flow measurements, pressure driven flow in pipes, dimensional analysis, and other open ended projects. Besides the obvious financial and logistic advantages that come from sharing of resources, the authors point to its pedagogical value and efficiency.

The second article by Pérez-Sánchez and López-Jiménez [9] in this collection highlights the use of PBL in a fluid mechanics course taught at a Hydraulic Engineering Department in Spain that caters to over 2000 students. Just as in an earlier paper [8], the project described in this article proposes the coordination of fluid-based labs in different subjects at both the bachelor's and master's degree levels. The paper discusses the improvement in student performance, as well as the new teaching approaches which the authors note to have "increased the student's satisfaction index".

The final article in this section by Zoupidis, Spyrtou, Pnevmatikos, and Kariotoglou [10] is aimed towards teaching the concept of 'floating and sinking' (FS) to elementary school children. The authors explain the value of a "density-based explanatory model . . . rather than the buoyancy-based" arguments typically used to explain FS phenomena, which is a conceptually challenging concept for children who instinctively associate floating and sinking with visceral experiences of 'lightness' and 'heaviness', respectively. The paper presents and evaluates the success of a novel instructional design paradigm founded on inquiry-based learning.

3. Experimental in Fluid Mechanics

The first article on experiments in FM by Wulandana discusses an impressive student driven project focused on building a recirculating flow tank [11]. Such tanks are an essential part of the collection of any fluid mechanics-based program and are very valuable due to their versatility and the ease with which many topics can be easily introduced. The only drawback of a prefabricated tunnel is its prohibitive cost. In this article, the author describes the design and fabrication of an open flow tank, built by students as part of their senior design project in a mechanical engineering program. The construction of this tank additionally provides opportunity for training in computer aided design (CAD) and computational fluid dynamics (CFD) as students perform comparative tests to validate flow structures in ideal and experimental conditions and also improve experimental designs so they meet expected flow conditions. The ideas introduced in this article can be replicated in any engineering program and also lead to interdisciplinary learning opportunities for students in physics and mathematics.

In the classic book, *A Splash of a Drop*, published by A.M. Worthington in 1895 [12], the world was introduced to the stunning visual world of droplet splashes. Worthington was certainly a strong influence on the movement in fluid dynamics scholarship to incorporate a visual element in order to understand the complex and beautiful structures that lie hidden behind the veil of transparency. Advancements and affordability of optical technologies makes is easier to introduce students to the fascinating world of flow visualization. The paper by Moghtadernejad, Lee, and Jadidi [13] introduces us to a course on multiphase flow where the instructors lead students on an experimental and theoretical investigation of splashes. Students also investigate the impact of temperature, wettability, impact velocity, droplet volume, shape, and relative humidity upon the splash dynamics. As the instructors note, this is an apt topic to introduce in a fluid mechanics course since it brings advanced knowledge into the classroom and also provides opportunities to discuss fundamental science and applications.

4. Computational Approaches

In this section we feature computation-based articles which cover both, articles of methodological nature and those that use computations to illustrate interesting physics of flows which can be introduced in any course on fluid mechanics.

The article [14] by Oz and Kara discuss computational methods relevant to 'Boundary Layer theory', a subject appropriate for introduction in an upper level undergraduate or graduate course. Relevant problems such as Blasius, Hiemenz, Homann, and Falkner–Skan flow equations are derived and numerically solved using the language, Julia. The codes have also been made freely available to the readers.

The article [15] by Ahmed, Pawar, and San provides a fundamental introduction to the mathematics and computational aspects of data assimilation methods which are fundamental to the study of climate science. Readers are exposed to the various methodologies through a series of Python modules which can be easily incorporated and adapted in an advanced course which treats such methods.

Mou, Wang, Wells, Xie, and Iliescu provide a survey of reduced order models which are computational models "whose dimension is significantly lower than those obtained through classical numerical discretizations" [16]. ROMs, in their various forms, have been found to be valuable in several complex computations involving uncertainty quantification, control, and shape optimization and in the numerical simulation of fluid flows. In this article, the authors summarize recent developments in ROM for barotropic vorticity equations, which are used to model geophysical flows.

In the article by Mongelli and Battista [17], the authors undertake a systematic study of pendulum dynamics by properly and fully accounting for the flow around a moving body which is not captured through the classical mechanical pendulum equations (see, also, another recent study that examines this issue through the lens of the least action principle [18]). The authors develop a computational fluid dynamics (CFD) model of a pendulum using the open-source fluid-structure interaction (FSI) software, IB2d. Comparisons with the results of the classical ODE model reveal very interesting and noteworthy results which ought to be discussed in any class which discusses pendulum dynamics.

Karlson, Nita, and Vaidya [19] discuss the interesting physics behind the vortex shedding phenomena. Computations using the program COMSOL are used to analyze the length of the primary vortex behind an elliptical body with varying eccentricities. The vortex length is used as a metric to understand and identify flow transitions from steady symmetric to asymmetric regimes which could potentially also be used as a noninvasive experimental strategy to distinguish flow regimes. The impact of the eccentricity of the body is seen to be particularly significant. While the physics itself is interesting and easy to follow and can even be discussed in an elementary FM course, such an example can be easily implemented in an advanced course in fluid mechanics or CFD course where students are exposed to a software for flow modeling.

5. Mathematical Fluid Mechanics

The final paper by Berselli and Spirito [20] is an extremely well written and much needed review of one of the most challenging mathematical problems of the last two centuries [21] and listed as one of the 'Millennium problems' in mathematics, namely the existence of solutions to the Navier–Stokes equation (NSE). While the history of this problem and various approaches is long and complex, the authors have done an excellent job in explaining and leading the readers through one aspect of this problem, namely the global existence of Leray–Hopf weak solutions to the NSE. I would strongly recommend that this article be made part of any course in an upper level undergraduate or even in an early graduate course in theoretical fluid mechanics or PDEs.

The papers in this volume, while selective and covering various different topics, showcase significant and cutting-edge knowledge of fluid mechanics in a manner that is easily adaptable for presentation in a course to undergrads, graduate students, and even in K12 settings. While the foundational materials traditionally taught in FM courses are important, our texts and curricula have not changed very much since the middle of the last century. The articles here provide templates for 'lesson plans' which can easily be implemented in our courses to make them more current and up-to-date. Many of the computation focused articles provide plug-and-play codes that can be implemented without much training and time and comprising the larger objectives of the courses. We hope that educators will take note and find these papers helpful in their own teaching efforts and also in encouraging their own efforts towards incorporating other newer results into their classroom discussions.

Conflicts of Interest: The author declares no conflict of interest.

References

1. Vaidya, A. Teaching and Learning of Fluid Mechanics. *Fluids* **2020**, *5*, 49. [CrossRef]
2. Gijbels, D.; Dochy, F.; Bossche, P.V.D.; Segers, M. Effects of Problem-Based Learning: A Meta-Analysis from the Angle of Assessment. *Rev. Educ. Res.* **2005**, *75*, 27–61. [CrossRef]
3. Sahin, M.; Yorek, N. A comparison of problem-based learning and traditional lecture students expectations and course grades in an introductory physics classroom. *Sci. Res. Essays* **2009**, *4*, 753–762.
4. Torp, L.; Sage, S. *Problems as Possibilities: Problem-Based Learning for K-12 Education*; Association for Supervision and Curriculum Development (ASCD): Alexandria, VA, USA, 1998.
5. Ahlfeldt, S.; Mehta, S.; Sellnow, T. Measurement and analysis of student engagement in university classes where varying levels of PBL methods of instruction are in use. *High. Educ. Res. Dev.* **2005**, *24*, 5–20. [CrossRef]
6. Blumenfeld, P.C.; Soloway, E.; Marx, R.W.; Krajcik, J.S.; Guzdial, M.; Palincsar, A. Motivating project-based learning: Sustaining the doing, supporting the learning. *Educ. Psychol.* **1991**, *26*, 369–398. [CrossRef]
7. Zyngier, D. Listening to teachers–listening to students: Substantive conversations about resistance, empowerment and engagement. *Teach. Teach. Theory Pract.* **2007**, *13*, 327–347. [CrossRef]
8. Garrard, A.; Bangert, K.; Beck, S. Large-Scale, Multidisciplinary Laboratory Teaching of Fluid Mechanics. *Fluids* **2020**, *5*, 206. [CrossRef]
9. Pérez-Sánchez, M.; López-Jiménez, P.A. Continuous Project-Based Learning in Fluid Mechanics and Hydraulic Engineering Subjects for Different Degrees. *Fluids* **2020**, *5*, 95. [CrossRef]
10. Zoupidis, A.; Spyrtou, A.; Pnevmatikos, D.; Kariotoglou, P. Teaching and Learning Floating and Sinking: Didactic Transformation in a Density-Based Approach. *Fluids* **2021**, *6*, 158. [CrossRef]
11. Wulandana, R. Open Water Flume for Fluid Mechanics Lab. *Fluids* **2021**, *6*, 242. [CrossRef]
12. Worthington, A.M. *The Splash of a Drop*; Society for Promoting Christian Knowledge: London, UK, 1895.
13. Moghtadernejad, S.; Lee, C.; Jadidi, M. An Introduction of Droplet Impact Dynamics to Engineering Students. *Fluids* **2020**, *5*, 107. [CrossRef]
14. Oz, F.; Kara, K. A CFD Tutorial in Julia: Introduction to Laminar Boundary-Layer Theory. *Fluids* **2021**, *6*, 207. [CrossRef]
15. Ahmed, S.E.; Pawar, S.; San, O. PyDA: A Hands-On Introduction to Dynamical Data Assimilation with Python. *Fluids* **2020**, *5*, 225. [CrossRef]
16. Mou, C.; Wang, Z.; Wells, D.R.; Xie, X.; Iliescu, T. Reduced Order Models for the Quasi-Geostrophic Equations: A Brief Survey. *Fluids* **2021**, *6*, 16. [CrossRef]
17. Mongelli, M.; Battista, N.A. A Swing of Beauty: Pendulums, Fluids, Forces, and Computers. *Fluids* **2020**, *5*, 48. [CrossRef]
18. Fitzgerald, K.; Massoudi, M.; Vaidya, A. On the modified least action principle with dissipation. *Eur. J. Mech. B/Fluids* **2021**, *89*, 301–311. [CrossRef]
19. Karlson, M.; Nita, B.G.; Vaidya, A. Numerical Computations of Vortex Formation Length in Flow Past an Elliptical Cylinder. *Fluids* **2020**, *5*, 157. [CrossRef]
20. Berselli, L.C.; Spirito, S. On the Existence of Leray-Hopf Weak Solutions to the Navier-Stokes Equations. *Fluids* **2021**, *6*, 42. [CrossRef]
21. Jaffe, A.M. The millennium grand challenge in mathematics. *Not. AMS* **2006**, *53*, 652–660.

Article

Large-Scale, Multidisciplinary Laboratory Teaching of Fluid Mechanics

Andrew Garrard *, Krys Bangert and Stephen Beck

Multidisciplinary Engineering Education, University of Sheffield, Sheffield S10 2TN, UK;
k.bangert@sheffield.ac.uk (K.B.); s.beck@sheffield.ac.uk (S.B.)
* Correspondence: a.garrard@sheffield.ac.uk

Received: 18 October 2020; Accepted: 9 November 2020; Published: 11 November 2020

Abstract: The nature of fluid mechanics makes experimentation an important part of a course taught on the subject. Presented here is the application of a novel, large-scale multidisciplinary model of practical education in a fluids engineering laboratory. The advantages of this approach include efficiencies through the economy of scale leading to better pedagogy for students. The scale justifies dedicated academic resources to focus on developing laboratory classes and giving specific attention to designing activities that meet learning outcomes. Four examples of applying this approach to fluid mechanics experiments are discussed, illustrating tactics that have been developed and honed through many repeated instances of delivery. "The measurement lab" uses a flow measurement context to teach identifying and managing general experimental uncertainty. In this lab, new students, unfamiliar with fluid mechanics, are guided through a process to gain understanding that can be applied to all future experimental activities. The "pressure loss in pipes" lab discusses the advantage of and process for sharing equipment and teaching resources between multiple cohorts. Here, the provision for students is adapted for context, such as the degree program or year of study. The "weirs big and small" lab provides a methodology for teaching the power of dimensional analysis to mechanical engineers using a field of fluid mechanics that is outside their usual theoretical studies. Finally, the "spillway design" lab discusses mechanisms for delivering independent, open-ended student experiments at scale, without excessive staff resource requirements.

Keywords: practical engineering education; fluid mechanics; learning and teaching; laboratories

---

1. Introduction

Laboratory practicals are often included as part of the scheduled delivery for courses teaching physical principles. They allow students an opportunity to understand the physical manifestation of underlying concepts and compare theoretical models to real world results and can cater for alternative learning styles. These justifications are pertinent for courses in fluid mechanics. The nature of the subject often involves the understanding of qualitative or counterintuitive concepts that are best understood through a tactile experience. It can be argued that a visceral instinct for the behaviour of fluids can only be obtained with sufficient experience of its application in the real world. In addition, teaching fluid mechanics usually involves understanding concepts and models to predict the behaviour of a fluid for specific scenarios, for example, flow in a pipe or around a wing. As these scenarios become more complex, models rely increasingly on the use of empiricism in order to overcome the inability of analytical methods to model the flow. Indeed, compared to in other engineering disciplines, the requirement to introduce empirical correlations into predictive models occurs for even relatively simple physical systems, such a turbulent flow in a straight, horizontal pipe. Understanding the value of experimental affirmation and validation is critical for the development of well-rounded students studying fluid mechanics. This has been clearly voiced in a recent publication, where the authors describe a blended approach to experimentation [1]. Within a university, each department is traditionally responsible for the delivery of laboratory teaching alongside a range of other teaching methods, such as lectures, tutorials, problem classes, design classes etc, which combine to provide students with courses in a particular subject discipline. The Faculty of Engineering at the University of Sheffield have adopted an alternative approach. One department, Multidisciplinary Engineering Education (MEE), is only responsible for the laboratory practicals of all 10 subject-specific degree programmes in the Faculty of Engineering, allowing the other departments increased time and resources to focus on classroom-based teaching methods and academic research. The volume of practical teaching delivered by MEE is consequently an order of magnitude greater than that which would typically be undertaken by departments offering individual degree programmes.

There are a number of advantages to this teaching model [2], such as the increased efficiency of infrastructure in common business processes, the reuse of teaching resources across similar activities and consistent experiences for students across the practical portion of their curriculum. The collective purchasing allows the possibility of buying many identical copies of experimental equipment and justifying the operation of large-capacity laboratories, with a related pedagogical advantage.

MEE is housed in the University of Sheffield's purpose-built Diamond building. Among the facilities is the Fluids Engineering Laboratory, which is used by Mechanical, Civil, Aerospace, Chemical, Bio and General Engineering students, in the order of 1000 students per year group. The laboratory has capacity for 80 students, typically working in groups of four, and is staffed by dedicated members of academic and technical staff. The laboratory is equipped with 20 identical copies of each of the pieces of experimental apparatus used for different aspects of fluid mechanics, including 20 hydraulic benches, on which a variety of internal flow experiments can be performed, and 20 wind tunnels, with which external flow experiments can be performed. The capacity provides three advantages of particular relevance to learning and teaching.

1. Large class sizes, with students performing the same experiment at the same time, result in laboratory teaching being temporally aligned with other classroom-based teaching, such as lectures and seminars. For example, if a cohort of 240 aerospace engineers required access to a single piece of experimental equipment, working in groups of 4, it would require 60 individual timetabled sessions. There is little chance of all sessions occurring at a specific time within the teaching calendar for the majority of students, which will impact on the effectiveness of their learning [3,4]. With 20 copies of the same equipment, all students can be provided with the same laboratory activity in three sessions. Thus, the design of the course structure, where and when topics will be taught, can be made by strategic design, rather than being constrained by timetable availability. Students can experience the reality of a practical laboratory simultaneously with being introduced to the physics in the classroom. In this context, a lecture can also be the introduction to an experiment that in turn shows the validity and application of the physics. An alternative approach could be to record practicals or use demonstrations, but this removes the engagement from the activity [5].

2. Laboratory sessions occurring in a predictable order due to Reason 1 makes it possible to sequence activities in the laboratory to be of increasing rigour and complexity, employing the principles of spiral learning. This would not be possible if all activities occurred in varying orders for different groups of students. It also makes assessment easier, as a single hand-in date is used for all students, and they all have the same amount of background to the topic.

3. The capacity of the laboratory and scale of the teaching load justifies a dedicated member of teaching staff focusing on laboratory education. With attention focused exclusively on the pedagogy of teaching fluid mechanics using experimentation, without the need to divert attention to other academic tasks such as the writing of exams, delivering lectures or providing feedback for assignments etc., significant effort can be placed in professionalising the laboratory activities. Similarly, the volume of teaching provides considerable opportunities to learn from and refine teaching methodologies to make them as effective as possible.

The Fluids Engineering Laboratory in the Diamond has been operating since 2015. The subsequent period of delivery has allowed a great deal of expertise to be developed in the teaching of practical fluid mechanics classes. Presented here are four examples of lab classes that have been honed through many repeated instances of running the activities.

2. The Measurement Lab

This laboratory is taught to all engineering students as one of the first tasks they perform when arriving at the University (typically in the first or second week). It is designed to equip students with a healthy scepticism for the results displayed on instrumentation and a toolkit for dealing with the uncertainty inherent in all forms of experimentation. A hydraulic bench is connected to a fluidic circuit containing a series of flow measurement devices. Although, at this stage in their programmes, students will be unfamiliar with almost all fluid mechanics concepts, including those that underpin flow measurement, it is explicitly stated and reinforced throughout the teaching that the activity is about understanding and managing general experimental uncertainty. Part of the intention is to imply the universality of error and uncertainty of techniques for any experimental set up, even ones for which the concepts under investigation or the outcomes are unknown.

Despite the lack of technical understanding of fluid mechanics principles, it is reasonably straightforward to explain the concepts of the conservation of mass and, for an incompressible flow, conservation of the volume flow rate. Students are aware that they will be studying fluid mechanics as part of the engineering programmes and are keen to understand these basic concepts early as well as be introduced to real-world instrumentation that they may not have previously been exposed to in schools. As water passes from one device to another in series through the hydraulic circuit, it is evident to the students that the volume flow rate through each device must be identical. The students are tasked with predicting if the various devices will all record identical readings for the flow rate.

Prior to starting the activity, students are given a briefing and watch an instructional video discussing the methods for capturing uncertainty for various pieces of instrumentation, how to record it and how it can be propagated when raw data are processed. The activity involves applying these principles to unfamiliar equipment.

Students record raw data from the instrumentation: the heights from water columns attached to a Venturi meter and an orifice plate, the flow rate from a calibrated rotameter measuring in litres per minute and the timing of water collection using a measuring tank. In order to compare the measured flow rates, the raw data from each device need to be converted. This requires students to consider that, in order to compare, the same measuring unit is required for each device and provides an opportunity to discuss the relative merits of the more commonly used litres/minute over the S.I. unit of meters cubed per second. While UK students are often trained to habitually convert into S.I. units, there is little justification for doing so in this case, particularly as it eliminates the need to convert the results of the rotameter.

Students need to process the height difference between the water columns of the Venturi meter and the orifice plate into a flow rate, the physics of which will not yet be known to them. For the Venturi meter, a pre-prepared spreadsheet is provided, which outputs the flow rate when the water column heights are input. The spreadsheet solves Equation (1), which is provided to students to allow understanding of the mathematical relationship, but the process of manual calculation is not required. The spreadsheet provides the opportunity to repeatedly calculate different answers quickly and determine the relative impact of the uncertainty for different parameters. The square root function within this equation makes this process more interesting.

$$Q = C_d A_1 \sqrt{\frac{2g\Delta h}{\left(\frac{A_1}{A_2}\right)^2 - 1}} \tag{1}$$

A calibration chart, shown in Figure 1, is provided to convert the raw data from the orifice plate. The process of reading and extrapolating results from the graph introduces additional uncertainty in processing the raw data, and students are encouraged to consider how to best incorporate this into their calculated flow rates. This facilitates an opportunity for students to become comfortable with a limited availability of precision in a calculation, which is a common occurrence in the application of practical engineering.

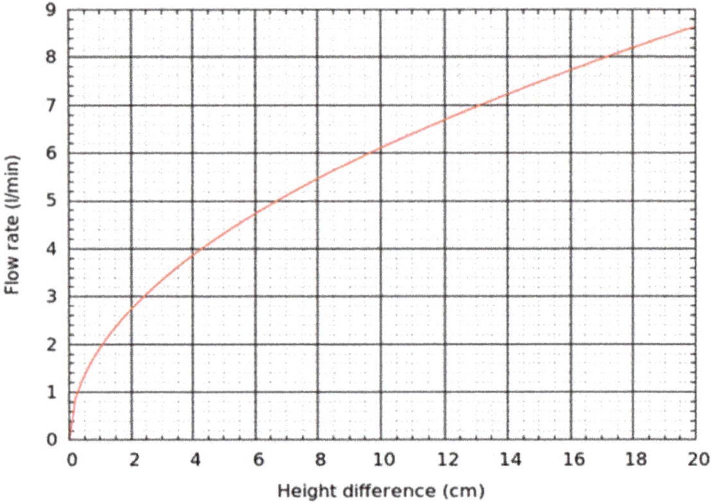

Figure 1. Calibration chart for an orifice plate provided to students.

Students are guided through a process of calculating flow rates and the associated uncertainty of their results for the four flow measurement devices and then through a process of presenting this information graphically with an introduction to the concept of error bars. If conducted correctly, the results show that all the devices will record different values of the flow rate but, when the error bars are considered, all the results overlap within a certain region. Further discussion can be had about the methods for reducing uncertainty in the raw data and the advantages of in-line flow measurement compared to the volume displacement of the measuring tank.

This lab has been designed to achieve the specific learning outcomes of introducing the concepts of, methods to record and process for handling error and uncertainty in experimentation. Students are clearly made aware of this expectation and achieve these explicit learning outcomes as a result of participating in the activity. The same learning outcomes could have been achieved with a paper-based exercise, delivered outside a laboratory. However, the act of learning through doing is more likely to result in the concepts being retained by students and provides a real-world context in which to apply these skills in an engaging form that enhances the student experience.

As this activity is delivered to all undergraduate engineers, typically, in excess of 1000 students per year, the investment of time to develop high-quality instructional material and training teaching assistants is easily justified and makes the activity very resource efficient.

3. Pressure Loss in Pipes Lab

The pressure loss in horizontal pipes is measured in an experiment run in the Fluids Engineering Lab for Mechanical, Aerospace, Civil and Chemical Engineering students. This is an important part of the engineering curriculum [6]. The reuse of teaching material and equipment for multiple cohorts results in efficient resource utilization. However, teaching material is adapted and contextualized for specific degree programmes. Subject-specific nomenclature or units should be used appropriately for different engineering disciplines. For example, civil engineers would measure, record and process pressure in the units of meters of head, whereas aerospace engineers would typically use Pascals. In addition, the level of academic rigour and expectation for the students is adapted depending on the placement of the activity within the degree programme. At the University of Sheffield, mechanical engineering students perform the experiment in the first year, and civil engineering students, in the second. For civil engineers, less prescriptive instructions are provided, and more independence is expected while conducting the experiment.

During the experiment, the students will collect raw data from manometers to measure the pressure drop and a measuring tank to measure the flow rate. A range of different-diameter and roughness pipes are available on each of the 20 hydraulic benches. The raw data are processed into Reynolds numbers and empirically derived friction factors, allowing students to generate their own Moody diagram that can be compared to a published version. The objective of the laboratory experiment is not to develop expertise in performing the mathematics. Students are provided with a "guided calculation", where the steps to process the raw data into a processed result are described in the instructions and executed by students on one piece of data, to ensure they understand the mathematical methods. Breaking each part of the calculation into defined steps makes the debugging of errors, by the students or teaching assistants, more straightforward, which is necessary when dealing with large class sizes. Once students have demonstrated they understand the process, a spreadsheet to automate the calculations on the remaining data points is released.

Mandating students to complete the hand calculation before using the spreadsheet opens the opportunity for discussion with students that perceive the activity to be a trivial task only required to access the spreadsheet. Performing the hand calculation and using a tool to perform the calculations allows a two-way validation of each process, by comparing the results from each. When using any tool that has been provided, it is wise to ensure it operates as expected. Articulating the general merit of a validation approach, and how it can be applied to a student's future engineering tasks, can be used to place value on performing the task. In addition, digital collection allows an individual student's data to be pooled into a larger dataset that can be shared with the cohort, for the purposes of error and reproducibility analysis.

Having multiple benches to support a large class size presents an opportunity to improve efficiency beyond the economy-of-scale issues previously described. In this experiment, to determine the influence of the pipe specimen (diameter and roughness) on the friction factor, multiple pipes should be investigated. With one hydraulic bench or a small number of hydraulic benches, pipes need to be installed and removed to test the full range. With as many benches as specimens, benches can be set up with particular specimens, and students can move around the laboratory to each piece of apparatus. Learning the procedure, executing it and performing the subsequent bleeding of air all consume student time and cognitive capacity in ways that do not directly relate to the intended learning.

As the students come into a session well prepared and the lectures on the subject are fresh in their minds, they do not require much assistance to conduct the laboratory. Hence, it can be taught by four staff (Academic, Technical and Teaching Assistants). These staff are able to spend the time discussing the work and providing feedback to the students, resulting in a much richer experience for everyone.

4. Weirs Big and Small Lab

One of the hardest fluid mechanics topics for students to understand is the importance and power of dimensional analysis. This is because there are a number of difficult concepts when contemplating scales, which can be simplified through the correct application of dimensionless numbers. These include the fact that experiments are needed to be able to obtain the constants for every geometry in a given situation. An understanding of dimensional analysis is needed for studying aerodynamics and heat transfer, but the teaching of it generally suffers from two deficiencies: firstly, students typically become very tangled up in the details of the subject rather than the method application, and secondly, almost all of the work and examples involve Reynolds numbers (as in the Pressure Loss in Pipes lab described above). Thus, an interesting approach to illuminating this topic has to have two requirements: not using Reynolds number and extracting some constants that are then applied to different scales of equipment. If it could be engaging, challenging and fun as well, that would be even better.

The hydraulic benches can be configured to allow students to perform open channel flow experiments with sharp-edged weir plates (square and triangular) and can measure the water height over the weir and water flow rate. The cohort for whom the lab was created are second year Mechanical Engineers whose curriculum does not contain free surface flow. This presents an ideal opportunity to introduce this topic to these students while showing the power of dimensionless groups. Students are shown, using the tools from their lectures (Buckingham π theory), that the dimensionless groups involved in this type of flow are the Froude number $\frac{V}{\sqrt{gH}}$ and length ratio $\frac{H}{b}$, as shown in Figure 2, and that the volumetric flow rate (Q, m^3·s^{-1}) can be derived from Equation (2):

$$Q = C_D b \sqrt{g} H^{\frac{3}{2}} \tag{2}$$

where C_D is the discharge coefficient and must be experimentally ascertained for the given geometry. The objective of this experiment is to experimentally determine the discharge coefficient for the small weirs and see how this scales to a geometrically equivalent, larger weir that is installed in the lab's 10 m flume. The main learning outcome of this activity is for students to be able to see both the power of dimensionless numbers and how extracting the constants experimentally is a required part of the process.

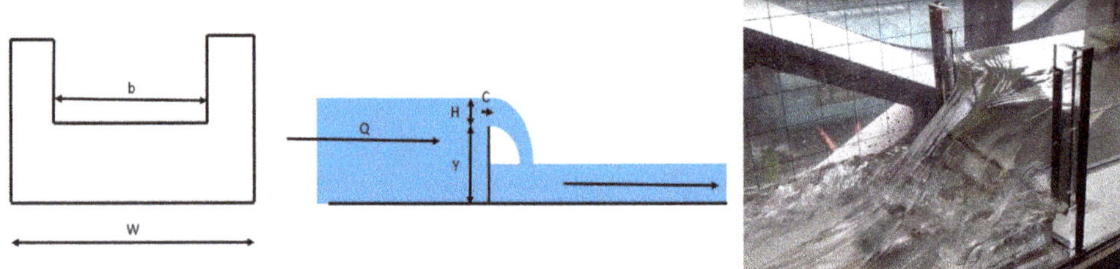

Figure 2. Dimensions used in weir calculations and the experimental apparatus.

In order to optimise the time students spend in the laboratory, a comprehensive pre-experimental activity was created, material from which can be seen in Figure 3. This consisted of a series of presentations. These were either recorded with overheads and voiceovers for theory (the flow over weirs, which Mechanical Engineers do not cover as part of their course). Another online lecture on fitting exponentials to a series of x and y data was created to help students understand one way of turning experimental data into equations, which they need to do to extract the constants in this experiment. Short quizzes created in the Virtual Learning Environment using adaptive release ensure that the students engage with each presentation prior to moving on to a subsequent section. The students are presented with three videos on using the large flume, operating the small flow rigs and reading the Vernier scale on the large flume. This culminates in a quiz on reading the Vernier and finally a compulsory test on Health and Safety issues relevant to the experiment and laboratory space. The use of adaptive release means that students cannot get to the final test without completing all the previous ones.

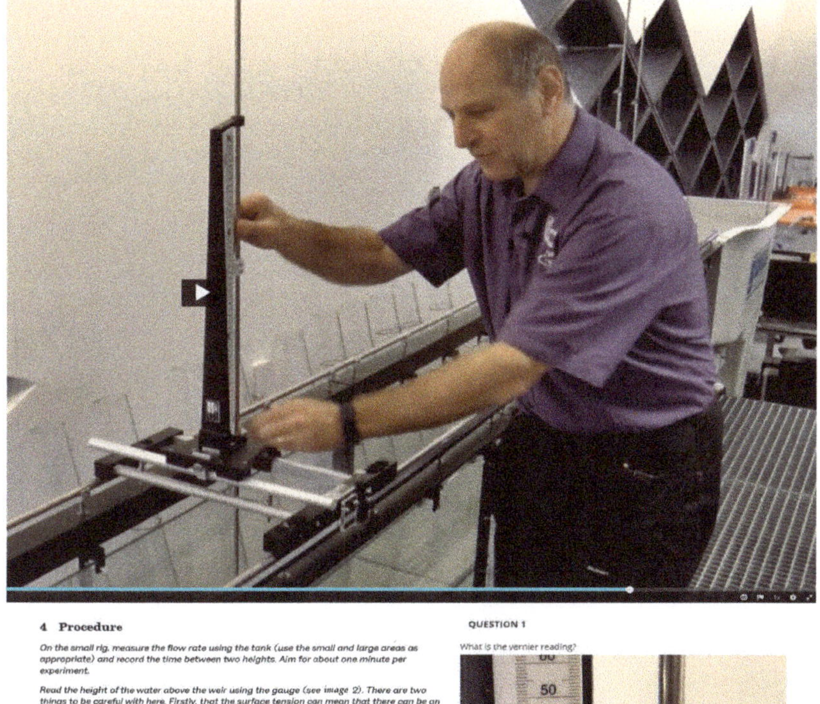

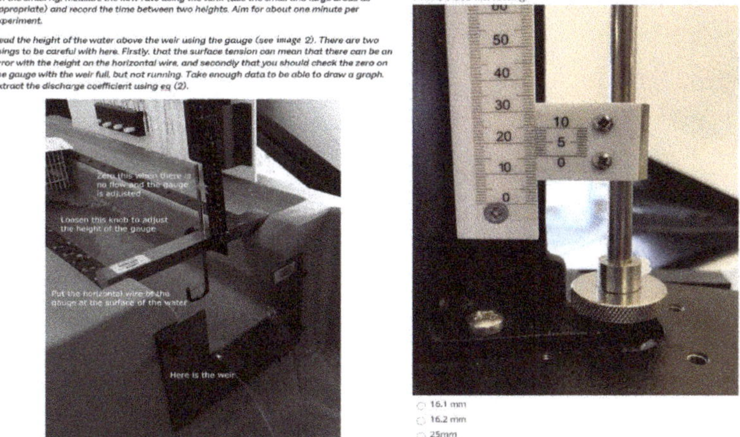

Figure 3. Examples of instructional material, including pre-experimental videos, provided to students.

This preparation works well. Students move on very quickly to the practical experiment. As required, they calculate the C_D on the small rigs from the flow and height measurements they recorded. For each session, the laboratory leader sets a different flow rate for the large flume, and the students predict the height above the weir that the water should reach before they measure it. This, in effect, gamifies the lab, as they are ascertaining their own experimental accuracy and are able to compare it with their peers'. They are thus able to grasp the power and value of dimensionless numbers and scaling in engineering. It shows them that methods such as the Buckingham π theory are merely tools to be used to solve real problems. It also illuminates the way that researchers need to use experiments to be able to create a set of results to identify and extract generalities. It is also a good way to use our facilities, as they only perform a single experiment on the large, unique rig but can perform a large number on the multiple small rigs.

By the end of the experiment, it was noted that the students appreciated how it was possible to scale between different sizes of models, but how one needed to be aware of all the important parameters. In a more general sense, the intellectual understanding that this journey provides through models and sizes acts as a conceptual bridge towards a better understanding of the dimensionless numbers that they will use in their fluid mechanics and thermodynamics careers and education.

5. Spillway Design Lab

Teaching large cohorts presents a tension: investing significant resources in activities that can be reused by many students is efficient and provides a high-quality, professional laboratory experience, but prescribed activities can limit the opportunity for students to explore open-ended activities and, for example, learn through failure. This tension can be partially overcome with the application of a multidisciplinary approach. Within MEE's portfolio of practical engineering education is manufacturing and fabrication. This provides a holistic integration of making that is available to staff and students.

With significant manufacturing capability, department-based workshop staff and tools, in-house builds of bespoke teaching equipment are feasible. Typically, engineering teaching equipment for use with students would have been bought from suppliers. The two significant downsides of this approach are that it is extremely expensive compared to an in-house build (if full-time staff time for design, fabrication and prototyping is excluded) and the equipment is not designed to achieve specific learning outcomes.

Unlike their Mechanical counterparts, second year Civil Engineering students study open channel flow and the design of flow control devices during their second semester. As they are reasonably advanced students, having been prescriptively taught the fundamentals of operating in a laboratory environment in preceding years, their practical activities are designed to be conducted independently, open ended and genuinely experimental, i.e., conducting empirical work to discover something previously unknown. To achieve these outcomes, a bespoke experimental rig was conceived, as shown in Figure 4.

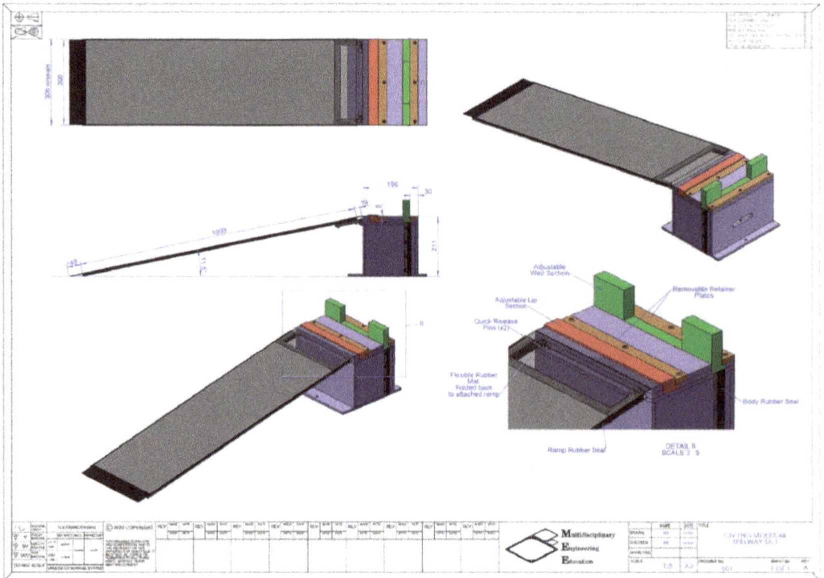

Figure 4. Bespoke rig for teaching spillway design.

The teaching of design principles is outside the remit of a fluid mechanics course, but a holistic approach to a programme can be used to apply previously learned design principles to the design of a weir and spillway to fulfil specific problem specifications. As such, the bespoke rig was designed to provide a constrained number of parameters that can be adjusted. Designated adjustable parts can be fabricated by students using readily available equipment, such as laser cutters or 3D printers, and inserted into the rig for testing in the fluids lab flume. Students are expected to design their adjustable parts based on theory delivered during lectures, predict the flow and test their predictions in the laboratory.

Prior to being given access to the flume, students are provided with extensive equipment and Health and Safety training to be allowed access to the laboratory without staff supervision. Compliance is established with online tests, and keys to the room/equipment are provided by reception staff who check for completion of the test. Students are able to book use of the equipment at a time convenient to them, and academic staff time input is minimised. This approach provides students with an opportunity to exert agency over their own learning.

Without a dedicated team of staff focused on providing students with an exemplary practical experience and the multidisciplinary team of academics and technicians working collaboratively, the development of practical teaching and bespoke equipment that is unobtainable from suppliers would be significantly more difficult for departments to justify resourcing.

6. Discussion and Conclusions

Historically, fluid mechanics laboratories have been run as a type of cottage industry, with lecturers specifying and delivering a couple of labs across the year on a single piece or possibly a couple of pieces of equipment. This meant that students could receive this laboratory at any time over a year. Each experiment had to be free standing; many students would conduct the experiment long before or after they were introduced to the theory, missing a crucial window for learning reinforcement. This meant that different students would, in effect, obtain different learning outcomes depending on their understanding of the background to the topic. The staff members responsible for the delivery of the lectures, tutorials, exams etc. set their own labs; they tended to be similar in difficulty, scope and assessment (usually a report). There was no coherence or progression along the course and, without the capacity to focus exclusively on the laboratory activities, very little in the way of designing teaching with constructive alignment towards the overall learning objectives.

MEE's multidisciplinary approach of professionalising and integrating the practical experience of the students allows many of these common issues to be obviated. The scale of the laboratories allows entire cohorts to perform an experiment in a short time period so that practical and theoretical work can be interwoven and used to support each other. In many cases, the formal lecture becomes the introduction to the laboratory. Many Electrical Engineering departments have rooms set up with multiples of equipment, but this approach is rare outside engineering. Having dedicated staff who deliver the only practical experience to a cohort, it is possible to *curate* an entire, progressive student experience starting from the closed and didactic (such as the Measurement Lab) and progressing to open-ended investigations such as the Spillway Design Lab. The result is that students receive an integrated and progressive learning experience culminating, after their first two years, in them becoming capable, reflective and autonomous experimenters ready to start independent project work.

As well as the efficient use of space and staff time, the experiments form a portfolio of work that can be renewed and repurposed as and when required. For example, within weeks of the creation (and delivery) of the Weirs laboratory, a lecturer from Civil Engineering asked if there was a laboratory for open channel flow for their MSc students. Not only was the answer "yes", but a version of all of the teaching and introductory material was ready for use. The laboratory sheet only needed updating to reflect the different approach to the theory and nomenclature used by a different discipline, but this was a minor investment of time and allowed the students to have an excellent practical experience to support their learning that would have been impossible under a different organisation.

There are, however, two potential drawbacks to the multidisciplinary approach, but these can be ameliorated if properly anticipated. Firstly, when the practical and theoretical teaching on a single module is delivered independently by different members of staff in different departments, the experience and messaging received by the students could become disconnected and incoherent. MEE overcomes these issues by setting up communication channels between the academics delivering classroom and practical teaching, allowing them to agree on how the labs are presented within the context of a module and ensure that the messaging to students is consistent. The tactics for achieving this include using material presented in lectures as part of laboratory tuition and vice versa. Secondly, there is a requirement for strong leadership within the faculty. The multidisciplinary model will only work if all the departments benefiting from the service agree to contribute to its resourcing. With any shared resource, issues of perceived value and equity for contributors can cause tension if not carefully managed.

Thus, in conclusion, there are a number of major advantages to teaching at scale in fluids laboratories, such as the efficiency, temporal proximity to lectures and scalability. Due to the integration and professionalisation of the practical teaching, it allows an integrated, progressive approach to student practical skills development to be implemented. Progressing from the usual method of teaching practical fluid mechanics to the new one demonstrated in the examples above is a difficult, long and potentially extremely expensive journey. We hope that we have shown you that the outcomes from it are worthwhile.

Author Contributions: Conceptualization, S.B., A.G. and K.B.; Methodology, S.B., A.G. and K.B.; writing—review and editing, S.B., A.G. and K.B. All authors have read and agreed to the published version of the manuscript.

Funding: This research received no external funding.

Conflicts of Interest: The authors declare no conflict of interest.

References

1. Chowdhury, H.; Alam, F.; Mustary, I. Development of an innovative technique for teaching and learning of laboratory experiments for engineering courses. *Energy Procedia* **2019**, *160*, 806–811. [CrossRef]
2. Garrard, A.; Beck, S.B.M. Pedagogical and cost advantages of a multidisciplinary approach to delivering practical teaching. In *The Interdisciplinary Future of Engineering Education Breaking Through Boundaries in Teaching and Learning*; Kapranos, P., Ed.; Routledge: Abingdon, UK, 2018.
3. Goodwin-Jones, J.; Nortcliffe, A.; Vernon-Parry, K. What does good engineering laboratory pedagogy look like? In Proceedings of the 44th Annual Conference of the European Society for Engineering Education—Engineering Education on Top of the World: Industry-University Cooperation, SEFI 2016, Tampere, Finland, 12–15 September 2016; SEFI: Brussels, Belgium, 2016.
4. Hofstein, A.; Lunetta, V. The role of the laboratory in science teaching. *Rev. Educ. Res.* **1982**, *54*, 201–217. [CrossRef]
5. Hesketh, R.P.; Farrell, S.; Stewart Slater, C. The Role of Experiments in Inductive Learning. In Proceedings of the 2002 American Society for Engineering Education Annual Conference & Education Conference, Session M4C, San Diego, CA, USA, 16–19 June 2002; p. 3613.
6. Krestae, S.M. Hands-on Demonstrations: An Alternative to Full Scale Lab Experiments. *J. Eng. Educ.* **2013**, *87*, 7–9. [CrossRef]

Publisher's Note: MDPI stays neutral with regard to jurisdictional claims in published maps and institutional affiliations.

Article

Continuous Project-Based Learning in Fluid Mechanics and Hydraulic Engineering Subjects for Different Degrees

Modesto Pérez-Sánchez * and P. Amparo López-Jiménez

Hydraulic and Environmental Engineering Department, Universitat Politècnica de València, 46022 Valencia, Spain; palopez@upv.es
* Correspondence: mopesan1@upv.es; Tel.: +34-96-387700 (ext. 28440)

Received: 6 May 2020; Accepted: 13 June 2020; Published: 15 June 2020

Abstract: Subjects related to fluid mechanics for hydraulic engineers ought to be delivered in interesting and active modes. New methods should be introduced to improve the learning students' abilities in the different courses of the Bachelor's and Master's degree. Related to active learning methods, a continuous project-based learning experience is described in this research. This manuscript shows the developed learning methodology, which was included on different levels at Universitat Politècnica de València. The main research goal is to show the active learning methods used to evaluate both skills competences (e.g., "Design and Project") and specific competences of the students. The research shows a particular developed innovation teaching project, which was developed by lecturers and professors of the Hydraulic Engineering Department, since 2016. This project proposed coordination in different subjects that were taught in different courses of the Bachelor's and Master's degrees, in which 2200 students participated. This coordination improved the acquisition of the learning results, as well as the new teaching methods increased the student's satisfaction index.

Keywords: outcomes competences; hydraulic engineering; hydraulic teaching; active methodology

---

1. Evolution of Teaching in the University

1.1. New Paradigms in Hydraulic Engineering Teaching

Hydraulics disciplines are in a higher number of degrees related to engineering topics [1]. Civil and environmental engineering courses are an example, although they are not exclusive. There are different Bachelors' and Masters' degrees, in which the fluid mechanics and hydraulic topics are present inside of the students' curricula, such as the mandatory or optative subject.

Teaching involves many methods to reach the learning results. Some of them are: master courses (i.e., a theoretical lesson taught by a professor); design projects; practical activities in the hydraulic lab; and informatic sessions, among others. All actions must provide students an integral and continuous vision of the hydraulic engineering (from fluid mechanics to environmental problems). However, the new students must experiment a necessary change in the new learning methodologies, allowing the students to reach the professional competences satisfactorily. Currently, the European Higher Education plans to decrease the credits to teach, increasing the required skills acquisition by activities, which are not an on-site class [2,3].

1.2. The Significance of the Learning Habilities

One of the problems that students must face is the complexity of numerous concepts in hydraulics subjects (e.g., fluid mechanics). The lecturer usually teaches theoretical matters and the student has to reach the learning results (e.g., master course, lectures, and exercises) using the professors' information (e.g., bibliography and exercises). Therefore, students must learn materials on their own with minimum guidance by professors. This methodology had good results in the last decades [4]. However, newer student generation demands the development of new teaching methods that rely on new tools. The use of active learning methods is highly recommendable since the students participate in the learning process actively [5]. These methods are based on student activities (e.g., 'playing and learning' using simulations, project-based learning, and role activities) they are a possible solution to improve the students' learning. Continuous project-based learning was proposed across different levels in 2016 as a part of these active strategies [6,7]. The current research shows the results, which were obtained by the coordination between bachelor and master matters from 2016 to 2018.

Currently, numerous researches show the professors should introduce the new learning tools and activities (e.g., simulations, experimental cases, and playing learning) using information and communication technology [8]. Using these tools engages students actively in the learning process. In this line, the Universitat Politècnica de Valencia (UPV) carries out the 'UPV generic students' outcomes 2015–2020' [9]. The main goal of this project is the introduction of 13 generic outcomes, which will improve the students' skills and their curricula [9].

To adapt the new strategic plans, an innovation and educational improvement project has been implemented between different professors of the Hydraulic and Environmental Engineering Department of the UPV since 2016. The main objective is to establish a transversal and vertical coordination in different subjects. The purpose is the acquisition improvement of the learning results by the students. Therefore, the project proposes an evaluation methodology using different rubrics according to the domain level (depending on the year and degree). Besides, the research compares the different subjects in different courses.

In this particular case, this proposal allowed students to start a hydraulic project draft in fluid mechanics topics (e.g., students sized a water branched network). Furthermore, they continued its development in hydraulic machinery matters (e.g., students designed a pump system) and finally, they designed a total project in their last matter fluid facilities (e.g., students sized fluid facilities in a hotel). The development of hydraulic projects, as a learning strategy, is a methodology that was proposed in other universities some years ago. When this method is used, the students must plan, implement, and evaluate complete works, which are applied in real case studies [8,10].

Project-based learning (PBL) allows students to acquire key knowledge and skills through the development of projects that respond to real-life problems [11]. The objective is to enhance students' autonomy. They become the main actor of their own learning process. This training evolves introducing new complex tasks each course using the same project [8]. In this learning process, professors guide and support the students throughout the entire project.

1.3. Hydraulic Engineering Learning Challenges

The stage of the studies of hydraulic engineering must be in constant evolution since the future professionals must face great challenges. These are aligned on terms of sustainability and optimization of the management. Therefore, hydraulics subjects at the university level cover many fields such as: urban hydraulics, watershed management, the pollutants dispersion, hydraulic machinery, river dynamics and restoration, water resources management, hydraulic works, aspects of flows to sheet free of charge involved in sanitation, the water-energy nexus and many other subjects that are being taught in different faculties masterfully. In all cases, the hydraulic engineering is present in the core subject, such as fluids mechanic and/or hydraulic machinery in the engineering bachelor's degree (e.g., electrical, mechanical, and chemistry).

Currently, the knowledge transfer requires the future students must be autonomous and capable. They have to develop skills, which allow them to solve the new challenges. The present manuscript shows the continuous project-based learning experience, which has been developing at UPV. This practice is focused on hydraulics subjects, which are teaching both the bachelors' and masters' degree. The proposed teaching project increases the development of real projects, decreasing the hours of lessons. This time decrease is complemented with online material (e.g., teaching video and laboratory tutorials) including workshops and specific conferences. These sessions are developed by companies or guest speakers in the university focused on students.

This research is a good example for engineering bachelors' and masters' degrees related to hydraulic and environmental topics at different levels. The manuscript summarizes teaching methodology and results, which developed a teaching project. The experience was carried out at Universitat Politècnica de València. Two thousand and two hundred students participated in thirteen hydraulic subjects, which were part of the teaching project and they were from different years. The students worked the hydraulic concepts using a methodology, in which they reached the learning results through the development of hydraulic projects. The strategy enabled to evaluate both specific and outcomes competences. Before this teaching project, the students were not evaluated of their skill competences and they did not use active methods. Previous to this project, the students' training was based on master courses and laboratory practices. The participation was up to 80% and the student's satisfaction was measured by surveys.

2. Materials and Methods

2.1. Structure of the Hydraulic Engineering for a Student of a Bachelor's and Master's Degree in the UPV

When the structure of the hydraulic engineering was analyzed at the UPV, there was a complete interweaving with other matters in their different bachelor's and master's degrees. These subjects (Figure 1), which were distributed throughout student training, were: (i) basis subjects in hydraulics and fluid mechanics; (ii) subjects related to hydraulic machines; (iii) subjects related to hydroelectric plants and wind power machinery; (iv) subjects related to hydraulic facilities; (v) materials in oleo hydraulic and pneumatic systems; (vi) matters in relation to the water-energy binomial; (vii) matters in computational fluid dynamics (CFD) modeling; (viii) matters in relation to the hydraulic aspects of wastewater treatment; and (ix) matters in relation to the dispersion of contaminants in receiving fluid media.

Figure 1. Structure hydraulic engineering subjects at Universitat Politècnica de Valencia (UPV; BD.—Bachelor's degree and MD.—Master's degree).

2.2. Learning Proposal Based on Learning Projects at Different Levels, Developing the Transversal Competence "Desing and Project"

The project-based learning (PBL) is a methodology focused on learning, research, and reflection. In this methodology, the students should reach the correct solution of a problem, using an autonomous and continuous learning. This problem was proposed by the lecturer once he/she teaches the theoretical concepts [12]. This methodology was included on the teaching project in the hydraulic and environmental engineering department [13]. It was applied on different matters, which were part of different courses and levels (i.e., Bachelor's and Master's). Therefore, the teaching project got a continuous project-based learning (CPBL) in the students' training [14]. Besides, the CPBL application at different training times of the student enables one to work and evaluate different transversal competences (e.g., time planning, permanent self-learning, and oral communication as well as design and project (DP)).

Time planning was proposed for each subject and it must be followed by both students and professors through the different phases. These steps (Figure 2) were divided on face-to-face and non-face-to-face lessons. The first phase allows students to know the theoretical concepts throughout the master course, the development of computer practices as well as the development of basic problems related to the taught issue. Once these are known, the learning results of each unit should be practiced in progressive development of the project. This practice is non-face-to-face and the students must use information from the UPV webpage. In this section, they have supplementary material. Along this phase, the students work on self-learning and the professor gives them help in group meetings. The collaboration between the professor and students improves the acquisition of the learning results.

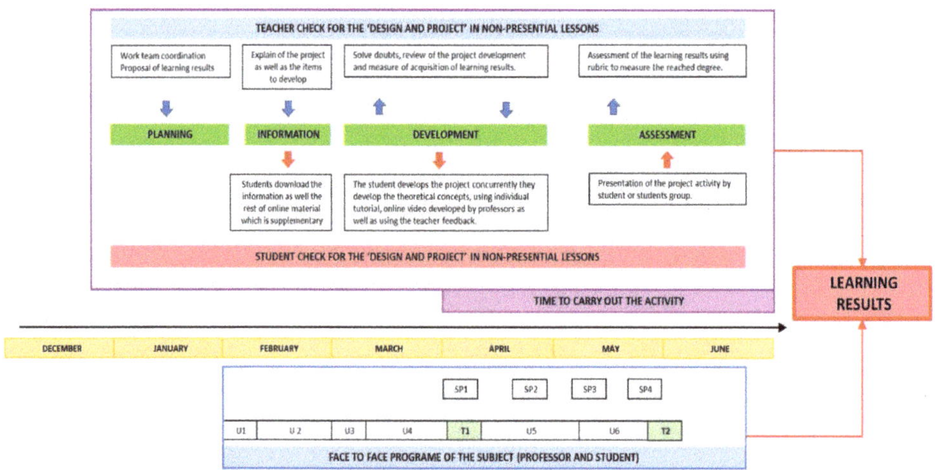

Figure 2. Example of temporal distribution to reach the learning results in a subject (U is unit, T is a test, and SP is simulation practice).

The different activities, their planning, and their dedication were defined using good practices sheets. This sheet was developed by lecturers, and it enabled to coordinate the subject between students and professors, and coordination improved when different teachers participated in teaching the subject.

2.3. Proposal of Rubrics

One of the objectives to develop an active learning is the definition of evaluation items. In this particular case, different rubrics were introduced to combine the evaluation of the student's competences (i.e., skills and concepts). These rubrics were composed of different indicators, which had four different descriptors for each one. The indicators measured the acquisition degree of the learning results. Table 1 shows the proposed indicators, which were used to do the proposed rubric (Appendix A). This rubric was used to evaluate the project in hydraulic machines. Other rubrics were published for different subjects: the wastewater network project [6], fluid mechanics [7], or fluid facilities in the chemistry industry [15]. The used rubrics were different in Master's and Bachelor's degrees, considering the reached level in the descriptor. In the Bachelor's degree, the pupils had to design a project with a level of draft. In contrast, the indicators were higher in the Master's degree, as the developed project should be more specific, and students must be more autonomous.

Table 1. Definition of weighted in the different indicators.

Indicator	Students' Actions	Weighted
I1.—The student bases the context and the need of the project	Define the need to develop the project	5%
I2.—The student formulates the objectives of the project coherently with regard to the needs detected in the context	Localize them and relate them with the taught concepts. Correct interpretation of the goals allows students to interpret the specific indicators of the follow group (iii) correctly	7.5%
I3.—The student plans the action to be developed effectively	The student has to propose and apply the solved methodology	20%
I4.—The student plans the actions efficiently	Design the proposed system. This group contains seven specific indicators	50%
I5.—The student identifies the risks and inconvenient of the project	Consider the negative and positive aspect of the project related to environmental and social concepts. This indicator is measured using two specific criteria.	7.5%
I6.—Review the results	Review, analyze, and critique with the obtained results, searching incoherent results.	10%

The attached rubric in Appendix A shows the new proposed rubric for hydraulic machines in which specific and skills competences were evaluated. The symbiosis between specific and transversal competences was developed using a matrix, which contained weights and ponderations. The first discrimination was done between 'not done' and 'developed task but the minimum is not reached'. When the student did not develop the descriptor, the numeric value was zero. If the student did the task, but it was not reached, the considered value was 3. If the descriptor was C, the numeric value was 5. When the descriptor was B, the considered value was 7 while the value was 10 when the descriptor was A. Each indicator had a specific weighted value, which was justified in Table 1.

Therefore, the numeric mark (NM) of the specific competence was obtained using Equation (1):

$$NM = 0.05I_1 + 0.075I_2 + 0.20I_3 + 0.50\frac{\left(\sum_{I=1}^{I=7} I_4\right)}{7} + 0.075I_5 + 0.10I_6 \tag{1}$$

This expression enabled one to get the numeric value through descriptors. If the NM was less than 4.5, the learning result was "No reach—D" for transversal competence. When the value was between 4.5 and 6, the learning result was "In Development—C". A learning result of "Good—B" was reaching, when the NM was between 6 and 8. Finally, 'Excellent—A' was reached when the NM was greater than 8.

3. Results

3.1. Students, Subjects, and Proposal of Projects

The project was developed between 2016 and 2019, although it continues currently. In these years, thirteen subjects were taught in Bachelor's (second, third, and fourth year) and Master's (industrial engineering and hydraulic and environmental engineering) degree at UPV. One thousand and one hundred students participated and 13 professors from the hydraulic and environmental engineering department collaborated in the project each year. The manuscript shows results for all subjects although main subjects of the students' curricula (i.e., fluid mechanics, hydraulic machines and fluid facilities) were described deeply in this research. Fluid mechanics was taught in a second-year course of the Bachelor's Degree. This subject was in chemical, electrical, and mechanical engineering degrees (in this particular case, the results were related to the mechanic engineering degree). The students were between 19 and 25 years. One hundred and seventy students were involved, who were divided into two groups for theoretical teaching and four groups for practical classes.

Hydraulic machines was taught in the third-year course of the mechanical engineering degree. The subject was focused on analyzing the pumps-operation principles (velocities triangle and Euler's equation) as well as the machines selection and their regulation according to demand. The students were between 20 and 26 years. One hundred and forty students were involved. These pupils were divided into two theory groups and four practical groups.

Fluid facilities was taught in the first-year course of the industrial engineering Master's degree, after students achieved their Bachelor's degree. The subject contained the analysis of the different types of the fluid facilities, that is, water distribution networks, gas networks, and waste-water networks as well as the facilities, which involved the comfort in society (ventilation and hot sanitary water). For each facility type, the normative, design, analysis, and regulation were analyzed, applying it to the real cases study. The students were between 23 and 30 years. Three hundred and fifty students were involved. These students were divided into seven theory groups and twenty-one groups in practical lessons.

In relation to the fluid mechanics subject, an elemental water supply network was proposed, in which students proposed different diameters for pipes, considering flow and pressure conditions. Once the system was sized, the students had to analyze it using Epanet software [16], considering the constrain conditions (e.g., demand, level node, minimum pressure, and maximum velocity). The sizing was developed as a function on demand over time, using the uniform hydraulic slope criterion. The students run an extended period simulation, analyzing the pressure and flow variations in the different lines. Finally, the students proposed a short budget, considering both length and the chosen material. In this case, the project was supervised by the lecturer. The initial information, which was available for students was: network topology, reservoir head, water demand in each point over time, modulation curve for the different consumption patterns, and the minimum operational conditions of the network as well as the material type and cost of the pipelines. The work was focused on establishing a methodology to develop hydraulic calculus, encompassing the Bernoulli's and continuity equations. The students compared the different studied scenarios as a function of demand pattern using Epanet software.

The evaluation was a formative type. The students did meetings with the professor and they show partial results. The professor verified the calculus and solutions, proposing improvements to students. There were two meetings. The first meeting included the proposal of the network. The second meeting addressed the sizing of the water system. The correction of the project was developed using rubric (Appendix A), and a third meeting was done to explain to students the errors in the project.

The proposed work in hydraulic machines was individual. The activity was focused on analyzing the energy consumption and regulation of a pumped system. This water network was supplied considering two options. Option A: the water network was supplied from a reservoir, which was filled using a pump station and Option B: the water was directly supplied using pump systems. Option A enabled one to analyze the influence of the reservoir volume in the pump selection (mainly pumped flow) when the energy cost was considered (i.e., schedule and operation time). Option B was focused on applying the similarity laws, regulating the operation curve. The students had to define the rotational speed of the machine as a function of the demanded flow over time. The students defined: the control rules, the operation costs, and the efficiency parameters for each pump system, trying to minimize the cost per cubic meter. The student only had two constrains: demand over time and energy cost, which was the current Spanish energy price.

Finally, when the student undertook the fluid facilities subject, the proposed work had a higher level than previous tasks developed in the Bachelor's degree. At this time, the students were more mature and the cases were near real buildings. Therefore, their training should be more intensive and closer to reality. In 2017, the proposed activity was to develop a complete project (summarize, calculus supplements, drawings and budget, defining the qualities, and normative for the different used materials). The project was related to a complex building (e.g., hotel, hospital, and school since for each students' group it is different). In this project, the students had to connect basic knowledge of fluid mechanics and hydraulic machinery with the new learning results, which are reached in the fluid facilities subject. The students designed the different pipelines and equipment, which were necessary to supply the building (e.g., cold and sanitary hot water system, pumps, and ventilation, among others). This work was developed by teams, composed of three or four students. Once the work was finished, the students had to explain it in an oral session.

In all cases, the students' doubts were attended by teachers. Generally, the questions were solved by face-to-face meetings. However, the doubts solution was also solved using mail and/or a video conference. Throughout the process, the student contacted the professor to validate the different items of the project in each one of the phases and stages.

3.2. Analysis of Results

3.2.1. Results

Figure 3a shows marks distribution in a students' group for hydraulic machines. Each indicator value can be observed for each student. The project mark was the upper 8/10 for 24 students while there were only six students who qualified below 5/10. Each indicator (from I1 to I6) is described in Table 2 and they are drawn in the Figure 3a.

Figure 3a shows the students worked really well I3 and I4 indicators. These were focused on the development of the simulations and the establishment of the control rules in the pumped systems to guarantee the hydraulic constrains (i.e., flow and pressure). In contrast, I5 was the worst developed indicator and it focused on the analysis and discussion of the results. However, the results were highly satisfactory.

Figure 3b,c shows the transposition from the mark to transversal competence in the different subjects that participated in the teaching project (Table 2). If observing the topic hydraulic machines (12659), 77% of students reached the A and B descriptors when the "Design and Project" competence was evaluated. Similar results were obtained in the rest of subjects shown in Figure 3b.

If all subjects were observed the satisfaction was higher, considering all students who participated in the teaching project. The participation in the project development was 82%, considering there were 1051 students in thirteen different matters in 2018 (1149 students were in 2017). When the "Design and project" competence was evaluated, 361 students reached an excellent degree (A). The B degree was reached by 286 students while 154 and 58 (6.75%) students obtained a C and D degree, respectively (Figure 3b).

Figure 4 shows there is a lineal relationship between exam and project marks in two years (i.e., 2016/2017 and 2017/2018). Therefore, the development of the activity helped students to acquire the hydraulic concepts as well as the methodology. Although there were no exams when PBL was applied, in order to compare the previous (traditional method using master courses) and new methodology

(CPBL), an exam was proposed. This improvement contributed to reaching the learning results favorably. This trend was observed in majority of the studied subjects. Besides, when the project delivery was after the exam, the test mark did not have a relationship between them. Therefore, there was a greater significance to establish the date delivery before the test. The final marks were compared with previous years. The score increased around the 1–2 point about 10, reducing the number of students who failed the subject (6% in 2016/2017 and 8% in 2017/2018).

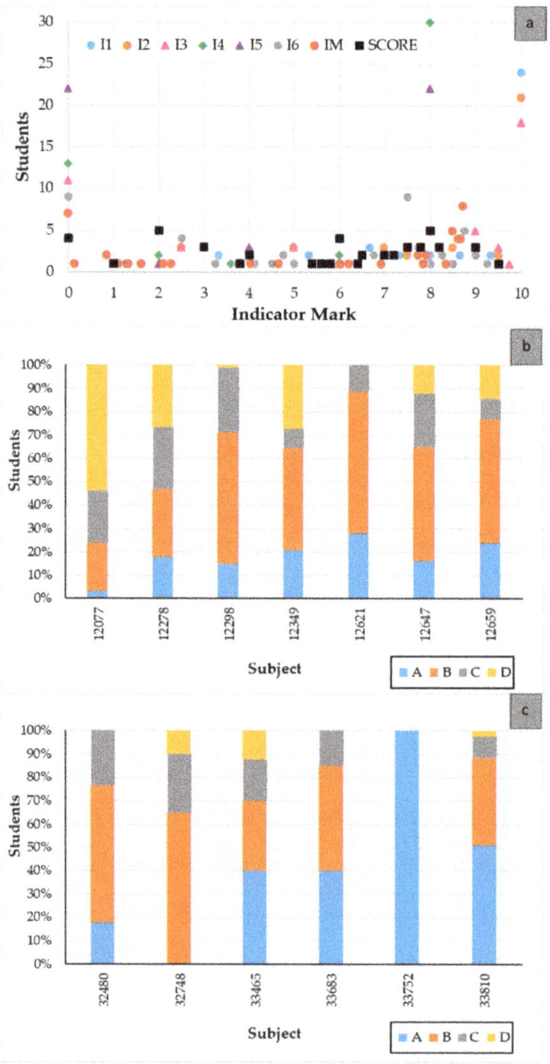

Figure 3. (**a**) Indicator in hydraulic machines related to Table 2; (**b**) results of the descriptors in the "Design and Project" competence for the subjects of the Bachelor's degree defined in Table 3; and (**c**) results of the descriptors in the "Design and Project" competence for the subjects of the Master's degree defined in Table 3.

Figure 4 shows the correlation between project and exam marks was strongly correlated when the students did not do the project correctly or they got a mark up to six. However, when the students developed an excellent project (mark between 7 and 9), they did not always get an excellent mark in their exams. It could be due to the student's collaboration and working on other skills such as analysis and resolution of problems. When they did the exam, the help between partners was not there, and therefore, they had to solve their doubts, which in some cases were not resolved correctly.

Table 2. Subjects of the teaching project.

Code	Subject	Bachelor's Degree	Master's Degree
12298	Hydraulic machines	Chemical Engineering	-
12621	Fluid Facilities in Building	Mechanical Engineering	-
12621	Fluid Mechanics	Chemical Engineering	-
12077	Fluid Mechanics	Electrical Engineering	-
12647	Fluid Mechanics	Mechanical Engineering	-
12349	Fluid Mechanics	Chemistry Engineering	-
12659	Hydraulic machines	Mechanical Engineering	-
33810	Fluid Facilities	-	Industrial Engineering
33752	Waste water treatment	-	Industrial Engineering
33465	Fluid Facilities in the chemical industry	-	Chemical Engineering
32478	Waste water networks	-	Hydraulic and Environmental Engineering
33683	Extension of Fluid Facilities	-	Industrial Engineering
32480	Analysis and modeling of water networks	-	Hydraulic and Environmental Engineering

Table 3. Questions related to planning.

ID	Question
Q1	Does the proposed activity allow you to apply the knowledge developed in theory classroom and practice lessons?
Q2	Does the temporary planning to develop the project design throughout the course allow you to start the activity well enough in advance to developing it properly?
Q3	Is the index developed by the teacher explaining the methodology and phases of the work, sufficiently clear and concise, to develop the proposed activity?
Q4	Did the project help you to acquire the knowledge, and to prepare other evaluations (e.g., tests and problems) of the subject?
Q5	Would you find it interesting that the development of the project proposed in this subject involved other subjects of your grade?

3.2.2. Surveys

Two different surveys were developed for each subject. First survey gave information related to the subject planning. This survey helped to analyze if the coordination between taught concepts and project development was correct. Related to this, five questions were proposed (Table 3). These questions were related to: (i) the application of the activity with the concepts, which were taught in the classroom (Q1); (ii) the synchronism between activity and taught concepts (Q2); (iii) if the index developed by the professor to explain the methodology was clear (Q3); (iv) if the project development helps student to reach the learning results and train to do the evaluations (Q4); and (v) if the student would be interested in development of a project considering different subjects of the degree (Q5).

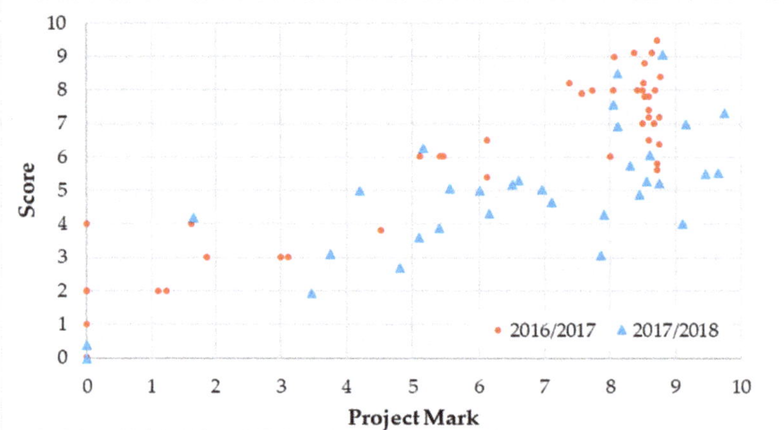

Figure 4. Relationship between the exam and project marks in hydraulic machines.

Figure 5 shows the results in the survey when it was done in hydraulic machines. The figure shows the results once 103 students (76%) answered it. There were 90% of students that positively agreed with Q1. This percentage was higher compared with other subjects in the UPV. This was a goal of the project, since it wanted to develop activities to increase the satisfaction in the students. These activities were focused on: (i) increasing the simulation lessons with software, (ii) visiting some buildings where the students can identify the studied facilities, and (iii) increasing the number of online videos in which they can visualize real solved case studies. In both years, the answer was similar between students. Therefore, they considered positive the use of this methodology to apply the teaching concepts.

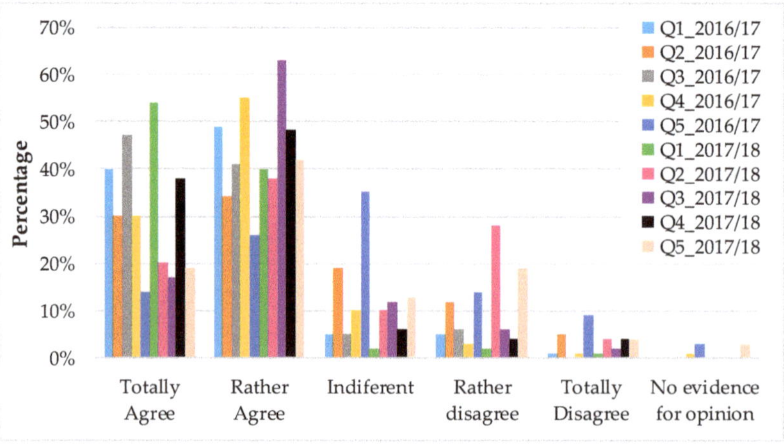

Figure 5. Survey to analyze the development of the project in the topic: hydraulic machines for the 2017 and 2018 years.

The rest of the questions were mostly approved. They showed the majority of students agreed to develop the teaching methodology. Similar results were obtained in fluid facilities (Figure 6).

The surveys analysis showed the student accepted this methodology although they had to invest more effort in the subject continuously. This learning obligated students to develop a planning to reach the objectives. In this case, results from only one year were presented, since the Master's students only undertook a course in the active methodology in 2017 (previously, they undertook a course in for their Bachelor's degree on the topic hydraulic machines using this teaching project).

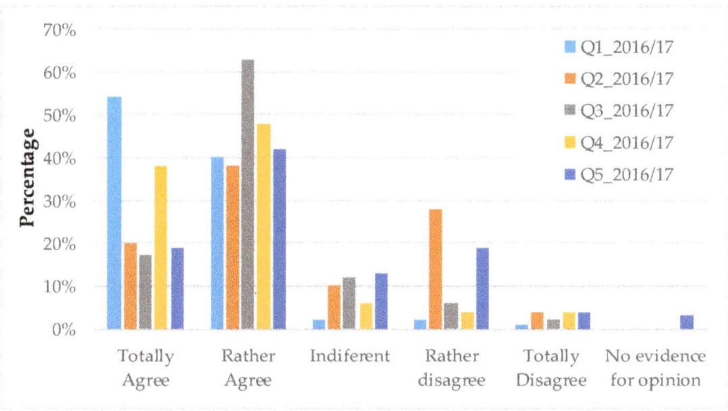

Figure 6. Survey to analyze the development of the project in fluid facilities.

Figure 7 shows the results of a survey, which had four questions. The survey was proposed to students in the second-year or third-year level (once the student studied fluid mechanics). The questions measured the vertical coordination between subjects. The questions (Table 4) were related to: (i) the developed project that helped students to improve the acquisition of competences (Q6); (ii) if the development of the project, which was developed on fluid mechanics in the previous year, helped to improve the development of the project in hydraulic machines (Q7); if the previous study of the hydraulic concepts helped students to develop the project (Q8); if the use of a similar methodology between the project developed both fluid mechanics and hydraulic machines that helped students to develop the project in the hydraulic machines topic (Q9).

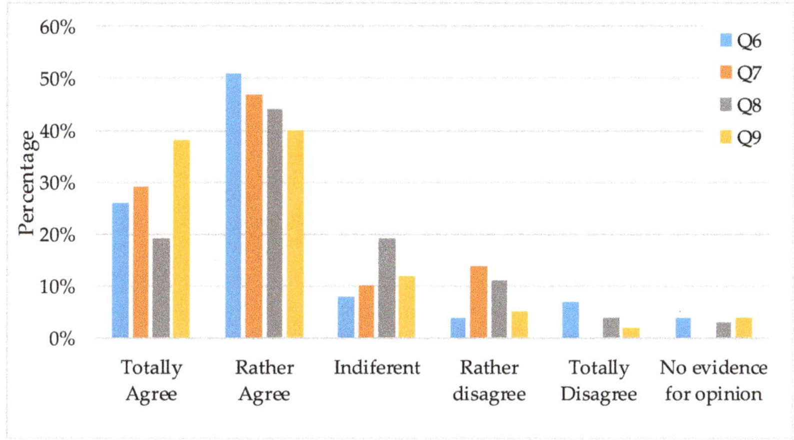

Figure 7. Survey to analyze the vertical coordination between subjects that are located on different courses and levels (Bachelor's or Master's degree).

Table 4. Questions related to coordination between years.

ID	Question
Q6	Does the development of a project that is related with studied subjects help you to improve the knowledge acquisition and competences in the 'design and project'?
Q7	Does the development of the project in fluid mechanics help you to understand and develop better the practical applications in hydraulic machines?
Q8	Does the study and understanding of common hydraulic concepts in different subjects help you to do the project?
Q9	Does the use of a similar methodology, which was used in fluid mechanics to do the project, give you autonomy to do the project in hydraulic machines?

If Figure 7 was analyzed, it shows that the majority of students (upper 60%) considered the application of this methodology positively and it had influence on achieving good results in the development to their competences.

The developed experience verified that the students improved the acquisition of the learning results in the different subjects when they were compared with the previous years. Therefore, the professors' experience joined to the students' opinion show the development of active methodologies increased the positive attitude of the students. This emotional state made the students show a greater interest in the subject, improving their efficiency. However, this effect cannot occur in some cases, in which students think they learn a lot in these scenarios, but when tested they really are not. This occurs when the students do not work in the activities continuously and correctly throughout the year.

Similar strategies are being developed currently in UPV and other universities to motivate students to develop continuous learning. Currently, the development of projects is being planned at an institutional level. The learning project includes subjects that are part of different years and are in different areas. This situation improves the integration of the subjects in the students' curricula. Besides, the students understand better the significance of the different subjects when there is a global learning project. It occurs even though the matters are studied in different years. The project existence allows students not to view the subjects individually, interrelating the different matters.

This methodology can be extrapolated to other knowledge areas or degrees, adapting the projects to the learning results of each subject. The success of this methodology is verified in other countries and universities [2,3,17]. The development of the good practices sheet [16] and the definition of the learning goals allow one to organize the active methodology for any subject.

4. Conclusions

A case study was described in this research, which joined different hydraulic engineering topics. The subjects were taught using continuous project-based learning. The implementation of this methodology was new at UPV to develop the skills competences in the students. The development of the methodology from basic subjects (i.e., fluid mechanics) enabled one to define the procedure, which can be applied on subjects at the upper level. The methodology allowed professors to establish a schedule in which face-to-face time and a non-face-to-face lesson fit perfectly. Therefore, the use of a good practice sheet allowed students and professors to know their activities for each time. The use of these sheets improved the synchronization of the teaching (i.e., theoretical concepts, practices lessons, and activities) between them. Besides, the development of the good practice sheet helped professors to organize subject learning. The good practice sheet contained different tasks, which should be carried out by professors and students, defining the data and time of their development.

The proposed methodology is crucial to give students an action strategy when they have to develop similar projects in matters of the hydraulic area. The strategy improves the vertical coordination in the Bachelor's or Master's degree. This organization maximized the reach of the learning results since it mixed a face-to-face class and online videos and material as well as real projects to apply the taught concepts according to the students' capacity.

The methodology included a rubric for each subject. These evaluation criteria enabled us to evaluate the acquisition of the learning results, joining both specific and transversal competence of the 'design and project'. The rubric, which was used in hydraulic machines was shown in this manuscript (Appendix A). It defined the specific indicators and the descriptors, which are necessary to develop the project. Besides, the used expression, which correlated the specific and transversal competences in the students' curricula, was presented.

Two surveys were proposed to students. These questions showed the students' satisfaction for the structure of the activity. Besides, they considered it necessary to improve the acquisition of the learning results. The students were grateful of the use of this methodology in other subjects related to hydraulic engineering topics.

Finally, the new challenge in teaching should be focused on:

- Professors need to establish active methodologies in which the students are involved, improving their learning results.
- The students' training should be coordinated in order to align the specific competences and outcomes competences as well as the sustainable development goals.
- Communication technologies (ICTs) joined to use software are tools, which must be incorporated in the teaching guides to improve the learning results and, therefore, the students' curricula.

Author Contributions: The author P.A.L.-J. wrote the introduction and she analyzed the results of both tests and rubrics. The author M.P.-S. proposed the methodology and he contributed with the analysis of the results. All authors have read and agreed to the published version of the manuscript.

Funding: This research received no external funding.

Acknowledgments: The described experience has been carried out as part of the work in the Innovation and Quality Education Teaching (EICE DESMAHIA) "Development of active methodologies and evaluation strategies applied to the field of Hydraulic Engineering" in the Universitat Politècnica de València.

Conflicts of Interest: The authors declare no conflict of interest.

Appendix A

Table A1. Specific indicators used to evaluate hydraulic machines.

Indicators (What Is the Analyzed Point?)	Descriptors			
	D. Not Achieved	C. In Development	B. Good	A. Excellent
1. The students bases the context and the need of the project	The student explains the need of the project but he/she does not justify	The student justifies the need of the project, using opinions that are not checked enough	The student justifies the need of the project correctly but it is incomplete	The student justifies the need of the project correctly and completely
Introduction and justification	There is not a definition of the goals	The student introduces the project to do but he/she doesn't justify its need or he/she does it incorrectly	The student introduces the projects but he/she does not justify the need	The student introduces the projects and he/she justifies the need
2. The student formulates the objectives of the project coherently with regard to the needs detected in the context	The student formulates the goals without considering the needs	The student formulates the goals but they are not coherent with the needs	The student formulates the goals and they are coherent with the needs	The student formulates the goals and they are coherent with the needs and these goals are operational
Goals	There is not a definition of the objectives	The student establishes the goals but these are ambiguous.	The student defines the objectives sufficiently	The defined goals are clear and operational
3. The student plans the action to be developed effectively	The student does not develop the justification of the action	The students develop the plan partially to reach the goal	The students develop the plan to reach the goal in their major points	The students develop the plan to reach the goal completely

Table A1. *Cont.*

Indicators (What Is the Analyzed Point?)	Descriptors			
	D. Not Achieved	C. In Development	B. Good	A. Excellent
For each section of the project	There is not a plan	The student does a short description or justification	The student describes and justifies the development only considering an academic point of view	The student describes and justifies the development, considering both the academic and technical point of view
4. The student plans the actions efficiently	He/she does not plan efficient actions	He/she plans efficient actions, although they are improvable	All actions are not efficient	He/she plans efficient actions completely
Setpoint curve	It is not calculated	It is calculated incorrectly	The result is correct but there is no discussion about this	The result is correct and there is an analysis of the result
Reservoir capacity	It is not calculated	It is calculated incorrectly	The result is correct but there is no discussion about this	The volume is correct and there is an analysis of the result
Pump selection	It is not developed	It is developed incorrectly	The result is correct but there is no discussion about this	The selection is correct and the student proposes alternatives (other types and manufacturers)
Pump selection when the network is pumped directly. Considering non-variable rotational speed	The new selection is not developed according to the setpoint curve	It is developed but it is incorrect	The developed selection is correct but it is not justified	The selection is correct. Besides, the student develops a justification and comparison, considering other solutions
Economic analysis when the rotational speed is fixed	The student does not develop the daily analysis	The student does the analysis but it is incorrect	The student develops an analysis correctly but the analysis is not justified	The student does a detailed analysis, developing indicators and comparing with others facilities
Pump selection considering variable rotational speed	The pump selection is not developed according to the new setpoint curve	It is developed incorrectly	The developed selection is correct but it is not justified	The selection is correct. Besides, the student develops a justification and comparison, considering other solutions
Economic analysis when the rotational speed is variable	The student does not develop the daily analysis	The student does the analysis but it is incorrect	The student develops an analysis correctly but the analysis is not justified	The student does a detailed analysis, developing indicators and comparing the values when the rotational speed is fixed
5. The student identifies the risks and inconvenience of the project	The student enumerates some risks but they are not analyzed	The student enumerates some risks but they are not analyzed deeply	The student enumerates some risks but they are analyzed but he/she defines constrains to solve the problems	The student enumerates risks, they are analyzed and solved for improving the project
Conclusion section	There is no conclusion	There is a conclusion, but the student does not discuss the results	Different results are discussed and compared	Results are compared, establishing the advantages and inconvenience for each solution
Language, format, and writing of the project	The presentation is poor, the writing and language style are not at a high enough level according to their academic status	The presentation is correct although the language is not at a high enough level since the student uses no technical words	Presentation, language, and writing are correct but the content exceeds the limit	Presentation, language, and writing are correct and the project is adjusted to the requirements established by the professors
6. Review the results	The student does not review the results	The student reviews the results but the review is not structured	The student plans the result evaluation (i.e., who, when, and how)	The student plans the result evaluation (i.e., who, when, and how), using indicators
Review the results of the facilities using EPANET	There is no evaluation	All results are not checked	All results are checked without doing comparisons	All results are checked. The student develops comparisons between a classmate or comparing values that are obtained from the bibliography

References

1. Chanson, H. Teaching hydraulic design in an australian undergraduate civil engineering curriculum. *J. Hydraul. Eng.* **2001**, *127*. [CrossRef]
2. Hotchkiss, R.H. Flow over a "Killer" Weir Design Project. *J. Hydraul. Eng.* **2001**, *127*. [CrossRef]
3. Novak, E.; Valentine, P. Teaching of hydraulic design at university of Newcastle upon Tyne. *J. Hydraul. Eng.* **2001**, *127*. [CrossRef]
4. Kelley, C.A.; Conant, J.S.; Smart, D.T. Master teaching revisited pursuing excellence from the students' perspective. *J. Mark. Educ.* **1991**, *13*, 1–10. [CrossRef]

5.   Pierce, R.; Fox, J. Vodcasts and active-learning exercises in a "flipped classroom" model of a renal pharmacotherapy module. *Am. J. Pharm. Educ.* **2012**, *76*, 196. [CrossRef] [PubMed]
6.   López Jiménez, P.A.; Andrés Doménech, I.; Pérez-Sánchez, M. Implementando metodologías de evaluación en proyectos de redes de saneamiento en el Máster Universitario en Ingeniería Hidráulica y Medio Ambiente. Caso de estudio. In Proceedings of the IN-RED 2017: III Congreso Nacional de Innovación Educativa y Docencia en Red, Valencia, Spain, 13–14 July 2017; pp. 1243–1255.
7.   Pérez-Sánchez, M.; Galstyan-Sargsyan, R.; López-Jiménez, P.A. The improvement of learning results in fluid mechanics topics through the transversal competence autonomous learning. In *3rd International Joint Conference Icieom-Adingor-Iise-Aim-Asem (IJC2017) Proceedings*; Poler Escoto, R., Mula Bru, J., Díaz-Madroñero, B., Francisco, M., Sanchis Gisbert, R., Eds.; Universitat Politècnica de València: Valencia, Spain, 2017; pp. 1–6.
8.   Savage, R.N.; Chen, K.C.; Vanasupa, L. Integrating project-based learning throughout the undergraduate engineering curriculum. *J. STEM Educ.* **2007**, *8*, 15–28. [CrossRef]
9.   UPV. Universitat Politècnica de València. Institutional Project of the Generic Outcomes. 2015. Available online: https://www.upv.es/entidades/ICE/info/Proyecto_Institucional_CT.pdf (accessed on 5 March 2020).
10.   Blank, W.E.; Harwell, S. Authentic instruction. In *Promising Practices for Connecting High School to the Real World*; University of South Florida: Tampa, FL, USA, 1997; pp. 15–21.
11.   Hadim, H.A.; Esche, S.K. Enhancing the engineering curriculum through project-based learning. *Front. Educ.* **2002**, *2*, 1–6.
12.   Martínez Martínez, A.; Cegarra Navarro, J.G.; Rubio Sánchez, J.A. Aprendizaje basado en competencias: Una propuesta para la autoevaluación del docente. *Profr. Rev. Curric. Y Form. Del Profr.* **2012**, *16*, 374–386.
13.   Miralles Martínez, P.; Guerrero Romera, C. Evaluación de la Competencia Transversal "CT-05 DISEÑO Y PROYECTO". Caso Estudio en Grado. In *Metodologías Docentes Innovadora en la Enseñanza Universitaria*; Universidad de Murcia: Murcia, Spain, 2018; pp. 289–300.
14.   Alptekin, S.E.; Deturris, D.; Macy, D.J. Development of a flying eye: A project-based learning experience. *J. Manuf. Syst.* **2005**, *24*, 226–236. [CrossRef]
15.   Rossman, L.A. *EPANET 2 Users Manual*; 2000epa/600/r-00/057; U.S. Environmental Protection Agency: Washington, DC, USA, 2000.
16.   Pérez-Sánchez, M.; Fuertes-Miquel, V.S.; Soriano-Olivares, J.; Gomez, E.; Gómez-Selles, P.A. Rubrics as a tool for evaluating hydraulic engineering projects in both bachelor's and master's degree. In *New Global Perspectives on Industrial Engineering and Management*; Springer International Publishing AG: Cham, Switzerland, 2019; pp. 343–350.
17.   Bell, S. Project-based learning for the 21st century: Skills for the future. *Clear. House* **2010**, *83*, 39–43. [CrossRef]

 *fluids*

Essay

Teaching and Learning Floating and Sinking: Didactic Transformation in a Density-Based Approach

Anastasios Zoupidis [1,*], Anna Spyrtou [2], Dimitrios Pnevmatikos [2] and Petros Kariotoglou [3]

[1] Department of Primary Level Education, Democritus University of Thrace, 68131 Alexandroupolis, Greece
[2] Department of Primary Education, University of Western Macedonia, 53100 Florina, Greece; aspirtou@uowm.gr (A.S.); dpnevmat@uowm.gr (D.P.)
[3] Department of Early Childhood Education, University of Western Macedonia, 53100 Florina, Greece; pkariotog@uowm.gr
* Correspondence: azoupidis@eled.duth.gr

Abstract: This essay synthesizes more than a decade of research, most of which has been published, on the teaching and learning of floating and sinking (FS) phenomena. The research is comprised of the iterative design, development, implementation and evaluation of a Teaching-Learning sequence (TLS) for the teaching and learning of density within FS phenomena. It was initiated within the frame of the European Community supported "Materials Science" project. Due to the many, different aspects of the project, each publication has focused on a particular part of the study (e.g., effectiveness and the iteration process). The didactic transformation for the teaching of FS phenomena is presented and discussed here. In doing so, it is essential to mention: (a) the students' ideas as the main cause of the scientific knowledge transformation, (b) the scientific/reference knowledge, and (c) the knowledge to be taught and its limitations. Thus, we intend to describe and justify the didactic transformation process and briefly synthesize the published (from previous papers) and unpublished results to show its effectiveness.

Keywords: inquiry-based instruction; science education; teaching-learning sequences; didactic transformation; primary level

1. Introduction

School children are familiar with floating and sinking (FS) [1], which is a main topic in the teaching of fluids in science education [2,3], especially at the primary and lower-secondary levels (10- to 15-year-olds). Although the topic is very common, and children have many everyday life experiences in FS phenomena, their interpretation is challenging, not only because of the difficulty of the scientific concepts and the respective explanatory models that are involved (e.g., density, buoyancy), but also because of these everyday experiences that students have and their subsequent ideas [4].

Research on FS has been extensive in the last few decades, both regarding students' ideas [5] and, consequently, about ways to effectively teach this topic [6]. Concurrently, Teaching-learning sequences (TLSs), i.e., medium-level curriculum unit packages, that include well-researched teaching-learning activities empirically adapted to student reasoning [7,8], are increasingly present in science education research, because they provide the opportunity to integrate teaching and learning theories and approaches, students' ideas about science concepts and explanations of natural phenomena, as well as the historical development of scientific concepts [7–10].

One of the most critical issues in the design and development of a TLS is the didactic transformation of the content, i.e., transforming the scientific knowledge into appropriate knowledge to be taught [7,11,12]. The choice of content and how it is transformed in order for it to be easily understood and readily adopted by students is crucial in the entire process of TLS development. Although this often takes place, none or very little of it is usually conveyed. In other words, even though the didactic transformation process is an essential aspect in every teaching effort, and especially in the design and development of a TLS, researchers rarely describe the process in an explicit and detailed manner, possibly due to space restrictions [8,13]. Thus, colleagues who wish to further investigate any such didactic proposal's effectiveness do not have all the necessary information to repeat its implementation.

To describe the didactic transformation process of certain content, in our case, FS phenomena, it is important to mention, among other things: (a) students' ideas about the phenomenon and the concepts related to it, as the main reason for the scientific knowledge transformation, (b) the scientific/reference knowledge, and (c) the knowledge to be taught, in its new form, following scientific knowledge transformation, including its limitations.

Students' alternative ideas have played a decisive role in the planning of teaching in science education in the last few decades [6]. Consequently, students' ideas about FS phenomena, and the difficulties they face in adapting interpretations to be consistent with the scientific ones, need to be taken into account in every teaching effort that is developed within the frame of the prevalent constructivist approach [14]. Moreover, the study of the historical development of scientific knowledge concerning FS interpretations could contribute to the didactic transformation process by revealing the difficulties scientists had come up against in understanding and interpreting those phenomena throughout the centuries [14–16].

The TLS entitled "Density of materials in floating and sinking phenomena: Experimental procedures and modelling", which was initiated in the framework of the European Community supported "Materials Science" project (FP6, SAS6-CT-2006-042942), has been described elsewhere [17–22]. However, because there were many different aspects of content to be taught, i.e., declarative (density and floating sinking), procedural (control of variables strategy), and epistemological knowledge (nature and role of models), the focus of the previous published papers has been other than describing and justifying the didactic transformation, which we hope to do here.

In this paper, we briefly describe elements of our developmental research that have already been published, focusing, however, on the didactic transformation of content, as this has not yet been thoroughly presented or discussed and which we consider to be of paramount importance. Therefore, our aim is to describe the didactic transformation of the content of the TLS concerning FS phenomena, to underline the factors which influenced the process of its development, and to present the limitations of the transformation. Specifically, we justify the reasons why we chose the density-based explanatory model for FS phenomena, rather than the buoyancy-based (see Section 3), as well as providing arguments for the didactic transformation of the concept of density, which is still an open issue for an effective approach to FS learning. Furthermore, selected essential aspects of the revised version of the TLS, which was adapted from an initial study, are also described [19]. Moreover, a short presentation of both our published (from previous papers) [20,21], and unpublished results in FS [22], on the implementation of the revised TLS in a real-class environment is given. In this sense, we consider that this work is an original sample of a developmental research description in the framework of Design-Based Research approach in science education [8,23,24].

2. Alternative Ideas and Difficulties in Explaining FS Phenomena

Students seem to perceive FS phenomena visually. That is to say that they decide whether an object is in a floating or sinking state based on the object's position relative to the surface of the liquid [1]. For instance, the majority of students in Joung's study [1] answered that an object was floating in the water when at least a part of it was above the surface, most of whom chose the case where the object touches the surface of the liquid and fewer chose the case where the object was half-submerged. Also, the majority of students considered that an object had sunk in the water, in the cases where it was below the surface, i.e., (a) at the bottom, or (b) in the middle (between the bottom and the surface of the liquid), with a decreasing frequency of occurrence, respectively. In the same research, students considered an object just below the liquid surface either to be floating or to have sunk. Few students recognized that when the object was between the surface of the liquid and the bottom, then it is in a state between floating and sinking, and therefore, was suspended and remaining at rest in the liquid at the same location [1,25].

In addition, it seems that students explain and describe the phenomena in relation to perception-based macroscopic natural properties, such as weight, length, and volume [5,26–28]. In other words, students formulate their estimation concerning the floating of solid objects in a liquid by taking into account: (a) the heaviness/size of the objects, (b) the existence of hollows, (c) the existence of holes, (d) the interface/edge, orientation, shape and/or texture of the floating object, (e) the dimensions of the tanks in which floating takes place, (f) the amount and/or depth of the liquid, and (g) the liquid stickiness [29]. Needless to say, one of the most prevalent alternative ideas that students hold is that of case (a), that is, students most often claim that an object floats because it is small and/or light, and it sinks because it is big and/or heavy [26,30].

Consequently, when interpreting FS phenomena, students tend to focus on the properties of the objects or the liquids. Additionally, they seem to merely use causal linear reasoning, referring only to an object's or a liquid's property, instead of causal relational reasoning, which involves comparing object and liquid densities in their interpretations [31]. However, this is not the only obstacle in students using the specific causal relational reasoning to explain FS phenomena. Researchers who have studied students' conceptions of density [26,32,33] have found that they had difficulty in understanding this abstract concept. Firstly, students find it hard to understand the ratio of two quantities [34], such as that of mass per volume, particularly when those quantities are changing simultaneously [35]. Secondly, the concept of density is a property that is not directly perceived through the senses but can only be understood through mental reasoning and/or calculations [33,36]. Thirdly, students' difficulty in understanding density is rooted precisely in an already developed conceptual framework about matter and material kind [37], which is composed of perception-based physical quantities where the raw scientific notions of weight, volume and density coexist undifferentiated [33]. Consequently, these students consider density to be proportional to the size of an object or the object's quantity of matter.

To fully understand the reasons why an object floats or sinks, one needs to comprehend that the concept of water pressure is an intensive property, while the concept of buoyancy is a force, and not, as is usually the case with students, a property of an object, within the framework of Newtonian mechanics [4,5]. However, students very often confuse the FS states with the explanatory model; that is, they equate buoyancy, a construct/force in an explanatory model, with the state of floating. They also seem to think that buoyancy is a property of an object opposed to the interaction between an object and its surrounding fluid, as they are unable to understand buoyancy as a force, i.e., the interaction between two entities. Furthermore, many students have the misconception that the buoyancy of an object is inversely proportional to its density, while others are not sure about the direction of buoyancy [4].

In sum, students confront severe difficulties in interpreting FS phenomena in the framework of both density-based and buoyancy-based explanatory models. The reasons for this difficulty, however, are not the same in both cases. In the former, the difficulty is mainly due to the non-differentiation of the concepts of weight, mass, and density, in contrast to the latter case, where it is mainly due to the students' inability to understand the concept of force as the interaction between two entities.

3. Floating and Sinking Teaching Approaches

The way educators approach the teaching of FS phenomena can be put into two broad categories, according to the central concept of the explanation of the phenomena (Figure 1). In the first category are those cases that provide density-based explanations, e.g., [29], following the so-called elimination of variables approach. This approach focuses on highlighting the variables that affect the FS phenomena in order to derive a prediction rule that will determine "which" body will float. In the second category are those cases that provide buoyancy-based explanations, e.g., [38], following the so-called scientific approach, that is, an interpretation using an equilibrium mechanism in order to explain "how" an object floats [15]. Several researchers provide both explanations concurrently [4,32].

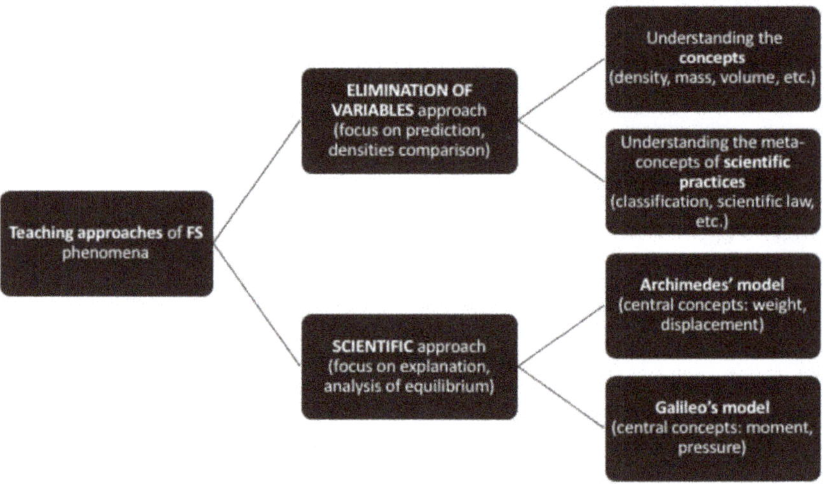

Figure 1. Categories of teaching approaches to floating and sinking phenomena; recreated by the authors according to [15].

The scientific approach is rooted in Archimedes' and Galileo's explanatory models [15,39]. In brief, Archimedes explained the floating of a solid in an infinite container, using only the concept of weight as a quality/property of objects [40] and comparing the weight of the solid to the weight of the fluid that was displaced by the immersed portion of the solid. He proposed that the weight of the fluid displaced would be equal to the weight of the solid. However, Archimedes' model was effective only for FS phenomena in infinite containers [40]. On the other hand, Galileo overcame Archimedes' inadequacy by explaining floating of a solid in a finite container. Galileo's model comprised important discriminations of new-defined concepts [15,40], such as the distinction of floating and surface tension phenomena [41] and the differentiation of the concepts of weight and specific weight in a qualitative manner [15]. Both models derive their floatation laws from an analysis of the conditions that result in equilibrium. However, Galileo's model consists of more delicate and abstract concepts and, subsequently, can explain a larger range of phenomena [15,39]. On the other hand, the elimination of variables approach is rooted in Inhelder and Piaget's work [42]. These researchers were the first to record children's explanations on the floating phenomenon, focusing mainly on the ability of children to (a) classify a set of objects according to whether they float or sink in the water, and (b) explain the criterion by which the classification was made [15]. Moreover, they were interested in testing children's ability to eliminate inconsistencies in their initial explanations, such as using the weight of objects in order to interpret FS phenomena, and at the same time, to formulate the predictive floating law; that is, objects float if their density is less than that of water, where density is defined as the ratio of weight to volume. Thus, contrary to the scientific approach, which focuses on the construction of an interpretive model that will explain "how" a body floats, the elimination of variables approach focuses on the construction of a predictive model for FS phenomena, with the latter being less complicated, and subsequently, an easier process for students [15].

Usually, the density-based approach is adopted by primary schools and junior high schools [6,15], while senior high schools and colleges/universities adopt the buoyancy-based approach [15] or a combination of both [4]. When one develops a curriculum for this topic, the chosen approach, whether density- or buoyancy-based, would also involve selecting different steps, practices and concepts in the teaching and learning process. Therefore, in the case of the elimination of variables approach being chosen, there would need to be an intermediate goal of the learning process, which would involve differentiating weight, volume, and density as this approach requires a clear understanding of these the concepts (Figure 1). Another intermediate goal would be explaining FS phenomena by using a partially correct explanatory model, that is, that an object's density as a property of materials determines its sinking or floating.

A correct explanatory model within the frame of the elimination of variables approach would be to use the concept of density in causal relational reasoning, that is, comparing the densities of an object and a liquid in order to come to a decision about the FS situation [31]. Contrastingly, in those curricula where FS is taught through an analysis of the equilibrium approach, it is not needed for students to discover or to be introduced to any intermediate/precursor concepts [29]. For example, sinking and floating can be explained as a result of the balance between gravity and buoyancy. In that case, the forces of buoyancy and gravity are the required concepts for the final learning goal. However, both buoyancy and gravity are scientific—rather than intermediate/precursor—concepts, which makes their understanding more difficult. Although the latter explanatory model is more potent than the first one, because it is capable of explaining more cases of natural phenomena, e.g., motion involved in FS phenomena, concurrently, it is more complex, and hence, more difficult to understand, especially for younger students [4]. This is one of the reasons why primary and junior high schools adopt the density-based approach, in other words, the elimination of variables approach. In this case, teachers are faced with the difficult choice of introducing density using one of the following three ways: (a) the mathematical ratio of mass per volume, (b) the particle theory of matter, or (c) a visual representation that emphasizes the qualitative aspect of density [15]. However, the first case, i.e., a mathematical introduction of density, has been proven to be ineffective, due to the fact that students find it hard to differentiate between the concepts of mass, volume and density, or understand that density is an intensive quantity [15,34]. In addition, the second case, i.e., using the particle theory of matter, would most probably create misconceptions or reinforce students' prior ideas, such as the non-differentiation of density and denseness [43]. It has been shown that introducing density through this approach is rather abstract, which makes it very difficult for 9- to 12-year-old students to comprehend [44]. It would, thus, appear that the last case is the most appropriate because, in contrast to the first two, the use of a visual representation also provides opportunities for differentiating the focal concepts [26].

In the last few decades, teaching approaches have been developed in the framework of inquiry, emphasizing both the content of science and scientific practices [29,45,46]. As can be seen in Figure 1, there is a need for the emphasis to be on the understanding of the meta-concepts of scientific practices, e.g., the process of developing evidence-based conclusions through an experimental procedure, which, from an early stage, had been the focus of the elimination of variables approach. By this comment, we do not claim that inquiry-based teaching is not suitable or feasible to be implemented in the scientific approach of the analysis of equilibrium for FS phenomena. Rather, in order for the results of the eliminating of variables, experimental procedures to be understood and adopted by students, they need to be aware of the reasoning behind the scientific practices involved [15,46]. We claim that the TLS described in the next section is an example of effective density-based implementation, concurrently aiming at declarative knowledge (FS and density), procedural knowledge (control of variables strategy), and epistemological knowledge (nature of models) [18].

4. Description of the TLS: Emphasis on the Content of FS and Its Didactic Transformation

4.1. The Design Principles of the TLS

A case example of the elimination of variables approach for teaching and learning FS phenomena is presented with the five-unit TLS entitled "Density of materials in floating and sinking phenomena: Experimental procedures and modelling". The design principles of the TLS have been discussed in detail in former publications, e.g., [19]. For the sake of clarity, a brief reference to the six design principles of the TLS is made here, focusing on those regarding FS didactic transformation and relevant activities. More specifically, the design principles of the TLS were as follows [19]:

1. The didactic transformation of content [11] concerning FS phenomena, that is, firstly, the decision to adopt a density-based approach to negotiate FS phenomena, and, secondly, the decision to introduce the concept of density in a qualitative way. Both decisions were driven by the difficulties that primary school students confront when prompted to negotiate FS phenomena.

2. The participatory design and developmental character [47] of the TLS includes teachers, together with researchers, in its designing, developing and evaluating processes. Teachers discussed the nature of the TLS activities with the researchers, how they understood the activities, the possible difficulties that students might face, and consequently the possible changes and/or specific teaching methods that would be suggested. Thus, the research was adapted to the particular needs of the school and the students.

3. The TLS's iterative process [8], which provided the opportunity to both researchers and teachers to evaluate together the initial implementation of the TLS and to propose improvement modifications [19]. Thus, the iterative evolution of the TLS contributed to the final version of the TLS and its didactic transformation.

4. The technological problem scenario, to provide a supportive context for learning [48]. The scenario was based on salvaging the Sea Diamond shipwreck. Discussion is initiated with an everyday problem (in our case, technological), which poses questions. These are answered through scientific knowledge that is eventually applied to solve the initial real-life problem. The combination of technological and scientific knowledge in teaching promotes active learning, improves students' performance and attitudes towards science, enhances positive interaction between teachers and students, and provides students with opportunities to participate in authentic exploratory processes, which are usually carried out by scientists [49,50].

5. The Inquiry-Based Science Education (IBSE) approach, by emphasizing the need to use scientific practices: (a) as a means of teaching and learning, e.g., investigating the variables that possibly influence the FS phenomena in groups, and (b) as an educational end, whose aim is to understand specific aspects of scientific practices [46,51,52]. Within this framework, learning is perceived as active and student-centered, due to pupils' increased interest and autonomy [53], and the intention is for them to gain ample practice in scientific reading and writing [54].

6. The use of digital tools, such as a simulation that was developed from scratch for the TLS [17]. Looking at existing educational software on FS phenomena, the proposals were inappropriate for our task, mainly because it was difficult to implement the inquiry-based activities and also because they tended to have a mathematical approach to the introduction of the concept of density. Therefore, a specially designed software package that followed the design principles of the TLS was designed from scratch. Furthermore, a simulated website on a local network was designed and developed for students to investigate information about materials, with the aim of becoming accustomed to scientific reading and writing skills.

4.2. The Didactic Transformation of the TLS

FS phenomena are not included in the Greek primary school curriculum. However, the concept of density is introduced in fifth grade (10–11-year-olds) as a property of materials, with a limited number of examples, including one task that negotiates the sinking of a real ship. Although the curriculum proposes a guided discovery approach for negotiating phenomena and concepts, and one of the aims explicitly referred to is for students to understand the scientific method, the majority of teachers implement traditional deductive teaching-learning practices, followed by experimental demonstrations, whereas group work is sporadic [19].

As mentioned, the TLS was developed within the "Materials Science" project and is proposed as part of a broader curriculum for primary and lower-secondary level students (10- to 15-year-olds). The general objective of this broader curriculum is to restructure students' conceptual framework as regards the concepts of matter and material kind [33], including fluids. The elimination of variables rather than an analysis of equilibrium explanatory model for FS phenomena was adopted for reasons to do both with students' difficulties and the project's characteristics. Students' difficulty in understanding and effectively implementing an equilibrium mechanism, such as the buoyancy-based model to explain FS phenomena, has already been documented [4,15] and is analytically discussed in Section 3.

Furthermore, the emphasis of the project, which, on the one hand, is on the properties of materials properties, and the other, on inquiry-based teaching and learning, appears to be more compatible with the elimination of variables approach, that is, the density-based approach for teaching FS phenomena [15] (see Section 3). By highlighting the variables of these phenomena in order to derive a prediction rule that will determine "which" body will float, instead of negotiating the forces that are acting on an object when it is immersed in a liquid, we believe is easier for students in this age range to understand, and thus more conducive to the teaching/learning process.

It was decided to introduce the concept of density through the visual "dots-in-a-box" representation (Figure 2) as a property of materials [26]. As students would have already investigated the variables that affect the FS phenomenon, this visual representation shows several variables in only the one diagram. Obviously, the "dots-in-a-box" representation depicts the weight of the object by the number of dots, and the cube represents its volume or size, while its kind of material is now assigned to the conjunction of the weight and volume and not to any realistic representation of its external appearance. This enables students to easily make comparisons of the densities of the different materials in order to predict and explain FS phenomena, and if possible, to grasp the "heavy for its size" intermediate/precursor concept of density. In this way, the usual introduction of density using the mathematical ratio mass to volume, which has been shown to be difficult for students of these ages to grasp such relationships, has been bypassed [34].

Figure 2. The visual "dots-in-a-box" representation of the density of several materials, reprinted by permission from Springer Nature Customer Service Centre GmbH: Springer, Iterative design of Teaching-Learning Sequences, by D. Psillos and P. Kariotoglou, 2016 and by permission from [20], http://earthlab.uoi.gr/tel/index.php/themeselearn (accessed on 20 March 2021).

With a view to preventing or eliminating any misconceptions, such as "all hollow objects float" or "objects with air always float", as reported in the study by Yin et al. [29], we also thought it best to first introduce students to homogenous objects in the variable of solids. In the TLS, density is introduced within the context of floating or sinking of various everyday objects. First, students are introduced to homogeneous and then to composite objects, such as an iron cube and a ship, respectively, using causal relational reasoning, in other words, by comparing the density of the object and liquid, one is able to interpret and/or predict the FS of each object [31].

In sum, the didactic transformation of the TLS consists of two core choices (a) selecting the density-based approach to teach FS, and (b) qualitatively introducing the concept of density through the "dots-in-a-box" representation. Reference is made here to the other two subsidiary but significant aspects of the didactic transformation: (a) integrating scientific and technological knowledge into a context-based approach (fourth design principle), and (b) incorporating guided discovery experimentation into the explicit teaching of procedural and epistemological knowledge within an inquiry-based approach (fifth design principle). The teaching/learning environment for the experimentation activities was initially structured, which was gradually decreased with the aim of enabling students to become more autonomously involved [19,53].

More specifically, FS phenomena are introduced in the fourth design principle through a technological-problem scenario based on salvaging the Sea Diamond shipwreck [19]. The scenario, which runs throughout the entire TLS, has a dual role, on the one hand, it forms the familiarization phase, and on the other, it involves the following scheme: "technological problem", "scientific investigation", and "return to the problem" with the aim of increasing students' curiosity and motivation leading to the solution of the problem [49,50]. In contrast to the traditional approach, which proposes only the scientific investigation of the phenomena and the related concepts, our context-based approach through the technological-problem scenario increases students' interest and succeeds in involving them in the entire teaching/learning process [48].

The content of the TLS within the fifth design principle, which is the IBSE approach, includes elements of the inquiry method, i.e., aspects of the control of variables method as well as the nature and role of the models [18,19]. In other words, students are explicitly taught that to test if a variable influences a phenomenon, e.g., FS, then only this variable should differ, and all the other independent variables should be controlled [55]. In addition, students are explicitly taught aspects of nature and the role of scientific models, for instance, that models are not an exact representation of reality and that they are used to describe, predict, or explain a phenomenon [19,56]. Our hypothesis that procedural and epistemological knowledge would positively affect the understanding/interpretation of FS phenomena has been confirmed [18]. Thus, our claim that inquiry as a teaching goal constitutes part of the TLS's didactic transformation has been reinforced.

Summing up, the didactic transformation of the TLS described in this paper is based on (1) the elimination of variables approach to teaching FS phenomena, and (2) the qualitative introduction of the concept of density using the "dots-in-a-box" visual representation [26]. The explanatory model proposed for students to use in order to predict and explain FS phenomena for both solid and composite objects is based on the "dots-in-a-box" representation, in conjunction with the causal relational FS rule, which is, if an object's density is smaller than the liquid's density, then the object will float, and if an object's density is greater than the liquid's density, then the object will sink. We maintain that it is easier for young students (primary and junior high school) to grasp a qualitative representation of density, as a property of materials, rather than the scientific knowledge of the specific content, which is traditionally presented as a mathematical ratio of mass per volume, or even weight per volume [26,27]. Understanding the concept of density as "heavy for its size", thus perceiving density to be related simultaneously to both weight and volume, was an implicit teaching goal of the TLS. In this sense, we claim that by using the "dots-in-a-box" visual representation in their explanations for FS phenomena, as a property of materials and not of objects, students can differentiate the concepts of density and weight and consequently come closer to an intensive perception of the concept of density.

Every didactic transformation of content in science education is characterized by limitations related to (a) the kind of explanation of the focal phenomena and/or (b) the range of the phenomena explained when the model that is related to the transformed content is being used [12]. Therefore, the explanatory model of FS presented here (visual representation of density in conjunction with causal relational FS rule) has some limitations in comparison to other more abstract explanatory models (e.g., the analysis of equilibrium model for the explanation of FS and/or mathematical ratio for the representation of density). These are:

1. Only static FS phenomena can be interpreted: It was decided not to study buoyant force as an alternative explanatory model of FS phenomena because we considered the analysis of equilibrium approach between buoyant and gravity forces, which is implied in the buoyant force model (see Section 3) is particularly difficult for primary school and some junior high school students. Additionally, the concept of displaced liquid, which is not necessary for the density-based model, was omitted as it might distract students from the expected learning outcomes of the TLS. This means that several aspects of FS phenomena regarding objects in fluids that are not at rest but in motion cannot be explained. For instance, the motion of an object that is initially sunk in the water and then is released when the density of the object is smaller than that of the water can only be explained if an analysis of equilibrium model is used.

2. Only a qualitative estimation of material density can be determined: We are aware that the qualitative "dots-in-a-box" representation does not accurately match the actual value of material density. It provides an approximate estimation of density that also enables an approximate estimation of the inequality of the relationships between the densities (larger-smaller). For example, the relationship between the density of oil and rubber, of course, is not equal to the ratio of one to two (1:2) (Figure 2); the diagram depicts only that the density of rubber is larger than that of oil. However, when it is necessary to determine the density of a composite object consisting of two materials of different densities, the only information the qualitative "dots-in-a-box" representation can give us is that the density of the composite object ranges between the densities of these two materials. In contrast, if we use the mathematical ratio of the concept of density, then we can precisely calculate the average density of the composite object. Another limitation of the qualitative introduction of density as a property of materials is that it cannot predict nor explain changes in density (especially of gases) under temperature and pressure fluctuations. Such changes could be explained microscopically, at an older age though, using the particle model of matter.

4.3. The FS Content of the TLS

The teaching and expected learning trajectory of the TLS implementation, along with a brief description and the activities in the five units, are presented here. The sequence of the activities is a fundamental element of the teaching design and implementation. In order to focus on the content that is directly relevant to FS phenomena, we do not include here any content to do with the control of variables strategy and models, which has, however, been described in previous publications [18,19]. The description follows the scheme of "main aim, content, and activities that students participated in" (Table 1). Students worked: (a) in groups of three to complete structured worksheets on both the real and simulated experiments that followed the POE (Predict–Observe–Explain) teaching strategy [57], and/or (b) in a whole class arrangement, following formative assessment activities [4].

In the first unit, our main aim was to provide students with a familiarization phase of FS through the technological problem of salvaging the shipwreck. Students participated in: (a) an introductory discussion about the variables that might influence floating and sinking, resulting in five independent variables: weight, material, and shape of the object, width of the container, type of liquid; and (b) a thorough discussion about the concept of a solid and homogeneous object, in contrast to a hollow object, in order to focus on solid objects. The teacher demonstrated an experiment in the POE approach to check if the first of the five independent variables, i.e., the weight of an object, influences the object's FS situation.

Table 1. Main aim, content and activities concerning FS in the five units of the TLS.

Unit	Main Aim	Content	Activities
1st	Familiarization phase through technological problem.	Technological problem of lifting shipwreck. Distinction of variables that possibly influence FS. Difference between solid and hollow objects.	Groups and classroom discussion. POE activities. Demonstration of experiments.
2nd	Test variables that influence FS.	FS of solid objects influenced both by the material of object and liquid. FS not influenced by other variables, e.g., weight of object.	Groups and classroom discussion. POE activities with gradual increase in openness.
3rd	Introduction of "dots-in-a-box" representation of density. Use of causal relational FS rule for solid objects in water.	"Dots-in-a-box" representation describes "heavier-lighter" relationship between different materials. Compare "dots-in-a-box" representations for solid objects and water to decide objects' FS in water.	Groups and classroom discussion.
4th	Generalization of causal relational FS rule. "Dots-in-a-box" renamed "density".	Density of a two-material composite object lies between the densities of these two materials. Compare composite object's and liquid's densities to decide object's FS in a liquid.	Groups and classroom discussion. POE activities.
5th	Lifting shipwreck.	Implementing generalized causal relational FS rule within the technological framework.	Groups and classroom discussion.

In the second unit, our main aim was for the students to understand that the FS of a solid and homogeneous object is influenced by the material of the object and the type of liquid. The POE activities that students participated in were characterized by the gradual decrease of scaffolding or the gradual increase of openness. In these activities, students were prompted to test the other four variables that might influence the FS of solid and homogeneous objects; we note here that the teacher has demonstrated the variable weight in the previous unit.

In the third unit, the main aim was the introduction of the "dots-in-a-box" representation of density and the use of this representation to predict and explain FS of solid and homogeneous objects, in conjunction with the causal relational FS rule, i.e., if an object's "dots-in-a-box" representation is smaller than the water's, then the object will float in water, and if an object's "dots-in-a-box" representation is greater, then it will sink in the water. Students: (a) searched and gathered information about the properties of several natural and artificial materials, such as glycerin and polyurethane, and the ways they can be used, (b) negotiated with cubes of the same volume but different material and were assigned with a task prompting them to express the "heavier-lighter" material relationship (Figure 3), and (c) completed a task in a simulated environment, using a balance in order to put cubes of the same volume but different materials in the order of heavier to lighter. The sequence of densities of the materials in Figure 2 resulted from this sequence of tasks.

Iron Rubber Wood

Figure 3. One of the students' proposals to represent the "heavier-lighter" relationship of cubes of different materials, noted on the relevant worksheet.

In the 4th unit, the main aim was to generalize the causal relational FS rule to explain FS in any liquid and for both solid and composite objects. First, students participated in real and simulated experiments of several homogeneous objects in glycerin instead of water, emphasizing once again the role of the type of liquid in the explanatory model of FS phenomena (Figure 4a). In addition, understanding that the density of a two-material composite object lies between the densities of these two materials was crucial, so students participated in real experiments of composite objects in water, emphasizing the role of the average density of an object in predicting and/or interpreting its FS. In this unit, the phrase "dots-in-a-box" was replaced with "density" to refer to the property of materials.

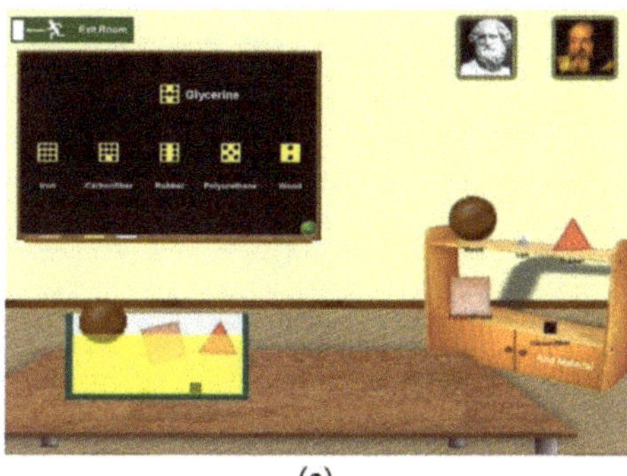
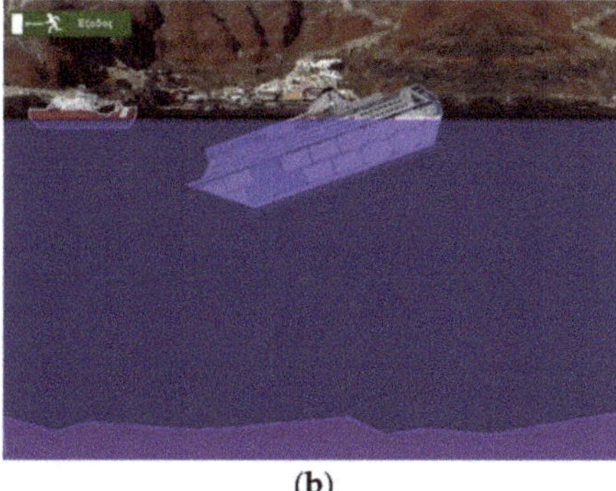

(a)   (b)

Figure 4. (a) A simulated experiment using glycerin instead of water, reprinted by permission from Springer Nature Customer Service Centre GmbH: Springer, Iterative design of Teaching-Learning Sequences, by D. Psillos and P. Kariotoglou, 2016; (b) simulated environment to investigate FS of the Sea Diamond shipwreck.

In the 5th unit, the main aim was to find the best solution for lifting the shipwreck, if possible, by implementing the generalized causal relational FS rule. The students had the opportunity to work in groups both in real and in a simulated environment (Figure 4b) so as to investigate the FS of the Sea Diamond shipwreck, studying the effect of excess water in the hold of the ship and discuss how it could be salvaged.

5. Selected Results

Here, a short description of the research method, and selected results on the learning of FS, of the implementation of the revised TLS are presented to show its effectiveness and its didactic transformation.

5.1. Method

The TLS described in this essay was developed, implemented and evaluated twice in an iterative evolution manner [19]. The initial and the revised versions were implemented on twelve and forty-one 5th graders, respectively, in a real-classroom educational context [20–22].

Data were collected using several sources (questionnaires, interviews, researchers' notes, etc.). For the sake of brevity, we focus on six Tasks of the questionnaire (see Appendix A) in order to evaluate the impact of the TLS implementation on students' FS explanations. The students answered the same questionnaire before, immediately after the TLS intervention, and seven months later (Pre, Post, and Delayed Post, respectively). More specifically, Tasks 1 and 2 were answered all three times, whereas Tasks 3–6 were answered in the Post and Delayed Post Tests. These tasks were not included in the Pre-questionnaire, either because students did not know the 'dots-in-a-box' representation before the intervention (Tasks 3 and 4) or because the questions were too complex to be attempted before the intervention (Tasks 5 and 6). While the results for Tasks 1–4 have been published in previous papers [20,21], the findings for Tasks 5 and 6 are presented here for the first time.

5.2. Results

The analysis results of students' responses for the six Tasks (Means and Standard Deviations) are presented in Table 2.

Table 2. Means and Standard Deviations of the students' responses in the six Tasks ($n = 41$).

	Pre		Post		Delayed Post	
	M	**SD**	**M**	**SD**	**M**	**SD**
Task 1	1.10	0.664	1.78	0.936	1.68	0.879
Task 2	0.78	0.613	1.39	0.862	1.37	0.859
Task 3	-	-	0.63	0.488	0.56	0.502
Task 4	-	-	0.73	0.449	0.54	0.505
Task 5	-	-	0.44	0.502	0.44	0.502
Task 6	-	-	0.51	0.506	0.46	0.505

Tasks 1 and 2 examined students' explanations about FS phenomena, whereas the 'dots-in-a-box' visual representation was not given. In both Tasks, the weight or the size of the object is the characteristic that could mislead students' responses to the alternative idea that "an object sinks because it is heavy" or "because it is big". The responses were classified thus: 0 for no or irrelevant answers, 1 for reference to the object's weight, 2 for reference to the material the object consists of, and 3 for reference to the causal relational FS rule, i.e., comparing the densities of the object and the liquid to predict and/or explain the object's FS. The study findings showed that there was a statistically significant improvement after the TLS intervention for both Tasks 1 and 2 ($z = 3.446$, $p < 0.001$ and $z = 3.801$, $p < 0.001$, respectively), which was retained seven months later. More specifically, for Task 1, the mean was 1.10 for the Pre-Test, which went up to a high 1.78 in the Post-Test, and which was maintained with a slight decrease seven months later in the Delayed Post-Test (1.68). The results were similar for Task 2, where the mean in the Pre-Test was only 0.78, which rose to 1.39 immediately after the TLS intervention in the Post-Test and was maintained almost at the same level (1.37) in the Delayed Post-Test.

Tasks 3 and 4 examined students' explanations about FS in a simulated environment, while the "dots-in-a-box" representation was also given. Students' responses were classified as 0 for causal linear reasoning, i.e., focusing only on one characteristic of the objects, and 1 for causal relational reasoning, i.e., a comparison of the densities of the object and the liquid. The results showed that most of the students were able to use the "dots-in-a-box" representation to successfully apply causal relational reasoning in their responses, immediately after the intervention and seven months later (Table 2). More specifically, for Tasks 3 and 4, the means for the Post-Test were 0.63 and 0.73, respectively, which decreased slightly in the Delayed Post-Test to 0.56 and 0.54, respectively. In addition, it appears that most of the students who were able to apply causal relational reasoning in Task 4 could also understand differences in float levels in relation to material density [20], a topic that had not been covered in the intervention.

The last two Tasks 5 and 6 examined whether students could effectively determine the position of an object, in relation to the surface of a liquid, in order to establish its density. In Task 5, in order for students to correctly decide the relationship between the densities of the two objects by applying the causal relational FS rule, they would have to disregard the size of the two objects, which could have been misleading. Students that did not, even qualitatively, differentiate between weight and density would intuitively think that the bigger object had a greater density. Students' responses were classified as 1 for using the causal relational FS rule and 0 in all other cases. The results in Table 2 show that a large number of students were able to successfully answer Task 5, with a mean of 0.44 in both the Post- and delayed Post-Tests, thus indicating that they, at least, qualitatively differentiated between weight and density. For Task 6, in order to successfully apply the causal relational

FS rule and come to the conclusion that the density of the object is equal to the density of the liquid, students had first to recognize that the object was in a state between floating and sinking, and therefore suspended, which meant that it remained in the liquid at the same location at rest. Responses were classified as 1 for recognizing that the object was suspended and for successfully applying the causal relational FS rule, and 0 in all other cases. The results in Table 2 show that a large number of students were able to successfully answer this question, with means of 0.51 and 0.46 for the Post- and Delayed Post-Tests, respectively. This finding indicates that the students were able to apply the causal relational FS rule in a new situation, that of the suspension of an object in a liquid, which had not been covered in the TLS intervention.

6. Discussion and Conclusions

In this paper, we describe and examine certain elements in the teaching of floating and sinking, with focus on the didactic transformation of content, which was part of a long-term developmental study. The didactic transformation process is an essential aspect in each teaching effort and especially in the design and development of a TLS.

In our research on the teaching of floating/sinking (FS) phenomena to 5th grade Primary school students, we adopted the elimination of variables approach (density-based) rather than an analysis of equilibrium explanatory model (buoyancy-based). The reasons for this decision were: (1) students find it difficult to understand and effectively implement the buoyancy-based model to explain FS, in contrast to density which is an easier concept to comprehend; (2) The emphasis of our developmental study project was on materials and their properties; and (3) Inquiry-based teaching and learning is more compatible with the elimination of variables approach, for the teaching of FS phenomena, as well as being more in line with the current Greek school curriculum.

The concept of density was introduced to students qualitatively through a visual representation called "dots-in-a-box". The qualitative method was chosen over (a) the mathematical ratio of mass per volume and (b) the particle theory of matter. The reasons for this choice are as follows: (1) the visual representation makes it easier for students to grasp this scientific concept, whereas the mathematical introduction of density has several times been shown to be ineffective, at least with primary and junior high school students [34]; (2) misconceptions would most likely arise with the particle theory of matter; and (3) introducing density with a visual representation also provides students with opportunities for differentiating the concepts involved in the interpretations of FS phenomena, e.g., density, weight and volume.

The study findings strongly suggest that our TLS and the didactic transformation of content that was developed had a significant level of success in the teaching/learning of FS to young students, which seems to have been maintained seven months later. Most of the students adopted explanations that were compatible with scientific ones and were able to overcome their prior alternative ideas, such as "heavy objects sink and light objects float". In addition, when given the "dots-in-a-box" visual representation of density, the students successfully implemented the causal relational FS rule, i.e., comparing the densities of the objects and liquid to predict and/or explain the FS state of objects. Finally, several students were able to implement the density-based model, which they had been taught, to explain situations that had not been covered in the TLS intervention, such as successfully predicting the floating level of objects made of different materials and applying the causal relational FS rule for objects suspended in a liquid. Both cases are considered difficult for students in this age range (10–15 y-o) to comprehend and explain [1].

We consider that this work is an original sample of a developmental research description, in the sense of the Design-Based Research approach in science education [8,23,24], and consequently an example of effective good teaching practice, that can help teachers to elaborate on their teaching and inspire innovative treatment of the topic of floating and sinking in science curricula for primary school physics. The teaching and learning intervention for floating/sinking phenomena, which is in itself a difficult conceptual science topic, to this young target population, was successful, we strongly believe, due to the contribution of the didactic transformation of content.

Author Contributions: The authors contributed equally to the preparation and writing of the present paper. All authors have read and agreed to the published version of the manuscript.

Funding: This research received no external funding.

Institutional Review Board Statement: Not applicable.

Informed Consent Statement: Not applicable.

Data Availability Statement: Not applicable.

Conflicts of Interest: The authors declare no conflict of interest.

Appendix A

The questionnaire tasks

Task 1

On a big ship, among other objects, you can find an anchor. Does it float or sink if we drop it into the sea? Justify your answer.
The anchor: floats sinks I do not know
Because:

. . . .

Task 2

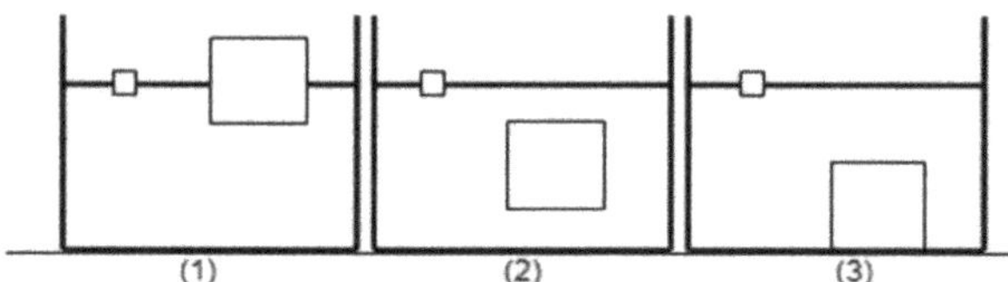

Costas drops a small piece of a particular material into a container filled with water, and he observes that it floats. Afterwards, Irene drops a bigger piece of the same material into the same container. In your opinion, at which point will the big piece stop moving? Circle which number: 1, 2 or 3 in the diagram you think represents the final position of the two bodies that Costas and Irene dropped into the container. Justify your choice.

Task 3

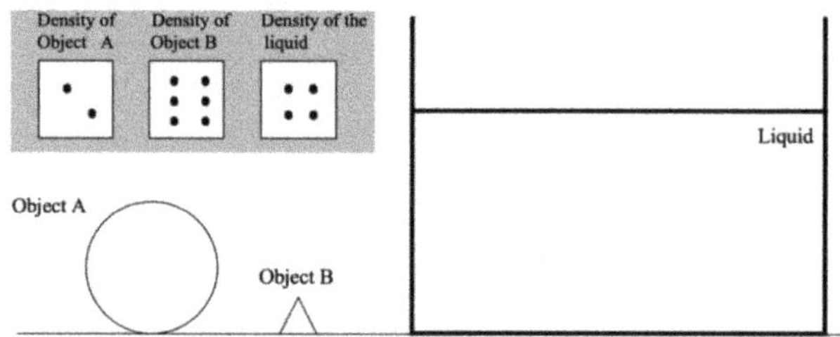

You are given two objects A and B, and a container which contains a liquid. The densities of the two objects and that of the liquid are given with the "dots-in-a-box" representation, as you can see in the gray box. If you drop objects A and B into the container with the liquid, what will their final position be? Draw objects A and B in their final position in the liquid. Justify your answer:

. . . .

. . . .

Task 4

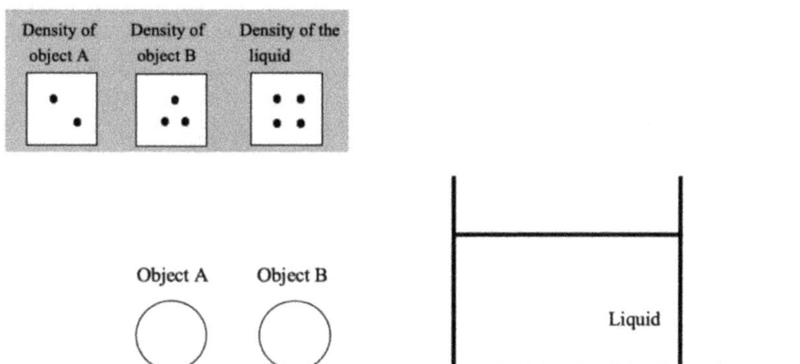

The densities of the two objects and that of the liquid are given with the "dots-in-a-box" representation, as you can see in the gray box. We drop objects A and B into the liquid. Circle which number: 1, 2 or 3 in the diagrams best represents the final positions of the two objects after we have dropped them into the liquid.

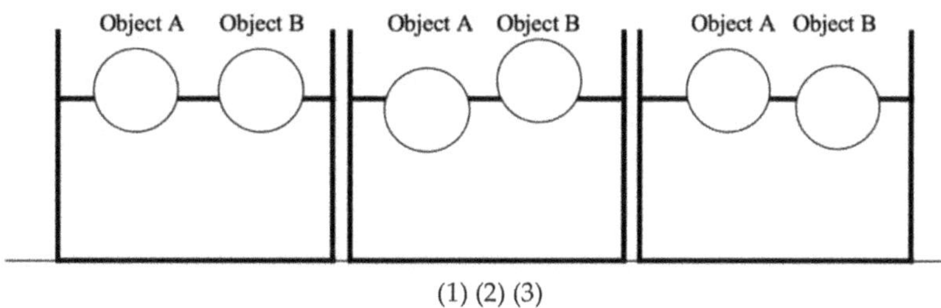

(1) (2) (3)

Justify your choice:

.. ..

.. ..

Task 5

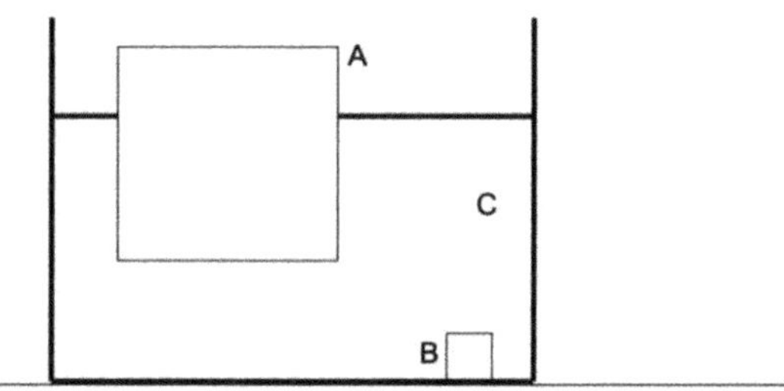

We drop the two Objects (A) and (B) in the liquid (C). Object (A) floats in the liquid, whereas Object (B) sinks in the liquid. Decide if the following sentence is correct or incorrect:

Object (A) has a greater density than Object (B).
It is right   It is wrong   I don't know

Justify your choice:

.. ..

.. ..

Task 6

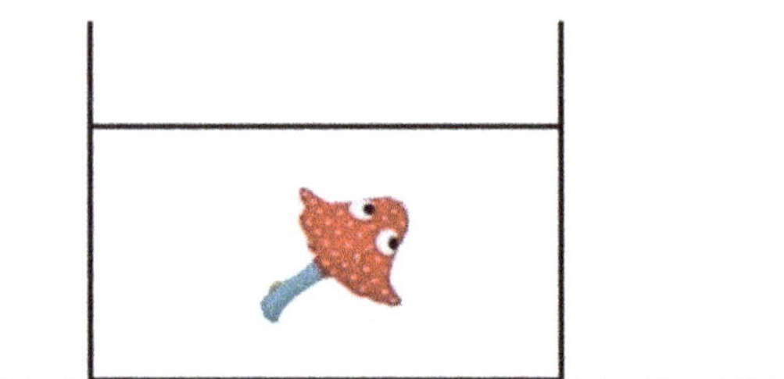

While Georgia, Petroula, and Sophia were playing with some toys, one of the toys accidentally fell in the container with the liquid that you can see in the picture. The girls noticed that the toy did not float up towards the surface of the liquid, nor did it sink to the bottom of the container. They wondered what the density of the toy could be, but they disagreed in their opinions:

Georgia says that this object has a greater density than the liquid.
Petroula believes that the object has a lower density than the liquid.
Sophia says that the object has the same density as the liquid.

Which of the girls do you agree with? With:

Georgia Petroula Sophia I don't know

Justify your choice:

....
....

References

1. Joung, Y.J. Children's typically-perceived-situations of floating and sinking. *Int. J. Sci. Educ.* **2009**, *31*, 101–127. [CrossRef]
2. Psillos, D.; Kariotoglou, P. Teaching Fluids: Intended knowledge and students' actual conceptual evolution. *Int. J. Sci. Educ.* **1999**, *21*, 17–38. [CrossRef]
3. Kariotoglou, P.; Psillos, D. Teaching and Learning Pressure and Fluids. *Fluids* **2019**, *4*, 194. [CrossRef]
4. Shen, J.; Liu, O.L.; Chang, H.Y. Assessing students' deep conceptual understanding in physical sciences: An example on sinking and floating. *Int. J. Sci. Math. Educ.* **2017**, *15*, 57–70. [CrossRef]
5. Yin, Y.; Tomita, K.M.; Shavelson, R.J. Diagnosing and dealing with student misconceptions about "Sinking and Floating". *Sci. Scope* **2008**, *31*, 34–39.
6. Ruiz-Primo, M.A.; Furtak, E.M. Exploring Teachers' Informal Formative Assessment Practices and Students' Understanding in the Context of Scientific Inquiry. *J. Res. Sci. Teach.* **2008**, *44*, 57–84. [CrossRef]
7. Méheut, M.; Psillos, D. Teaching-Learning Sequences: Aims and tools for science education research. *Int. J. Sci. Educ.* **2004**, *26*, 515–535. [CrossRef]
8. Psillos, D.; Kariotoglou, P. *Iterative Design of Teaching-Learning Sequences: Introducing the Science of Materials in European Schools*; Springer: Berlin/Heidelberg, Germany, 2016.
9. Tiberghien, A.; Vince, J.; Gaidioz, P. Design-based Research: Case of a teaching sequence on mechanics. *Int. J. Sci. Educ.* **2009**, *31*, 2275–2314. [CrossRef]
10. Viennot, L. Physics education research and inquiry-based teaching: A question of didactical consistency. In *Designing Theory-Based Teaching-Learning Sequences for Science Education: Proceedings of the Symposium in Honour of Piet Lijnse at the Time of his Retirement as Professor of Physics Didactics at Utrecht University*; Kortland, K., Klaassen, K., Eds.; Utrecht University: Utrecht, The Netherlands, 2010.
11. Chevallard, Y.; Bosch, M. *Encyclopedia of Mathematics Education*; Lerman, S., Ed.; Springer: Dordrecht, The Netherlands, 2014.
12. Duit, R. Science education research internationally: Conceptions, research methods, domains of research. *Eurasia J. Math. Sci. Technol. Educ.* **2007**, *3*, 3–15. [CrossRef]
13. Duit, R.; Gropengießer, H.; Kattmann, U.; Komorek, M.; Parchmann, I. The Model of Educational Reconstruction—A Framework for improving teaching and learning science. In *Science Education Research and Practice in Europe: Retrospective and Prospective*; Jorde, D., Dillon, J., Eds.; Sense Publishers: Rotterdam, Netherlands, 2012; pp. 13–37.
14. Chinn, C.; Brewer, W. The Role of Anomalous Data in Knowledge Acquisition: A Theoretical Framework and Implications for Science Instruction. *Rev. Educ. Res.* **1993**, *63*, 1–49. [CrossRef]
15. Snir, J. Sink or Float—What do the Experts think? The Historical Development of Explanations for Floatation. *Sci. Educ.* **1991**, *75*, 595–609. [CrossRef]

16. Macbeth, D. On an actual apparatus for conceptual change. *Sci. Educ.* **2000**, *84*, 228–264. [CrossRef]

17. Spyrtou, A.; Zoupidis, A.; Kariotoglou, P. *The design and development of an ICT- Enhanced Module concerning density as a property of materials applied in floating-sinking phenomena. In Physics Curriculum Design, Development and Validation*; Constantinou, C.P., Papadouris, N., Eds.; University of Cyprus: Nicosia, Cyprus, 2008; pp. 391–407. Available online: http://lsg.ucy.ac.cy/girep2008/papers/THE%20DESIGN%20AND%20DEVELOPMENT%20OF%20AN%20ICT-ENHANCED.pdf (accessed on 20 March 2021).

18. Zoupidis, A.; Pnevmatikos, D.; Spyrtou, A.; Kariotoglou, P. The impact of procedural and epistemological knowledge on conceptual understanding: The case of density and floating–sinking phenomena. *Instr. Sci.* **2016**, *44*, 315–334. [CrossRef]

19. Zoupidis, A.; Spyrtou, A.; Malandrakis, G.; Kariotoglou, P. The evolutionary refinement process of a Teaching Learning Sequence for introducing inquiry aspects and density as materials' property in floating/sinking phenomena. In *Iterative Design of Teaching-Learning Sequences*; Psillos, D., Kariotoglou, P., Eds.; Springer: Dordrecht, The Netherlands, 2016; pp. 167–199.

20. Zoupidis, A.; Spyrtou, A.; Pnevmatikos, D.; Kariotoglou, P. Explicitly Linking Simulated with Real Experiments for Conceptual Understanding of Floating/Sinking Phenomena. *Themes eLearning* **2018**, *11*, 35–52.

21. Spyrtou, A.; Lavonen, J.; Zoupidis, A.; Loukomies, A.; Pnevmatikos, D.; Juuti, K.; Kariotoglou, P. Transferring a Teaching Learning Sequence Between Two Different Educational Contexts: The Case of Greece and Finland. *Int. J. Sci. Math. Educ.* **2018**, *16*, 443–463. [CrossRef]

22. Zoupidis, A. Teaching and Learning through the Use of Scientific and Technological Models: The Case of Floating and Sinking Phenomena. Ph.D. Thesis, University of Western Macedonia, School of Education, Florina, Greece, 2012, unpublished work.

23. Hjalmarson, M.; Lesh, R. Engineering and design research: Intersections for education research and design. In *Handbook of Design Research Methods in Education*; Baek, J.Y., Lesh, R.A., Eds.; Routledge: Abingdon-on-Thames, UK, 2008; pp. 96–110.

24. Kelly, A.E.; Lesh, R.; Baek, J. *Handbook of Design Research Methods in Education: Innovations in Science, Technology, Mathematics and Engineering Learning and Teaching*; Routledge: Abingdon-on-Thames, UK, 2008.

25. Biddulph, F.; Osborne, R. Pupil's ideas about floating and sinking. *Res. Sci. Educ.* **1984**, *14*, 114–124. [CrossRef]

26. Smith, C.; Snir, J.; Grosslight, L. Using Conceptual Models to Facilitate Conceptual Change: The Case of Weight-Density Differentiation. *Cogn. Instr.* **1992**, *9*, 221–283. [CrossRef]

27. Kawasaki, K.; Herrenkohl, L.; Yeary, S. Theory Building and modeling in a sinking and floating unit: A case study of third and fourth grade students' developing epistemologies of science. *Int. J. Sci. Educ.* **2004**, *26*, 1299–1324. [CrossRef]

28. Havu-Nuutinen, S. Examining young children's conceptual change process in floating and sinking from a social constructivist prospective. *Int. J. Sci. Educ.* **2005**, *27*, 259–279. [CrossRef]

29. Yin, Y.; Tomita, M.K.; Shavelson, R.J. Using Formal Embedded Formative Assessments Aligned with a Short-Term Learning Progression to Promote Conceptual Change and Achievement in Science. *Int. J. Sci. Educ.* **2014**, *36*, 531–552. [CrossRef]

30. Fassoulopoulos, G.; Kariotoglou, P.; Koumaras, P. Consistent and Inconsistent Pupil's Reasoning about Intensive Quantities, The Case of Density and Pressure. *Res. Sci. Educ.* **2003**, *33*, 71–87. Available online: http://www.ingentaconnect.com/content/klu/rise/ (accessed on 20 March 2021). [CrossRef]

31. Perkins, D.N.; Grotzer, T.A. Dimensions of causal understanding: The role of complex causal models in students' understanding of science. *Stud. Sci. Educ.* **2005**, *41*, 117–166. [CrossRef]

32. Hardy, I.; Jonen, A.; Moeller, K.; Stern, E. Effects of Instructional Support Within Constructivist Learning Environments for Elementary School Students' Understanding of "Floating and Sinking". *J. Educ. Psychol.* **2006**, *98*, 307–326. [CrossRef]

33. Wiser, M.; Smith, C. Learning and teaching about matter in grades K-8: When should the atomic-molecular theory be introduced? In *International Handbook of Research on Conceptual Change*; Vosniadou, S., Ed.; Routledge: Abingdon-on-Thames, UK, 2008; pp. 205–239.

34. Rowell, J.; Dawson, C. Teaching about Floating and Sinking: An Attempt to Link Cognitive Psychology with Classroom Practice. *Sci. Educ.* **1977**, *61*, 245–253. [CrossRef]

35. Smith, C.; Maclin, D.; Grosslight, L.; Davis, H. Teaching for understanding: A study of students' preinstruction theories of matter and a comparison of the effectiveness of two approaches to teaching students about matter and density. *Cogn. Instr.* **1997**, *15*, 317–393.

36. Xu, L.; Clarke, D. Student difficulties in learning density: A Distributed Cognition Perspective. *Res. Sci. Educ.* **2011**, *42*, 769–789. [CrossRef]

37. Vosniadou, S.; Vamvakoussi, X.; Skopeliti, I. The framework theory approach to the problem of conceptual change. In *International Handbook of Research on Conceptual Change*; Vosniadou, S., Ed.; Routledge: Abingdon-on-Thames, UK, 2008; pp. 205–239.

38. Radovanović, J.; Sliško, J.; Stepanović Ilić, I. Active learning of buoyancy: An effective way to change students' alternative conceptions about floating and sinking. *J. Phys. Conf. Ser.* **2019**, *1286*, 012011. [CrossRef]

39. Bonera, G. Distinguished papers series: Signor galileo galilei's scales. *Int. J. Sci. Educ.* **1996**, *18*, 881–887. [CrossRef]

40. Galili, I. Weight versus gravitational force: Historical and educational perspectives. *Int. J. Sci. Educ.* **2001**, *23*, 1073–1093. [CrossRef]

41. Bormashenko, E. Surface tension supported floating of heavy objects: Why elongated bodies float better? *J. Colloid Interface Sci.* **2016**, *463*, 8–12. [CrossRef]

42. Inhelder, B.; Piaget, J. *The Growth of Logical Thinking from Childhood to Adolescent*; Psychology Press: London, UK, 1958.

43. Hewson, M. The Acquisition of Scientific Knowledge: Analysis and Representation of Student Conceptions Concerning Density. *Sci. Educ.* **1986**, *70*, 159–170. [CrossRef]

44. Strauss, S.; Globerson, T.; Mintz, R. The influence of training for the atomistic of the development of the density concept among gifted and non gifted children. *J. Appl. Dev. Psychol.* **1983**, *4*, 125–147. [CrossRef]
45. National Research Council. *A framework for K-12 Science Education: Practices, Crosscutting Concepts, and Core Ideas*; The National Academies Press: Washington, DC, USA, 2012.
46. NGSS Lead States. *Next Generation Science Standards: For States, by States*; The National Academies Press: Washington, DC, USA, 2013.
47. Couso, D. Participatory approaches to curriculum design from a design research perspective. In *Iterative Design of Teaching-Learning Sequences*; Psillos, D., Kariotoglou, P., Eds.; Springer: Dordrecht, The Netherlands, 2016; pp. 47–71.
48. Krajcik, J. Supporting Science Learning in Context: Project-Based Learning. In *Portable Technologies: Science Learning in Context*; Tinker, R., Krajcik, J., Eds.; Kluwer Academic/Plenum Publishers: New York, NY, USA, 2001; p. 92.
49. Benett, J.; Lubben, F.; Hogarth, S. Bringing science to life: A synthesis of the research evidence on the effects of context-based and STS approaches to science teaching. *Sci. Educ.* **2007**, *91*, 347–370. [CrossRef]
50. Waight, N.; Abd-El-Khalick, F. The impact of technology on the enactment of "inquiry" in a technology enthusiast's sixth grade science classroom. *J. Res. Sci. Teach.* **2007**, *44*, 154–182. [CrossRef]
51. Abd-El-Khalick, F.; BouJaoude, S.; Duschl, R.A.; Hofstein, A.; Lederman, N.G.; Mamlok, R.; Niaz, M.; Treagust, D.; Tuan, H. Inquiry in science education: International perspectives. *Sci. Educ.* **2004**, *88*, 397–419. [CrossRef]
52. Bybee, R.W. Scientific inquiry and science teaching. In *Scientific Inquiry and Nature of Science*; Flick, L.B., Lederman, N.G., Eds.; Springer: Dordrecht, The Netherlands, 2006; pp. 1–14.
53. Taber, K. *Progressing Science Education. Constructing the Scientific Research Programme into the Contingent Nature of Learning*; Springer: New York, NY, USA, 2009; Volume 45, p. 310.
54. Chamberlain, K.; Crane, C. *Reading, Writing & Inquiry in the Science Classroom*; Corwin Press: Thousand Oaks, CA, USA, 2009.
55. Boudreaux, A.; Shaffer, P.; Heron, P.; McDermott, L. Student understanding of control of variables: Deciding whether or not a variable influences the behavior of a system. *Am. J. Phys.* **2008**, *76*, 163–170. [CrossRef]
56. Treagust, D.F.; Chittleborough, G.; Mamiala, L.T. Students' understanding of the role of scientific models in learning science. *Int. J. Sci. Educ.* **2002**, *24*, 357–368. [CrossRef]
57. White, R.; Gunstone, R. *Probing Understanding*, 1st ed.; Routledge: Abingdon-on-Thames, UK, 1992. [CrossRef]

Article

Open Water Flume for Fluid Mechanics Lab

Rachmadian Wulandana

Mechanical Engineering Program, SUNY New Paltz, New Paltz, NY 12561, USA; wulandar@newpaltz.edu

Abstract: Open water flume tanks with closed-loop circulation driven by centrifugal pumps are essential for hydro experimentation in academic settings as well as research centers. The device is also attractive due to its versatility and easy-to-maintain characteristics. Nevertheless, commercial open flume systems can be expensive and become less prioritized in engineering schools. This paper describes the design and fabrication of an affordable, medium-size water flume tank, suitable for education purposes. The central piece of the system is a transparent observation chamber where fluid experiments are typically conducted and observed. The expected maximum average water speed in the observation chamber of about 60 cm per second was achieved by the inclusion of a 3 hp centrifugal pump. The size and capacity of the current design were constrained by space limitation and available funds. The educational facility was assigned as a two-semester multi-disciplinary capstone senior design project incorporating students and faculty of mechanical, electrical, and computer engineering programs in our campus. The design process provides a training platform for skills in the area of Computer Aided Designs (CAD), Finite Element Analysis (FEA), Computational Fluid Dynamics (CFD), manufacturing, and experimentation. The multi-disciplinary project has contributed to the improvement of soft skills, such as time management, team working, and professional presentation, of the team members. The total material cost of the facility was less than USD 6000, which includes the pump and its variable frequency driver. The project was made possible due to the generous sponsor of the Vibration Institute.

Keywords: open water tank; education; fluid mechanics

check for **updates**

Citation: Wulandana, R. Open Water Flume for Fluid Mechanics Lab. *Fluids* **2021**, *6*, 242. https://doi.org/10.3390/fluids6070242

Academic Editor: Mehrdad Massoudi

Received: 31 May 2021
Accepted: 26 June 2021
Published: 3 July 2021

1. Introduction

Open water flumes provide effective hands-on multidisciplinary learning tools for students matriculated in engineering programs, as well as in the physical education program [1]. Senior students participating in the design and fabrication process of the flume obtained the opportunity to reinforce fundamental engineering concepts and to master valuable technical skills [2–4]. The reported open flume was motivated by the need of a test chamber for an ongoing research on bladeless turbines, as well as the desire to have such device in the fluid mechanics laboratory. The current design was inspired by a small scale commercial water flow tank used in the investigation of vortex-induced autorotation and oscillation of straight cylinders [5,6], as well as the energy potential from such vibration modes of symmetric geometries [7,8]. Open channel tanks are paramount for various hydrodynamic research areas, such as designs of hydrokinetic energy harvesters, investigation of flow characteristics in the presence of obstacles, drag and lift of objects exposed to fluid flows, etc. The vortex-induced vibration (VIV) of objects exposed to flow has emerged as attractive potential sources of renewable energy. Additionally, the versatility of open channel flumes has allowed plethora of vibration modes to be experimented. Sun et al. used a 40 cm wide and long rectangular channel to convert vortex-induced vibration and galloping of blunt bodies into electricity via piezoelectric strips [9]. In a small scale water tank, Cao et al. attempted to amplify the vibration amplitudes of piezoelectric strips by means of magnet fields [10,11]. Arionfard and Nishi utilized a small (30 cm × 1 m) water channel to investigate power harnessing potential of pivoted cylinder exposed to highly turbulent flow [12]. The utilization of much larger size channels allows modeling of near-realistic flow and full-scale energy harvesters. Rostami and Fernandes studied the energy harvesting from the torsional galloping, fluttering, and autorotation modes of flat and S-shape plates using a 1.4 m, wide 22 m long water channel [13–16]. Furthermore, a 1 m wide water flume tank is used in the development of VIVACE (Vortex-Induced Vibration of Aquatic Clean Energy), which exploits the transverse-to-the-flow (lateral) oscillation mode of VIV at large range of Reynolds numbers using magnets attached to cylindrical bars as harvesters [17–19]. An open channel with similar width but five times longer (about 42 m long) was recently used in the attempts to utilize the galloping modes of triangular cylindrical bars at much higher velocity [20,21]. Generally, interests in the designs of innovative hydrokinetic energy turbines have been well facilitated by open water flumes. The facility is adaptable for various types of turbines and its blades, as well as a range of measurement needs. Barber et al. studied the power and thrust improvement of axial turbines after the utilization of carefully designed adaptive pitch composite blades [22]. A small commercial flow loop water tank was used in the investigation of wake behind the propellers of wind turbines [23]. Open water flume experimentations of vertical axis hydrokinetic generators, such as Savonius and Darrieus turbines, have been intensified lately. These turbines are attractive renewable energy harvesters due to their simple designs, low-cost manufacturing, easy maintenance, and independence of flow direction. Talukdar et al. demonstrated that the performance of two-bladed semi-circular Savonius turbines is better than two-bladed elliptical and three-bladed semi-circular designs [24]. On the other hand, Sarma et al. pointed out that the three-bladed Savonius model performs better in water than in air flow [25]. It is interesting to note that placing an upstream obstacle can improve the performance of a modified two-bladed Savonius turbines [26] and Darrieus turbines [27]. Attempts to assemble and test the two turbines into one single hybrid unit have shown promising outcomes [28,29]. In the hydrology area, large size tanks facilitate investigations of particle sedimentation carried by water flow [30], improvement of hyporheic zones due to riverbed restoration [31], and effects of flood waves on river banks [32] among many other studies. The open channel allows examination of flow qualities, such as head loss and turbulence, by placing out obstacles in its test section. Investigation of the head loss due to the presence of submerged baffle-posts [33,34] and examination of water turbulence due to rib roughness [35] are examples of important studies to be carried out for better irrigation systems. Examinations on the interaction between vegetation and water flows shed light on the life quality of river. Here, the vegetation under investigation is carefully arranged along the base of the test chamber and its effects on water waves are studied using camera [36]. Lastly, we would like to mention flume experimentations to test the drag and lift of 3D-printed models of swimmers [1] that show the capability of water flume to accommodate large range of multidisciplinary applications. The transparent chamber of the flume certainly allows flow measurement by either direct observation [33,34] or flow visualization technology such as dye injection [31], hydrogen bubble [37], particle image velocimetry (PIV) [23,37,38], planar laser-induced fluorescence (PLIF) [39], stereoscopic digital particle image velocimetry (SDPIV) [40], laser doppler velocimetry (LDV), and acoustic doppler velocimetry (ADV) [30] methods. It is obvious that the versatility and easy-to-maintain characteristic of the open flume systems makes the device very attractive to be included in fluid mechanics and hydrology labs.

The presented work discusses the design and fabrication of a small size (15 cm wide and 75 cm long test chamber) open water flume with closed flow loop driven by a 3 hp centrifugal pump. The designs must consider limited funds, available space in the lab, and mobility requirement. The overall size of the system, the materials for the structural frame and the chambers, as well as the pump selection, can be determined by the flow requirement in the observation chamber and experiments that would be conducted. In this project, the maximum fund was set to be USD 10,000 and the overall size is governed by the size of the freight elevator in the school. The overall length of the system must be kept to about 2.5 m or shorter, so that it can fit in the school's freight elevator. Reported in this paper are the pump selection process, structural analysis of the supporting frame, computational fluid dynamic analysis of the flume, manufacturing process, and conducted experimentations on the flow visualization using dye injection and investigation of drag forces on submerged objects. A brief discussion on the budget is presented before the paper is closed with a discussion and conclusion sections.

2. Basic Design and Pump Selection

This water flow tank is designed to provide straight uniform flow with an average speed of as much as 60 cm/s through its transparent observation chamber. The cross-sectional dimension of the observation chamber is planned to be at least 15×15 cm^2. The length of the observation chamber was designed to be 60 cm. This length is needed to provide enough observation room for vortex trail and wakes behind objects exposed to water flow. The final length of the test chamber is 75 cm. The selection of the maximum speed and chamber dimension was based on a commercial water flume used in experiments on vortex-induced oscillation of cylinders [41]. An upstream manifold in a shape of converging chamber of about 44 cm is added to provide room for the water to reduce its turbulence and complex characteristics prior to enter the observation chamber. The height of the observation chamber was designed to be ~1.2 m, a little less than the average height of human eyes of ~1.4 m. The top side of the observation chamber is expected to be open to allow direct physical access to the flow and, more importantly, easy placement of objects exposed to the flow. Tight covering of the top side to achieve high speed of water flow is possible, but the pressure increase must carefully be calculated. The flow tank was designed to be mobile, so that it can be easily moved when needs arise. The overall length and width of the flow tank therefore is constrained to size of the school's freight elevator of approximately 2.16 by 2.44 m^2. This size put limitation on the total length of the equipment to be about 2.5 m. The overall budget of this device is set to be less than USD 10,000, based on the maximum fund by the sponsor for this project: Vibration Institute. As the project is scheduled for a senior design project, the design and construction must be finished in two semesters.

The design process begins with a preliminary calculation for the pump specification, which would determine the pipe diameters, the dimension of supporting frames, and required power outlet. Detail calculation would require flowrates, height of the observation chamber, lengths, diameters, and materials of various pipes involved in the designs, as well as connections and pipe bends. In the absence of many of these parameters, the needed pump power can be calculated using the required flowrates and maximum elevation that needs to be overcome by the water. Other parameters need to be estimated. Figure 1 panel (a) shows an example of simple conceptual sketch of the water flow system drawn by students involved in this project. The flowrate can be determined from the demanded speed $\overline{V}$ of ~0.6 m/s and cross section area of the observation chamber of $A = 0.15 \times 0.15 = 0.0225$ m^2, through the equation for flowrate $Q = \overline{V} \times A = 0.0135$ m^3/s, or about 12,800 gph or 214 gpm. Figure 1 panel (b) depicts a simple open flow diagram to represent the design of the water tank. The pump must overcome the height of the observation chamber and resistance by all pipes and chambers (major losses) and various connections (minor losses) along the loop. Note also that the effects of gravity on the returning channel from the observation chamber is not included in the calculation, as the power discount may not be significant. The flow diagram shown in Figure 1 guides us in determining the Bernoulli equation to be used.

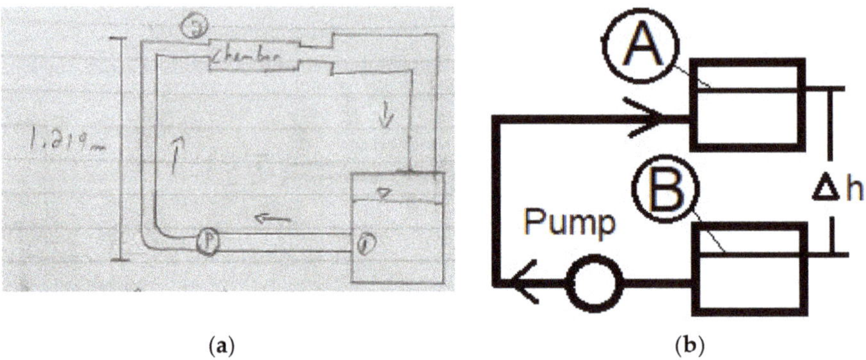

(**a**)　　　　　　　　　　　　　　　(**b**)

Figure 1. (**a**) Sketch of 1st design of the flow tank loop given by students and (**b**) the basic flow diagram for the calculation of pump head needed to determine the pump power.

The total head h_p for the calculation of pump power can be obtained from the Bernoulli equation easily found in fluid mechanics text books [42,43];

$$h_p = \Delta h + \Sigma\, f \frac{L}{D} \frac{\overline{V}^2}{2g} + \Sigma\, K \frac{\overline{V}^2}{2g} \tag{1}$$

The Δh represents the elevation difference between two end points in the flow diagram. In our case, we would take this as the largest height difference available in the water tank system, which is the height of the observation chamber. The second and third terms in the equation are known as the major and minor losses, losses that are caused by the wall friction along the pipes and channels in the system, and losses that are caused by any other obstacles in the system, respectively. For the major loss, the Darcy's friction factor f is known to depend on the Reynolds numbers of the flow and roughness of the pipe. The L and D are length and diameter of pipe, respectively. In the third term, the constant K represents resistance factors for various obstacles along the loop, such as bends, junctions, entrances, etc. The Reynolds number is defined as

$$Re = \frac{\rho \overline{V} D}{\mu} \tag{2}$$

where ρ and μ are the water density and absolute viscosity, respectively. In calculating the major and minor losses, the average speed $\overline{V}$ of the water should be taken at each section along the pipe where the loss is calculated, and the D represents hydraulic diameter of that particular section. As the detail geometry of the flume is not yet known, we assume the contribution of these losses to be simply double the elevation difference, hence the total head is $h_p = 1.2 + 2 \times 1.2 = 3.6$ m. It will be shown later that this estimation is larger than an alternative estimation based on possible major and minor losses in the flume. The required pump power can be calculated using [42,43]

$$W_p = \frac{\rho\, g\, Q\, h_p}{\eta} \tag{3}$$

In estimating the needed pump power, we took the density (ρ) and absolute viscosity (μ) of water at atmospheric pressure and 20 °C to be 998 kg/m^3 and 0.001 Pa.s [43,44], respectively, while the gravitational constant $g = 9.81$ m/s^2. Not knowing the pump to be purchased yet, we estimated the pump efficiency η from commercial pump examples presented in textbooks [43,44]. Based on these examples, for the required flow rate of 214 gpm, the efficiency was found to be about $\eta \approx 60\%$. Using these data, we can estimate the required pump power to be only $W_p = 793$ W or about 1.06 hp. Note that the power calculation here is based on high flow rate but very low head loss, due to the very short pipe and minimal height involved in the current design. Consequently, such a pump, with such combination of low pump power and high flow rate, may not be available in the market. Here, students need to realize that the market availability of the pump puts a constraint in the design.

Alternatively, the major and minor losses could be estimated based on the possible materials and components to be used in the designs. This detail calculation process provides valuable realistic exercise for the students involved in this project. Table 1 shows list of pipes and channels, as well as estimated major head losses. The mean velocity for each section is estimated from the flowrate of 0.0135 m^3/s based on the required flow speed of 60 cm/s. The major loss is defined as $h = f \frac{L}{D} \frac{\overline{V}^2}{2g}$ and the friction factor for each section is determined from the Moody chart based on the associated Reynolds number and given roughness factors of the pipes [42–45]. Due to the short pipes involved in this design, this calculation renders net head much less than 1 m, well below the estimated head mentioned before. As expected, the required head is dominated by the height of the water column that must be overcome.

Table 1. List of friction and major head loss of pipes and channels involved in the designs.

Materials	Length (m)	Diameter (m)	Roughness (mm)	$\overline{V}$ (m/s)	Re	f	*h* (m)
Plexiglass	0.6	0.15 × 0.15	0.0015	0.6	100,921	0.018	0.001
Stainless Steel	4.0	~0.4 × 0.3	0.002	0.11	49,339	0.021	<<0.001
Plastics	1.0	$\pi \times 0.05^2$	Smooth	1.7	95,315	0.017	0.05

Similarly, the minor losses can be estimated by considering all possible pipe components that are to be included in the system and evaluate its contribution. Table 2 shows the list of possible components to be installed in the system and its corresponding "*K*" values obtained from a textbook [42,45]. The total *K* value is about 4, and this results in minor head loss of about 0.2 m, after assuming velocity of 1 m/s. Combined with the major head loss calculated above, the total would still be less significant compared to the head by elevation difference.

Table 2. In this table, we list of possible minor loss components to be included in the design.

Components	Amount	*K*
Ball valve	1	0.05
Couplings	7	0.08
Entrances	6	0.5
Sudden Expansion	1	0.5

Three (3) pumps for minimum 190 gpm flow rate are selected from McMaster–Carr [46]; (A) High-Efficiency Circulation Pumps for Water, Coolants, and Oil, (B) Harsh-Environment Self-Priming Circulation Pumps for Water and Coolants, and (C) High-Flow Inline Circulation Pumps for Water. Shown in Table 3 are features of the three pumps considered for the system. The price range is approximately USD 1000 to 2000 (2018 price). The specification of these pumps shows that the flow rates and power requirement slightly above our design requirement and the prices are within the budget range.

Table 3. List of potential centrifugal pumps available in McMaster–Carr and its working fluid, maximum flow rate, power requirement, maximum head, and unit price.

Pump #	Product Name	Working Fluid	Max. Flow Rate (gpm)	Power (hp)	Maximum Head (m)	Price (USD, 2018)
A	High-efficiency Circulation Pumps	Water, coolants, oil	190	5	19	~1800
B	High-Flow Harsh-Environment Circulation Pump	Water, coolants	375	3	20	~1105
C	High-Flow Inline Circulation Pumps for Water	Water only	240	3	14.6	~2100

In selecting the pump, students may be asked to setup a simple decision matrix. The matrix allows a group of students to quantify their opinions on factors that weigh the buying decision such as price, capacity, pump power, and type of fluids that can be handled by the pump. In using the matrix, each student in the design team gives score between 1 to 3 (as there are three pump candidates), indicating the level of preference, for each category. For example, a student who prefers to buy pump B as they think that the price is the most reasonable, not necessarily the cheapest, should score 3 for pump B under the category of "Price". If the student considers that they prefer another pump for the power, then they can put a score of either 2 or 1 under the "Power" category for pump B. Students involved in the pump selection process collect their matrix, and the pump with highest total score should be selected. Table 4 shows an example of such decision matrix performed by one student. This example shows that the student preferred to purchase pump B over the other two pumps.

Table 4. An example of decision matrix for pump selection process. Each student involved in the project may participate in the selection process by filling out this table.

Product	Working Fluid	Flow Rate	Power	Maximum Head	Price	Total Score
A	1	1	3	2	2	9
B	3	2	2	3	3	13
C	2	3	1	1	1	8

The selected 3"-self-priming pump is manufactured by AMT, a Gorman–Rupp Company based in Mansfield, OH, USA [47]. This is a 3-phase pump that draws 10 Amps at 208–230 V. The stainless-steel impeller is encased in cast iron and capable of delivering 375 gpm maximum capacity of water at its 3450 rpm. The pump is selected due to its availability, capability in delivering 200 gpm, and affordability. Testing of the pump after the complete assembly indicates that the flow speed reaches 60 cm/s at the maximum 50 Hz input.

3. Static and Dynamic Analysis of Supporting Frame

Structural analysis is paramount to determine the integrity of the flow tank and its supporting frame. Figure 2 panel (a) shows the final Computer Aided Design (CAD) drawing prepared by students and its major components. The final product of the water flow tank is shown on panel (b). The final total length of the water tank system is 2.51 m, which includes the 75 cm long observation chamber (1) and inlet manifold into the observation chamber (5) with enough entry length. In the discussion below, only the structural analysis of the supporting frame is presented.

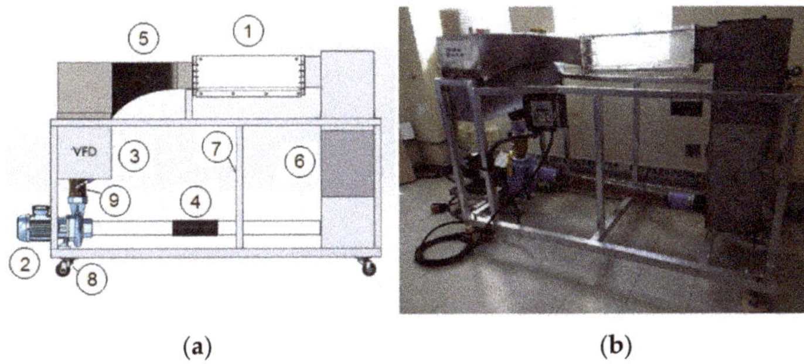

(a)   (b)

Figure 2. Left panel (**a**) shows the drawing of the water tank and its major components; 1. observation chamber, 2. centrifugal pump, 3. variable frequency drive or pump controller, 4. return pipe to the pump, 5. inlet manifold into the observation chamber, 6. holding tank, 7. structural frame, 8. wheels, and 9. ball valve. On the right, panel (**b**) shows the final product of the water tank supported by frames and wheels.

The material chosen for the supporting frame is ASTM A513, which has a yield strength of 220.6 MPa (32,000 psi). Considering a safety factor of 1.3, the maximum yield strength for the system is σ_y = 169.72 MPa (24,615 psi). The beam profile used is 5.08 × 5.08 cm^2 hollow square tubing with a thickness of 2.11 mm. The static analysis involves calculations of dead loads acting on the system, which includes the weight of the water and the tank. Table 5 lists all possible static load items and its estimated amount. The list was prepared by students as part of the full static and dynamic structural analysis of the supporting frame. Several possible designs were analyzed and results from one design is presented here.

Table 5. The table shows main loads to be considered in the design of the supporting frame.

Components	Weight (kg)	Description
Water	362.87	Total weight of water at full capacity
Observation chamber	11.16	This section is made out of acrylic
Pump	43.54	Self-priming centrifugal pump
Piping	4.72	Total weight of steel channels
Manifolds	29	Inlet and outlet chambers, made out of stainless steel
Steel strucure	4.99	Constructed out of 2 × 2 square tubings
Total dead load	456.28	

The maximum deflection of 0.0254 mm was discovered on one of the cross beams directly supporting the observation chamber. This value is considered very minimal, and it indicates that the selected channel is sufficiently strong. In this model, the frame is assumed to be supported by pins on its four bottom corners. Static loads, representing items listed in the Table 5 are applied as point loads on several locations on top beams. The water weight is distributed throughout several points on the frame. Figure 3 displays the deformation of the frame in the full 3D static analysis. As is expected, the bottom beams show minimum deformation due to the supports, while the top beams show large deformation due to the application of load.

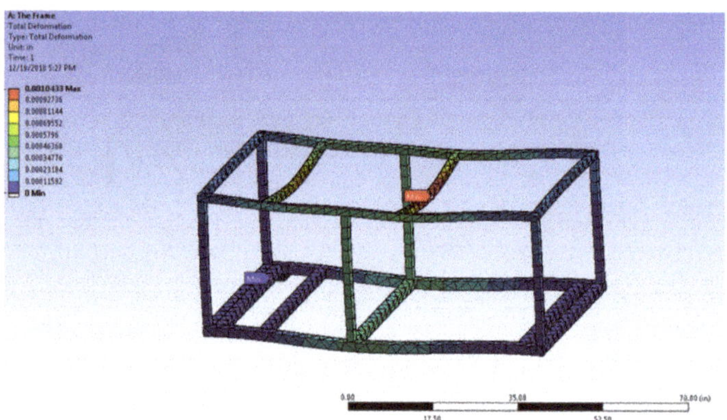

Figure 3. The deformed shape of the supporting frame due to the dead loads indicate maximum deflection of about 0.001 inch occurs on the cross beam that supports the observation chamber. The 3D static analysis in ANSYS is performed by members of capstone senior design team.

Additionally, a dynamic modal analysis was performed using ANSYS (ANSYS Inc., Canonsburg, PA, USA) to check the mode shapes and natural frequencies of the supporting system. Outcomes from the modal analysis indicate possible resonance and amplification of deformation when the pump's operating frequency is in the vicinity of the natural frequency of the support system. Due to the simple vibration load that is being considered, the modal analysis was set to find only the first six (6) natural frequencies. The resulting natural frequencies of the frame are presented in Table 6. These natural frequencies were used in harmonic response analysis to determine the maximum deflection of the frame. The maximum deformation of 5.51 mm occurs for Mode 5 corresponding to the frequency of 73 Hz. Other frequencies result in maximum deformation of less than 2.54 mm. The first three (3) modes correspond to the operating frequencies of the pump of 0 to 60 Hz. Nevertheless, due to the rigidity of the structure, the amplification was found to be very minimal. The outcome suggests that the structure and material selection for this flow tank is sufficiently strong to sustain the vibration load by the pump. The study also shows that only the first 3 (or at most 4) modes need to be considered as the maximum operating frequency of the pump is only 60 Hz. Table 6 shows the six (6) first natural frequencies of the supporting frame. The maximum deformation from the harmonic analysis, based on the given static load, is also shown in this table. A maximum deformation of 5.08 mm can potentially occur for Mode 5, corresponding to ~73 Hz. Other modes show maximum deformation of less than 2.54 mm.

Table 6. The first six natural frequencies of the supporting frames are listed in this table, along with the maximum deformation associated with the harmonic analysis.

Modes	1	2	3	4	5	6
Freq. (Hz)	23.53	36.21	39.28	70.91	73.18	86.59
Max δ (mm)	2.108	2.489	2.108	2.007	5.512	2.108

The first, second, and third modes represent predominantly lateral deformation of the top beams due to the rigid support of the bottom beam. The lateral deformation can occur harmoniously between the parallel top beams (Modes 1 and 3), but it can also occur non-harmoniously as in Mode 2. Mode 3 shows lateral deformation in the long direction of the frame. While the dynamic analysis shows sufficient stiffness and unlikeliness for large amplification, rubber footings were placed underneath the pump's platform to damp the vibration.

4. Computational Fluid Dynamics

Computational fluid dynamics (CFD) analysis suggests useful information for the design of various channels of the flume. Nevertheless, the CFD simulation and analysis provide hands-on exercise materials that allow students to better understand the dynamics of flow phenomena through visualization and parametric studies [48]. Possible analysis to be performed ranges from two-dimensional to full three-dimensional study of either unsteady or steady water flow. Examples of topics to be studied include studying the flow characteristics in the observation chamber, pressure losses across the flume, undesired circulation zones, areas of turbulence, extreme stresses caused by fluid flow, etc. The computation domains of such studies can be either developed from scratch or imported from the structural mechanics CAD designs. In this section, results from a full 3D steady state analysis of water flows in a semi-loop model of the tank are briefly presented. The study is performed using the commercial package COMSOL Multiphysics 5.5 with CFD module (COMSOL Inc., Burlington, MA, USA). Most of the computations are performed using HP ProBook 640 G2 Notebook PC (Hewlett-Packard Company, Palo Alto, CA, USA). operated using 64-bit Windows 10 Enterprise version 1909 with a total capacity of ~465 GB. The computer has an installed RAM of 16 GB and is equipped with Intel® Core™ i7-6600U CPU @ 2.60 GHz. One model is run using a desktop HP Compaq Elite 8300 CMT computer equipped with Intel® Core TM i7-3770 CPU at 3.40 GHZ and 16.0 GB RAM.

The computation domains can be differentiated into several parts, as shown in Figure 4 panel (a). The pipe inlet domain represents the vertical cylindrical thick PVC pipe channel that delivers water from the pump's exit into the main part of the water tank. This channel is 8 cm in internal diameter and the length is 40 cm. The pipe is extended into the diffusing chamber before it splits into two horizontal branches with the same internal diameter. The resulting T-pipe allows the water to be distributed in the diffusing chamber. The T-pipe also prevents the water to "shoot up" and put high pressure on the ceiling of the diffusing chamber. The next important domain is the converging chamber that functions to collect the water and channel it into the observation chamber. A flow straightener could be included in the converging section, however this part is not modeled due to its large amount of mesh. This simplification is justified by the computation results that indicate straight uniform flow in the observation chamber (Figure 6). The observation chamber is a 60 cm long prismatic pipe with a square cross section of 15×15 cm^2. At the end of the chamber, the water is collected in a rectangular drainage tank of $60 \times 30 \times 40$ cm^3. The CFD model does not include a horizontal transparent plastic pipe that returns the water back from the drainage chamber into the centrifugal pump.

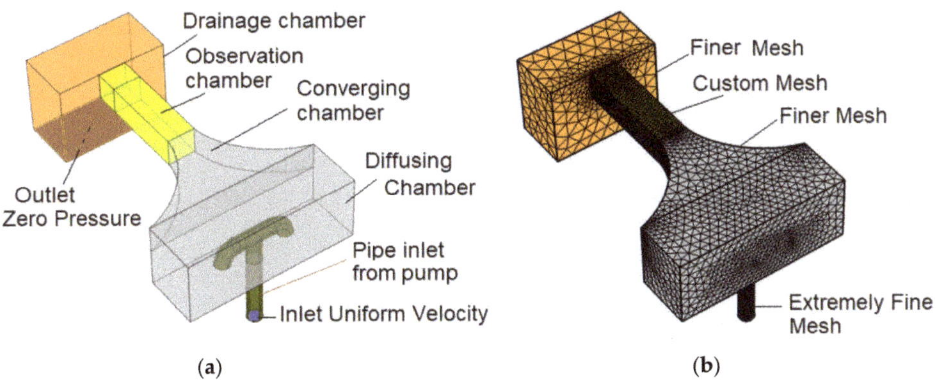

Figure 4. Panel (**a**) shows the computational domains involved in the analysis and panel (**b**) shows the non-uniform mesh generation implemented for the analysis.

Several simplifications are implemented in the CFD model. For instance, the actual drainage chamber is converging downward, but in the model only the top 40 cm section is included. It is assumed that the inclusion of this part is sufficient to recreate necessary backflow effects that occur in the observation chamber. The converging chamber attached to the distal face of the observation chamber is made to be plain horizontal, parallel to the observation chamber. The actual configuration of this curved chamber includes a slight decline in the vertical direction into the entrance of the observation chamber. Lastly, in the CFD, the observation chamber is assumed to be a closed rectangular pipe. The actual observation chamber is open on its top side to allow direct access to the water.

The inlet boundary condition is located at the free end of the inlet pipe that represents the conduit between the centrifugal pump and water tank body. Not knowing the exact velocity profile produced by the centrifugal pump, a uniformly distributed velocity profile is assumed at the inlet. The exit boundary condition is located at the bottom of the drainage chamber and a zero-pressure outlet is assumed here. The actual pressure level required by the centrifugal pump to produce the flow can be obtained by adding known hydrostatic and dynamic pressures occurred at the exit level. The internal wall is assigned to the wall of the T-pipe domain, as this pipe is located inside the diffusion chamber. The physical T-pipe has a non-zero pipe thickness that has been ignored in this CFD modeling. This important feature certainly leads to a convenience and efficient mesh generation. The remaining walls are assumed to have zero velocity. Figure 4 shows the complete model, domains, boundary conditions, and partial mesh generation that results in non-uniform mesh size throughout the domains.

CFD analysis requires a delicate balance between demands for accuracy of the outcomes versus the computational cost that are the direct results of both mesh generation and complexity of the selected flow model. Table 7 lists CFD models studied in this work and their mesh generation scenarios. Models As and Bs are all meshed using "Global Mesh" method, where the same mesh size or scheme is applied uniformly on all domains of the model. However, models As are computed by means of "Laminar" flow model, while models Bs are executed using "Turbulent" flow model. The C, D, E, and F models are discretized using "Partial Refinement" method and are executed using "Turbulent" flow model. Effects of the two flow models will be explained later in this section. All models are computed using the HP ProBook laptop, except model D that is executed using the desktop computer. The Global Mesh method is straightforward, but for three-dimensional models this method may result in astronomical element numbers and computation time with possible less-than-sufficient accuracy at places of interest. In the current work, we later show that partial mesh refinement on places of interest should be practiced as a choice of practice in CFD modeling, particularly for three-dimensional modeling. In our work, the "Extremely Coarse" mesh option of COMSOL Multiphysics (COMSOL Inc., Burlington, MA, USA) for our model results in 21,011 elements. Increasing the accuracy by selecting the "Normal" mesh generation option multiplies the element number to 414,735 elements, almost 20 times more than "Extremely Coarse" option. However, the computational time needed to run the model using turbulence k-ε scheme is multiplied from 6760 s to almost 286,538 s, 40 times longer. The partial refinement mesh generation, used for models C, D, E, and F, is performed by selecting domains or regions of interest (ROI) to be discretized finer than other domains. Here, the observation chamber, T-pipe, and inlet pipe are selected as the ROI so that the velocity profiles and other important aspects in the region can be studied more accurately. This method aims to reduce the amount of mesh and computational time, and to avoid unnecessary data collection in non-ROI parts such as the diffusing chamber, drainage chamber, and converging chamber. The Partial Refinement method certainly opens numerous possible mesh arrangements. To limit the study, we focus on mesh refinement of the ROI only. The refinement of the observation chamber sub domain will enhance accuracy of the velocity profiles in the region needed for possible two-dimensional modeling. On the other hand, the refinement of the inlet pipe and the T-pipe is expected to give accurate information on the amount of pump pressure needed in the design. These three domains will be refined using "Extremely Fine" mesh or customized mesh, while the remaining body will be meshed using either "Normal" of "Finer" scheme. For Model C, the ROI are meshed using "Extremely Fine", while the remaining domains are meshed using "Normal". For Models D and E, the mesh for the test chamber domain of Model C is further refined by reducing the "maximum element size" from 0.0318 to 0.015 m. Model D is the only model that is computed using the HP desktop computer. It should be noted that the "Normal" size used in the non-ROI domains of models C, D, and E employs "maximum element size" that is about triple than that of the "Normal" size used in Model B4 with global mesh method. Consequently, although the ROI domains of Models C, D, and E, are finely meshed, the total number of elements of these models are less than that of Model B4. The difference in the "maximum element size" is due to the different "Calibration" option offered by COMSOL (COMSOL Inc., Burlington, MA, USA). The mesh configuration for Model F is the finest among models with partial refinement mesh. Here, the mesh configuration of the ROI used in Model D is maintained, but the remaining domains are meshed using "Finer" mesh. This refinement results in 355,730 elements, and the computation time is 145,860 s, only half of the time needed by model B4.

Table 7. This table lists CFD models used in this project and their associated number of elements and computation times. Only models A1 and A2 are executed using Laminar fluid model. All other models are executed using Turbulent model—RANS K-ε. All models are run in the HP ProBook, except Model D.

Model #	Mesh Method	Fluid Model	Number of Elements	Computation Time (s)
A1	Partial Refinement	Laminar	91,162	3630
A2	Partial Refinement	Laminar	289,796	21,787
B1	Global Extremely Coarse	Turbulent	21,011	6760
B2	Global Extra Coarse	Turbulent	43,753	17,861
B3	Global Coarser Mesh	Turbulent	76,983	34,030
B4	Global Normal Mesh	Turbulent	414,735	286,538
B5	Global Fine Mesh	Turbulent	9,797,088	Failed to converge
C	Partial Refinement	Turbulent	113,624	25,626
D	Partial Refinement	Turbulent	160,774	20,680
E	Partial Refinement	Turbulent	160,378	35,046
F	Partial Refinement	Turbulent	355,730	145,860

Shown in Figure 4 panel (b), results of Partial Refinement strategy where high mesh density ("Extremely fine mesh") is applied on the observation chamber, pipe inlet, and T-junction domains. The remaining domains are discretized using "Finer" mesh. The minimum and maximum element sizes employed in the observation chamber are 5 mm and 15 mm, respectively. The application would result in at least 10 computation nodes across the 15 cm wide observation chamber, which should be enough for accuracy.

The volumetric flowrate (Q) and mean velocity ($\overline{V}$) in the observation chamber in the inlet pipe and observation chamber will be used to measure the accuracy of the solution. The volumetric flowrate across planes perpendicular to the long direction of the observation chamber must equal to that of the inlet. This flowrate can be obtained from the uniform inlet velocity V_{in} provided at the pipe inlet; $Q = V_{in}\frac{\pi}{4}D^2$, where D is the pipe diameter. The mean velocity in the observation chamber can be easily estimated using $\overline{V} = \frac{Q}{A_{ch}}$, where the cross-section area of the chamber is known as $A_{ch} = 0.15^2 = 0.0225$ m^2. As the outlet at the base of the drainage chamber is assumed to have a pressure drop, the flowrate at this location should also be verified.

Second only to the mesh generation, the selection of appropriate physical modeling of the fluid is the most crucial aspect of CFD modeling. The selection between Laminar and Turbulent models is often determined by Reynolds numbers involved in the problem. The laminar threshold for fully developed flow in long pipe with circular cross sections is typically around 2300, while for an open channel flow is 500. Students have to understand that the change from laminar to turbulent flow is a gradual transition, and therefore it is not practical to point out a single number to differentiate the two regimes. As pointed out by Lowe, the confusion over the critical Reynolds number is not new in the Fluid Mechanics community [49]. Nevertheless, smaller Reynolds numbers should be treated as turbulent when the complex geometry of the domains include circulation zones that are best captured using turbulent models. In the current water tank design, the maximum expected mean velocity occurring in the 15 × 15 cm square observation chamber is 60 cm per second, which results in the Reynolds number of 90,000, which puts the model under turbulent category. A long list of turbulent models is available in COMSOL Multiphysics software (COMSOL Inc., Burlington, MA, USA). For this project, we selected the standard k-ε model to serve our purpose. This suggested model is selected due to its popularity for industrial applications, easy convergence, and low memory requirement [50]. Alternatively, the analysis on such water flume can be performed using k-ω SST model [51]. However, a comparison study performed on a water flume with 5 m wide and 17 m long test section suggest that the outcomes from the two different Turbulent models do not show significant differences [52]. Heyrani et al. compare the performance of seven (7) turbulent models used in steady state modeling of a venturi flume. It was discovered that the k-ε model performs slightly better than the k-ω SST model [53].

Figure 5 shows the accuracy and computation times that are plotted against the mesh number for different models. The graph shows that Turbulent model with Partial Refinement (rectangular markers) produces the best accuracy among the three models (Laminar model with Partial Refinement, Turbulent model with Global Mesh, and Turbulent model with Partial Refinement). Moreover, its computing time is less than the Turbulent model with Global Mesh (circular markers). The computing time of the Laminar model with Partial Refinement (triangular marker) is the lowest, but the accuracy is also very low. The laminar model should only be used to make sure that the CFD model can be properly executed. High accuracy can be easily achieved by the Turbulent model with Partial Refinement, even with a low amount of mesh.

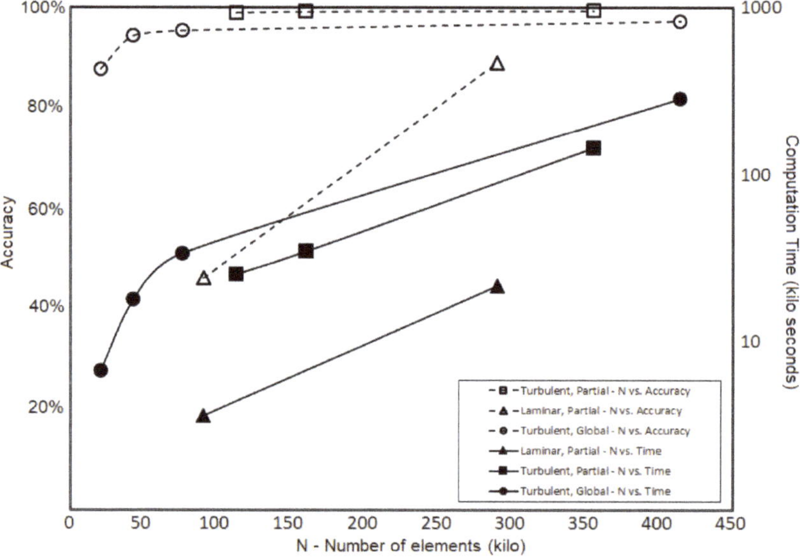

Figure 5. Convergence studies—data with dark labels indicate relationship between number of mesh and computation time, while the white markers indicate the relationship between the number of mesh and accuracy. The Laminar model shows the lowest accuracy and computing time. The Turbulent models can take 10 times longer time than the laminar flow, but its accuracy is very high.

Streamlines in the water tank when the mean inlet flow is 3 m/s are shown in Figure 6. The streamlines are obtained using "standard point controlled" method and using "Number of Points" entry of 80 points. The streamlines on the left and right are obtained when the computation modules used are Laminar model and Turbulent k-ε model, respectively. The streamlines from the Laminar model are parallel and straight in the observation chamber, but they are clearly absent from the circular flows in the drainage chamber, as well as in the diverging chambers that are demonstrated by the Turbulent model.

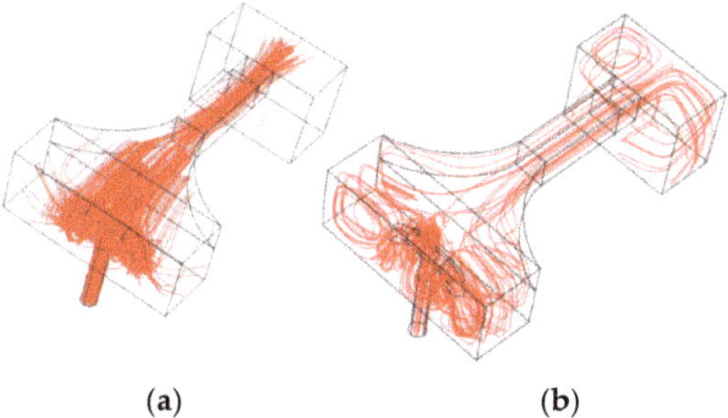

(a)   (b)

Figure 6. Three-dimensional streamlines obtained from an (**a**) Laminar model and (**b**) Turbulent model. The Turbulent model captures well the twin circulation zone occurring in the drainage tank and the collecting chamber. The Laminar model fails to capture the circulation zone despite of the use of large number of elements.

Observing the axial velocity profiles along the observation chamber, the Laminar model shows reasonable parabolic shapes (data are not shown here), however the amount of flow rate and mean velocity show deviation from the correct values. The profile of axial velocity along the observation chamber obtained from the Turbulent models are depicted in Figure 7. Each velocity profile is obtained at the same horizontal mid-section of the observation chamber, but at different distances "Y" from the starting of the chamber (taken as the interface between the converging chamber and observation chamber). The development of the velocity profile can be observed in this figure as the profile changes into a more parabolic shape as the Y is increasing. The velocity profiles demonstrate the typical characteristic of turbulent profile consisting of a turbulent core in the middle and laminar sublayer near the wall [45]. Note, however, that for the Model B4 (global mesh), the laminar sublayer is pronounced, while for Model F (partial refinement), it is very difficult to observe due to its small thickness. The accuracy of Model F in predicting the expected flowrate and mean velocity however is higher than that of Model B.

The velocity map and streamlines, viewed from the top of the flume, can be seen in Figure 8. The flow direction goes from top to bottom (diffusing chamber to the drainage chamber). Panel (a) on the left shows the velocity map of laminar model, while panels (b) and (c) are from the turbulent model, but with different mesh schemes. All models present reasonable straight parallel streamlines in the observation chamber. The Laminar model shows gradual increase in speed as it enters the observation chamber, as can be observed from the changing color. On the contrary, the Turbulent models both show consistent average velocity along the observation chamber. The Laminar model fails to produce the twin circulation zones that occur in the drainage chamber. These circulation zones are physically observed during real experimentation. The two Turbulent models clearly show these circulation zones with slight difference. The circulation zones by the Model B4 (Figure 8 panel (b)) are shown to be less chaotic than that of Model F (Figure 8 panel (c)).

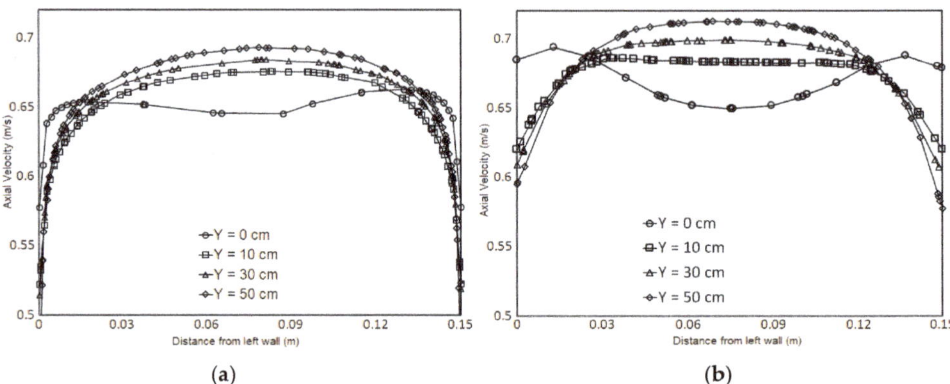

Figure 7. Axial velocity profiles obtained at the horizontal mid-section of the observation chamber taken at different distances Y from the entrance obtained using (**a**) Global Mesh refinement and using "Normal" scheme (Model B4) and (**b**) Partial Refinement method (Model F).

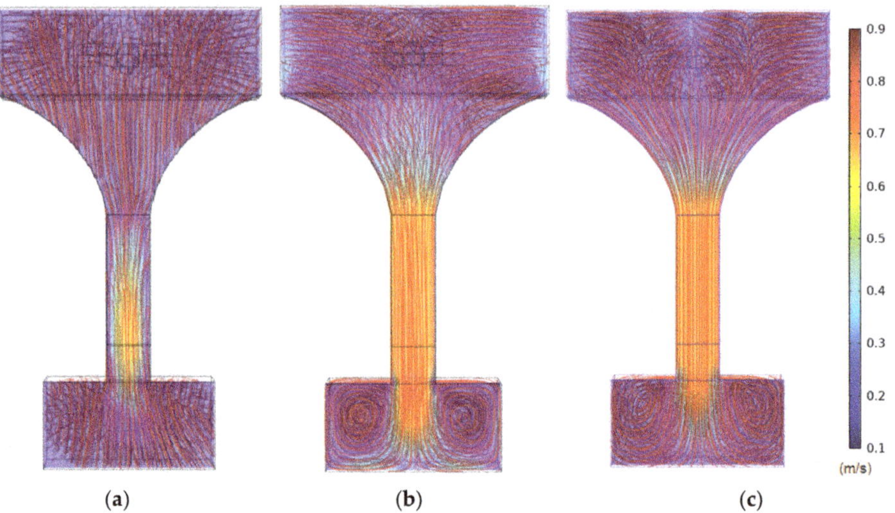

Figure 8. Streamlines in the observation chamber and the diffusion chamber as well as in the collecting chamber produced by (**a**) Laminar model, (**b**) Turbulent model with Global Mesh "normal" refinement, and (**c**) Turbulent model with Partial Refinement method. The mean velocity of these models at the inlet is 3 m/s.

Vorticity is a local measure of the rotational motion of fluid particle relative to its own centroid, while the circulation can be considered as the measure of the rotational motion in global sense, relative to a distance reference point. After passing the observation chamber, water enters a short drainage chamber that is three times wider than the observation chamber but much deeper in the vertical sense. Here, due to the high speed in the observation chamber, the water shoots into the back wall of the chamber, but it also starts to fall due to the gravity. Hence, the stream splits into a pair of almost identical spiraling flows downward. Shown in Figure 9 are streamlines projected on a two-dimensional plane cut across approximately the middle height of the drainage chamber. On each panel, the streamlines demonstrate a pair of spiraling circles running in opposite directions. Note that the flow is coming from the observation chamber located, according to these pictures, relatively above the rectangular areas shown in the panels. The color map describes local vorticity values, and the scales indicate that, on average, the two pairs have a similar amount but in opposite directions. In this figure, each panel represents a circulation zone in the chamber from the same Reynolds numbers, around 72 K (uniform inlet velocity of 3 m/s), but each is computed using different mesh density and element types used in the drainage chamber domain. On panel (a), the fluid model is laminar, and the elements used in the drainage chamber is tetrahedral. On panels (b), (c), and (d), the k-ε Turbulent model is used, but the number of elements used (only in the drainage chamber) are 5498, 14,681, and 61,052, respectively. On panel (b), only tetrahedral elements are used, but various elements are used for results shown in panels (c) and (d). It can be seen that the utilization of various elements results in smooth vorticity map and streamlines that are less discrete. The symmetry of the circular flow is captured when the mesh density is very high but, generally, panel (c) shows that essential flow structure is sufficiently captured. The maximum and minimum values on the scales indicate the amount of vorticity details that can be captured by the respected mesh density. On panels (b), (c), and (d), while the maximum and minimum vorticity are different, the average values inside and around the cores of the circulation area is around 4 to 5. On panel (c) and (d), zones with large vorticity up to 20/s can be captured, but these areas are concentrated near the walls. As is expected, the large vorticity occurs near the walls where large shear strain is expected to occur. On panel (a), the Laminar model results in a pair of seemingly symmetric circulation zones. The zones, however, appear to be shifted to the side walls. Moreover, the vorticity map shows some level of local spins, but the distribution is different from that shown in other panels. The laminar model certainly has failed to produce the desired outcomes, despite of the high mesh density that has been employed.

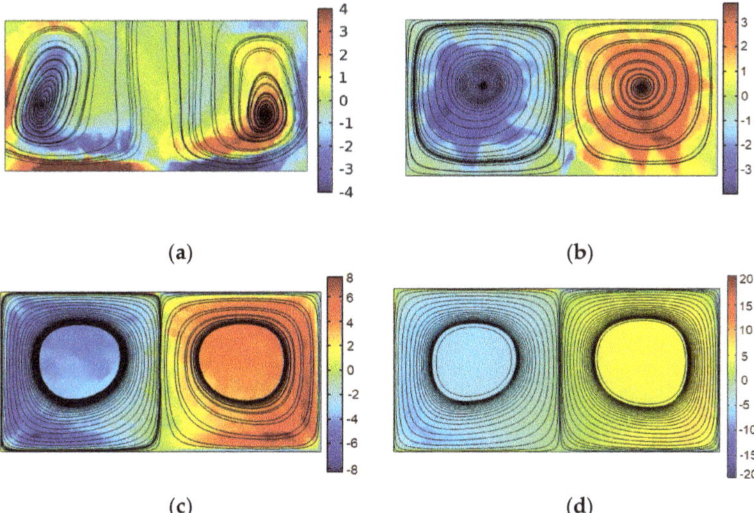

(a)

(b)

(c)

(d)

Figure 9. Panel (**a**) shows the circulation zone captured using laminar modeling; (**b**) Partial Refinement results in 5498 tetrahedron elements in the drainage tank, a minimum element quality of 0.6541, and an element volume ratio of 0.001275; (**c**) Global "Coarser Mesh" mesh with Turbulent model with 14,681 mixed elements, a minimum element quality 0.1307, and an element volume ratio 0.0021; and (**d**) Global Mesh refinement with "Normal" mesh with Turbulent model with 61,052 mixed elements in the drainage tank, a minimum element quality of 0.1251, and an element volume ratio of 9.8×10^{-4}.

5. Fabrication and Manufacturing

Welding was the most challenging process of the manufacturing activity in this project. All fabrication (cutting, folding, welding, assembly, etc.) had to be performed mostly off-site at a private workshop due to the space limitation for hot works in our campus. Particularly, the welding of individual manifold was the primary challenge. The inlet and outlet manifolds needed to be designed, cut, welded, and assembled without the benefit of a fully equipped fabrication facility. Furthermore, the limitations of the available workshop dictated that tolerances could be held only down to 1/4 " (~6.35 mm) for most features of size and 1/2 " (~12.7 mm) for some larger features. The choice of 18 GA (0.049" or 1.27-mm) Austenitic Stainless Steel for the manifolds dictated that TIG (Tungsten static electrode Inert Gas shielded) welding was necessary to accomplish production. Alternative processes were ruled out for manifold production due to heat transfer concerns that would lead to excessive warpage of the fabricated sheet metal structures. Additional requirements of the design included the need for flush weld joints which would be less easily accomplished than other welding processes. Shown in Figure 10 one of the completed holding tanks (left) prior to being assembled with the inlet manifold (top right). The student who performed the welding used many cardboard templates (such as the one shown on panel (a) of Figure 10) to determine the cutting of the steel plate.

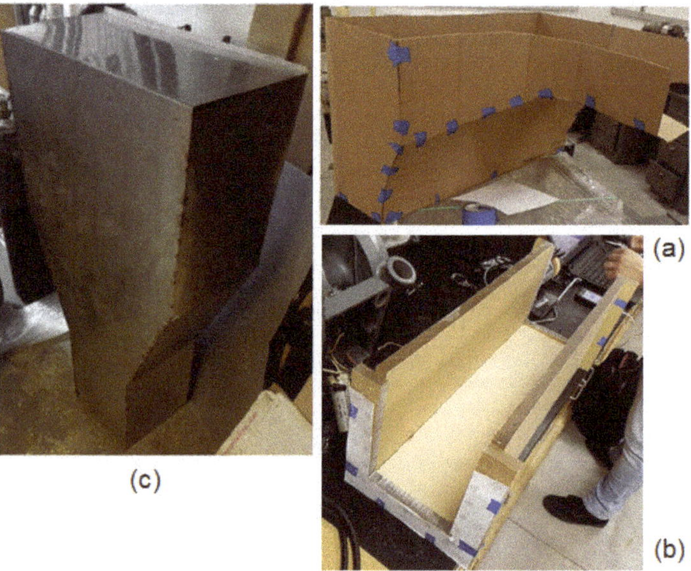

Figure 10. Panel (**a**) shows the cardboard model used in the making of the diffusion channel. Panel (**b**) shows the observation chamber being assembled, and panel (**c**) shows the final product of drainage chamber before being assembled into the main body.

MIG (Metal wire feed electrode Inert Gas shielded) welding was chosen for frame production. Mild Steel Square Tube with 1/16" wall thickness (~1.59 mm) was chosen for the frame and could withstand MIG welding with minimal concerns for warpage. The design did not require the welded joints to be flush with the welded materials. The steel square tubes that were welded together required a coat of rust protective paint due to the material properties that would cause rusting. A Miller Multi-Matic 220 multi-process welder (Miller Electric Manufacturing Co., Appleton, WI, USA) was used for all welding processes, selected due to its versatility of process selection and advanced user interface and arc control systems. The frame had undergone minor modifications from the original design to acclimate for the weight of the pump, transportation, and vibrations that would pass through the system. Jack screws, similar to stands, allowed the system to elevate off the wheels to limit vibration through the floor, level the entire system and to relieve stress from the wheels, as their specifications stated that they would only withstand 544.31 kg.

The frame that the pump was set onto had its own set of jack screws that allowed it to be lifted off the frame for the rubber insulators to absorb vibrations produced by the pump. The $15 \times 15 \times 60$ cm^3 observation chamber was made of 0.635 cm thick transparent plexiglass. The corners are secured using both glue and bolts. To prevent leaking, rubber gasket is applied throughout. Finally, several holes are drilled on the upper edge of the chamber wall (not shown in the figure) to allow experiment platform to be secured onto.

6. Experiment Results and Discussion

The following experiments were conducted using the water tank after its fully built to test its capability:

1. Velocity Measurement of flow in the observation chamber;
2. Tests on flow straightener designs;
3. Flow visualization using dye;
4. Measurements of drags.

6.1. Velocity Measurement of Flow in the Observation Chamber

The speed of the water flow can be controlled by adjusting the rotational speed or frequency of the pump's propeller using the programmable speed controller. In the first experiment, we wish to establish the relationship between the frequency of the propeller and the resulting water velocity in the observation chamber. The average velocity in the observation chamber is measured using a flow meter by Vernier (Vernier, Beaverton, OR, USA) (product number FLO-BTA) shown in Figure 11 panel (a). The sensor is designed for external flow measurement such as flow in open channels and rivers. The measurement range of this affordable sensor is up to 4 m/s, with a typical resolution of ± 1.2 mm/s, and an accuracy of about 1 percent. A data acquisition card, LabQuest Mini, also by Vernier (product number LQ-MINI), is used to collect the data and transfer the data to a laptop through a USB cable. The dynamic data can be displayed by Graphical Analysis GUI software available for free by Vernier. The data can be exported to either Excel, Matlab, or Phyton, for further analysis and post processing. Typical data sampling in our experiment is 1 Hz.

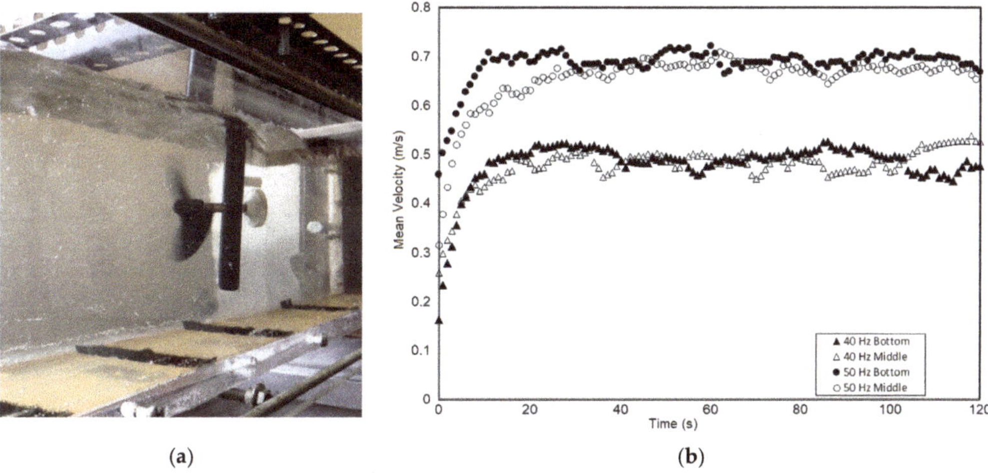

(a) (b)

Figure 11. The panel (**b**) displays the velocity reading by Vernier flow meter versus time when the pump operates at 40 and 50 Hz. The propeller is placed either near the floor of the chamber ("Bottom") or approximately in mid-height ("Middle"). Data suggests minimum difference by the two methods. The panel (**a**) shows the propeller being lowered in the "Middle" position in the observation chamber.

During one set of data collection, we seek to check the speed variability across the chamber's depth. Here, we perform the flow measurement by placing the sensor's propeller near (about 3 cm above) the chamber's floor and approximately at mid-height above the chamber's floor. The data collection for each trial was done at least for 40 s. On each trial, the frequency of the pump is increased from 10 Hz to 60 Hz using intervals of 10 Hz. This data collection results in insignificant variability of the time-averaged speed between the two locations. Examples of data collected over a period of 120 s for pump frequencies of 40 Hz and 50 Hz are shown in Figure 11 panel (b). The transient regime for about 25 s at the beginning of data collection can be easily identified in both data. The steady state regime maintains a constant average speed and indicates small fluctuations. Data with solid labels are taken for measurement at "Bottom", while data with open labels are from "Middle". Data indicate that the placement of samples in between these two locations would result in more-or-less identical outcomes.

6.2. Tests on Flow Straightener Designs

The second set of experiments involved velocity measurement in the presence of a flow straightener. The flow straightener functions to dissipate the turbulence water flowing into the observation chamber. The three (3) designs by students are shown in Figure 12; Models 1, 2, and 3 displayed in panels (a), (b), and (c), respectively. The cross section of all the models is 15×15 cm. Model 1 was constructed by PVC tubes of the same diameter (12.7 mm) cut at the same length (~10 cm) that are glued along its long sides. These cut pipes were then arranged in parallel, and bonded together on its long sides to form a channel. This model has the largest area of holes. Model 2 is a 3D-printed PLA plastic rectangular block of $15 \times 15 \times 5$ cm^3, furnished with 12-by-12 holes, each about 1.2 cm in diameter. Model 3 is a combination of parallel 0.5″-PVC (12.7 mm in diameter) pipes that are hold together using 3D-printed blocks with holes on their ends. Among the three models, Model 1 is the cheapest, but is very tedious to make, as each pipe has to be glued one by one. During the bonding process, the pipes can be placed in between two parallel walls. Model 2 is the most expensive, when the printing time is included in the calculation. While the SolidWorks (Dassault Systèmes SolidWorks, Inc., Waltham, MA, USA) CAD model for this model is easy to make, it took more than 12 h to print the model using a regular desktop 3D-printer. The thin walls separating the holes require the model to be printed with high density (high infills), and this prolongs the printing process. While involving 3D-printed blocks, Model 3 is inexpensive, as the holes on these blocks are well separated from one another, requiring less printing density compared to Model 2. Nevertheless, the arrangement of the pipes into these blocks is quite tedious. In terms of durability, it was quickly found that Model 1 disintegrated quite easily after several tests in the water. Model 2 also shows signs of plastic fibers detached from the bulk structure after several tests. Model 3 is the most durable, but its effectiveness is the lowest.

The data of flow velocity, taken when the pump's speed is 10 Hz, versus time for different models of flow straighteners, are presented in Figure 13. The blue, red, black, and green lines indicate water velocity without flow straightener, with Model 1, with Model 2, and with Model 3, respectively. On average, the velocity data for Model 1 (red line) is very similar to the case where no flow straightener is installed (blue line). Some degree of deviation from the flow with no straightener (blue line) is reflected by data using Model 2. Compared to other models, Model 3 shows the largest deviation from the flow with no flow straightener (blue line). Clearly, the Model 1 produces the least flow disturbance in terms of time-averaged velocity outcome. The expected outcomes can be attributed to the difference in the cross-section areas of the holes of each model. The largest area of holes for Model 1 is contributed by the holes by the pipes and the holes in between these pipes. Nevertheless, the Model 2 can be seen to produce a more stable velocity (black line) across the time compared to Model 1 and Model 3.

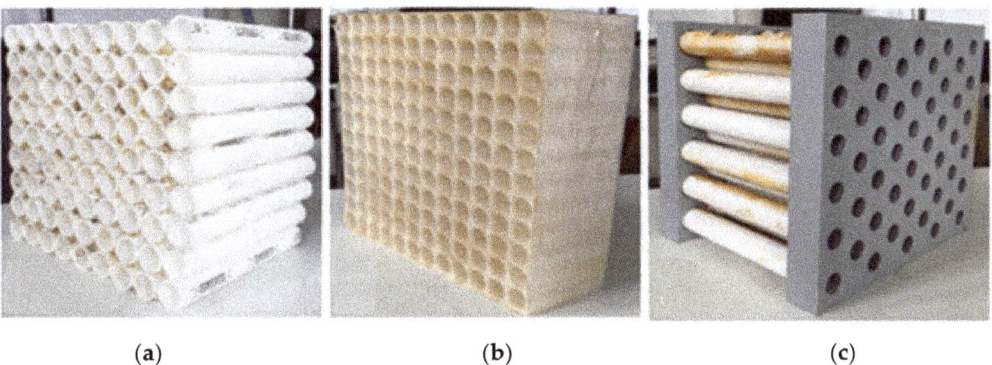

(a)   (b)   (c)

Figure 12. Models of flow straightener used in the flow experimentation—(**a**) Model 1 is made of cut PVC pipes that are glued together in parallel arrangement. (**b**) Model 2 is a full 3D-printed apparatus designed in SolidWorks. (**c**) Model 3 is an assembly of PVC pipes fitted into 3D-printed block with holes.

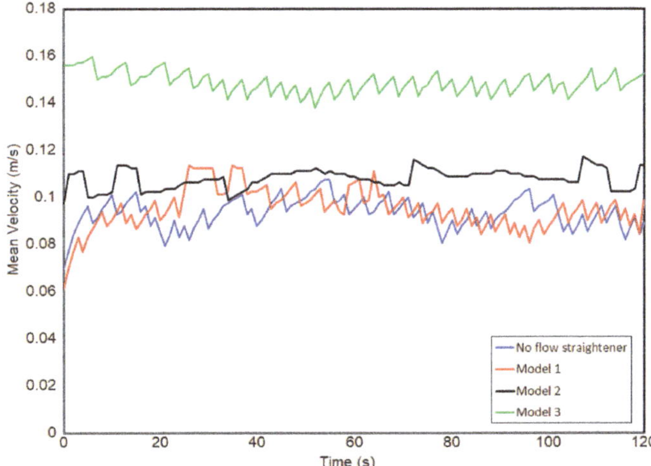

Figure 13. This figure shows mean velocities from the flowmeter for various flow straightener models taken for a period of 120 s when the pump speed is 10 Hz. The blue line is the velocity of water flow without flow straightener. The velocity after the installation of Model 2 shows stability compared to other models.

Time-averaged (over the steady state period) water flow velocity obtained from the flow meter for various pump's frequencies are presented in Figure 13. The data points are labeled using white circles (no flow straightener), white square (Model 1 of flow straightener), and white triangle (Model 2 of flow straightener). A linear relationship of

$$\overline{V} = 0.014 \times Fr - 0.042 \left(R^2 = 0.9999 \right) \tag{4}$$

was discovered for data obtained without the flow straightener. Here, $\overline{V}$ (m/s) and Fr (Hz) are the mean velocity and pump's frequency, respectively. The non-zero intercept suggests the minimum frequency (associated with a minimum water velocity) needed to turn the propeller of the flow meter of about 3 Hz. The known linear relation between the pump's frequency and average velocity establishes the formulation of $\overline{V}$ *vs.* *F.* for future experimentation. Velocity data obtained using the flow straightener Model 1 shows similarity to the data obtained without the flow straightener. On the other hand, Model 2 produces velocity data that deviates at high pump's frequency. The linear relationship is also quantified as $\overline{V} = 0.0166 \times Fr - 0.0663 \ (R^2 = 0.9951)$.

On a separate, simple experiment, two students measured the water velocity by means of a Lagrangian approach. Here, the time needed by a ~1 × 1 × 2 cm^3 cork to travel for 50 cm is measured using a manual stopwatch. This measurement results in smaller velocity compared to ones measured using the propeller. These Lagrangian velocities are displayed in Figure 14 using dark cross legends. The lower estimation can be attributed to the drag produced by the weight of the cork, as well as accuracy of the manual data collection. Nevertheless, the manual measurement using cork provides a simple and quick validation of the Eulerian method employed in the automatic measurement using digital flow meter.

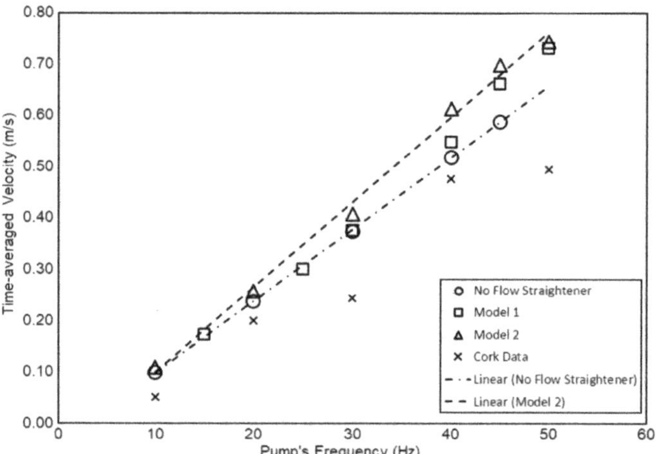

Figure 14. This figure shows the comparison of time-averaged velocity data versus pump's frequency. The data are obtained using Models 1 and 2 of the flow straighteners. The data from measuring the water velocity using floating cork are presented using cross symbols. Data from Model 1 shows minimal deviation from the reference data. The cork data underestimates the reference data, but it gives good estimate for a quick verification of the flow meter.

6.3. Flow Visualization Using Dye

The effects of placing the flow straightener in the water stream was examined by means of dye injection. While this method is simple and affordable, students performing the tasks can learn valuable lessons. For example, students realize that the fluid density had to be carefully chosen, as a fluid that was too light would rise to the top (float) and the fluid mixture was too dense then it would sink to the bottom. The fluid was also mixed with food coloring to make it easier to observe. After running several tests using hand soap, oil, sugar, and honey, it was concluded that honey diluted with a small amount of water gave the best results. These tests were run by using a syringe with the mixed fluid. Figure 15 shows the difference in flow behavior through the utilization of injected dye. The left picture (panel (a)) shows the flow turbulence when there is no flow straightener inserted in the chamber. Panels (b) and (c) show the resulting flow behavior when Model 2 is used. Here, the flow is almost straight and uniform with minimum turbulence characteristic.

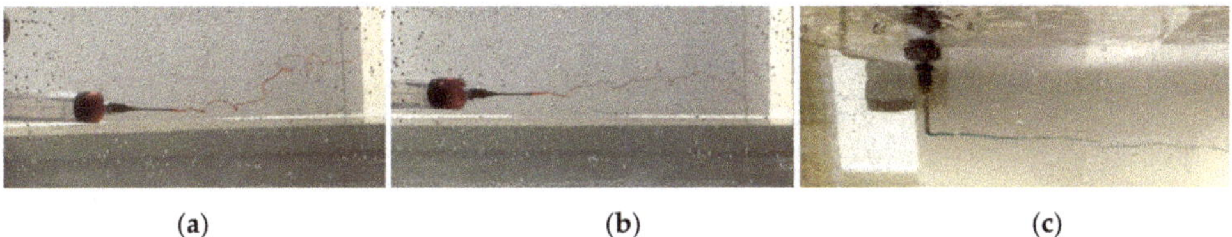

(a)   (b)   (c)

Figure 15. The flow behavior at about 10 cm/s. was observed using color dye injected by a syringe. The left panel (**a**) shows the dispersion of the dye when no flow straightener was used. Turbulence characteristic can be seen from the dyes. Panels (**b**,**c**) shows the flow characteristic when Model 2 is used as flow straightener. Straight streamline is easily observed in this case.

6.4. Measurements of Drags

Objects exposed to fluid flow experienced both drag and lift forces. Being able to demonstrate this phenomenon for objects in the water flow is one important objective of the water flume project. Experimentation on drag reported below was performed by students in Thermal Fluid Lab course on Fall 2019. The drag force acting on objects, such as long prismatic cylinders and bars, is registered using a force sensor from Vernier (Go Direct $^\circledR$ Force Sensor—GDX-FOR) that uses strain gauge to measure as small as ± 0.1 N up to ± 50 N of axial force, sufficient range for our fluid application. A sampling rate of 50 Hz, out of a possible maximum of 1000 Hz, is used in our application. This device can be either directly connected to a laptop using a USB cable, or wirelessly using a Bluetooth to a mobile device. The stream data can be displayed on a laptop and recorded using a software Graphical Analysis available for free from Vernier. The force sensor can also be connected to the data acquisition card by Vernier—LabQuest Mini. This allows simultaneous measurement of multiple sensors with other data such as water speed, temperature, and lift. The $\sim 7 \times 4 \times 5$ cm^3 sensor device is placed on $\sim 15 \times 15 \times 1.5$ cm^3 (1.36 kg) aluminum sled supported by four smooth wheels (rollers) that rest on the walls of the observation chamber (Figure 16). The sled was machined to include a hollowed pin on its top to secure the force sensor. A hole was drilled at the center of the sled to allow samples to be secured using bolt and nut. The other end of the bar is free. The force registration is performed by aligning the sensor's hook to the flow direction and securing it to a reference point fixed to the chamber's wall. The four smooth wheels reduce the contribution friction caused by the sled's weight and maximize the registration of the hydrodynamic drag by the sensor.

(a)   (b)

Figure 16. The figure on the left panel (**a**) shows the force sensor supported by the aluminum platform (sled) with its wheels placed along the wall of the observation chamber. The panel (**b**) on the right shows the complete setup of the drag sensor secured on the observation chamber with the flow meter placed upstream to allow simultaneous measurements of flow speed and drag forces. The water flows from left to right.

The drag measurement has been conducted for several objects: plastic cylinders, 3D-printed cylinders, solid aluminum cylinders, and square bars, as well as rotating bladeless turbine models. A preliminary attempt using a 3D-printed fixture to hold hollow plastic cylinder samples and 3D-printed samples resulted in promising data. However, it also indicated the need for rigid and sturdy support and samples that do not float. The force data is converted into Drag Coefficient, C_D using the following formulation [43,44]:

$$C_D = \frac{2\overline{F}_D}{\rho \overline{V}^2 A}$$ (5)

where $\overline{F}_D$, ρ, $\overline{V}$, and A are the time-averaged drag force, water density, time-averaged water velocity, and frontal area of the cylinder sample, respectively. The frontal area only accounts for surface perpendicular to the water flow. Only data from the 16 mm diameter aluminum cylinder and square aluminum bar are presented in this report.

Shown in Figure 17 is an example of raw data of the force obtained from the force sensor for a period of 120 s when the data is taken at 50 Hz sampling frequency. The average velocity of the water flow here is about 59.4 cm per second. The oscillation of drag force is expected to be caused by the unsteadiness of the flow and the vortex shedding that occurs when fluid flow passes the object. The average force is about 0.38 N, and its standard deviation is ±0.072 N.

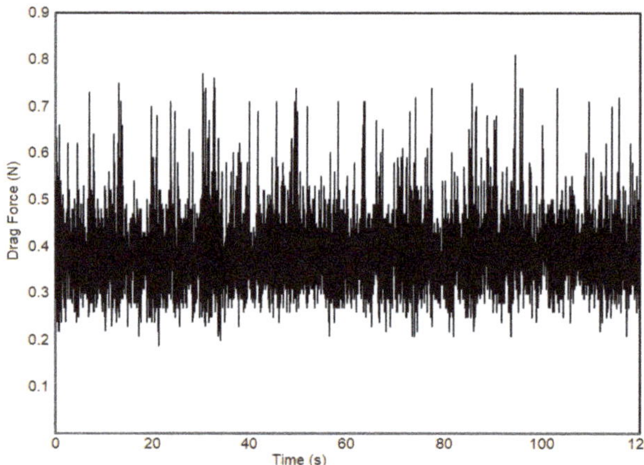

Figure 17. This graph shows typical force data collected during a test. For this data, the time-averaged velocity is 0.594 ± 0.01 m/s (not shown), while the time-averaged force is 0.386 ± 0.072 N. The sample is 16 mm cylinder bar using data sampling of 50 Hz.

The drag force data for each sample was collected for four different pump speeds: 0.35, 0.50, 0.60, and 0.66 m/s. All measurement was obtained when steady state had been reached. Figure 18 shows monotonic and non-linear increase in the drag forces for the samples in the range of the applied flow speed. The two samples demonstrate a very similar trend. The error bars show the standard deviation over the data population of speed and force measurement. The error percentage presented by the flow meter is less than that by the force sensors. A consistent increase in the deviation from the mean of the force data is observed when the flow speed increases. This is expected, as the force should be proportional to the square of flow speed. Data from the flow meter indicate consistent deviation throughout the range of the flow speed.

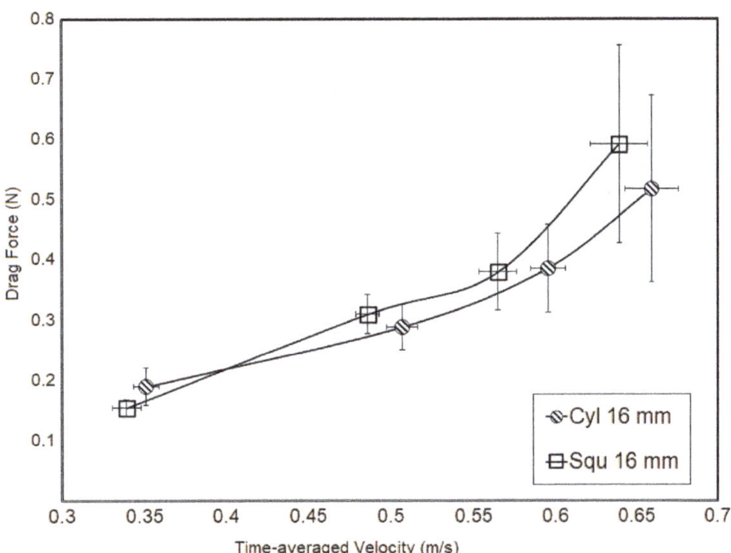

Figure 18. The mean velocity vs. mean force shows monotonic increase in the drag force with the increase in applied velocity. Error bars show standard deviations that increase with the velocity. The errors from the flow meter are relatively smaller compared to that from the force sensor.

Figure 19 shows the plot of the average drag coefficients for increasing Reynolds numbers. The average $\overline{C}_D$ here is calculated using the time-averaged velocity and time-averaged force taken at a specific pump's frequency. The error bars indicate the "maximum" and "minimum" values of the drag coefficients, calculated using the maximum and minimum velocity and force data reading. The large fluctuation of force data at high velocity is reflected here as large range of "maximum" and "minimum" coefficients. Nevertheless, the average drag coefficients for both cases remain relatively constant in the presented range of Reynolds numbers. The average drag coefficients, over the given Reynolds numbers, are 1.48 ± 0.25 and 1.58 ± 0.13 for the cylinder and square bar, respectively. The plot of C_D for cylindrical bar as a function of Reynolds numbers for $0 \leq Re \leq 10^4$, proposed by Tang et al. [54], is also shown in this figure. The expression is stated here for clarity:

$$C_D = 10^{\left(0.0041 \times (\log{(Re)})^3 + 0.0853 \times (\log{(Re)})^2 - 0.65 \times \log{(Re)} + 1.05\right)} \tag{6}$$

The function is obtained by regression analysis of C_D data on cylinder bars obtained from various data collection. The dash line in Figure 19 shows the constant C_D number for a long cylinder that is typically published in textbooks. The presented comparison shows that the C_D values obtained in our lab can be considered reasonable. For C_D of the cylinder bar, the deviation from both published C_Ds is large when the Reynolds number is low, but the error is minimized as the Reynolds number increases. The outcome certainly warrants further refinement of the experiment apparatus and more data collection at various Reynolds numbers.

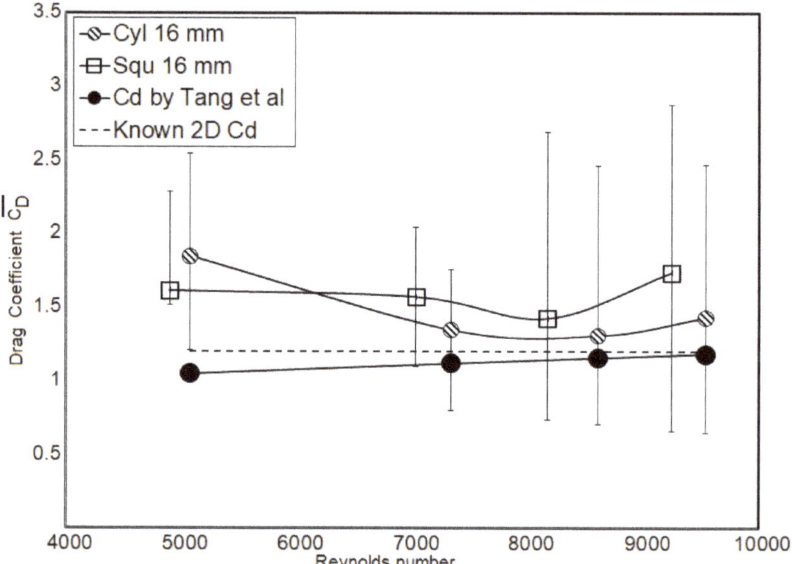

Figure 19. Average drag coefficients ($\overline{C}_D$) for the circular cylinder and square bar for various flow speeds (Reynolds numbers). The error bars indicate the "maximum" and "minimum" C_D values calculated using the extreme values of speeds and forces obtained during the sampling period of 120 s. The average C_D values across the range of Reynolds number for the cylinder and square bar are 1.48 and 1.58, respectively. The data marked by dark circles are obtained using formulation proposed by Tang et al. [54], and the dash line represents the constant known C_D typically presented in textbooks.

Alternatively, the drag coefficients can be obtained by first plotting the drag force data against the square of the time-averaged speed, as shown in Figure 20. Using standard linear regression analysis, the linear relationship between F_D and $\overline{V}^2$ can be obtained. For the presented samples, the following relations are obtained $\overline{F}_D = 1.1623\,\overline{V}^2$ and $\overline{F}_D = 1.3449\,\overline{V}^2$ for the cylinder bar and square bar samples, respectively. The high R^2 values for the two samples indicate a strong fit in the regression process. It also indicates that the linear relation is appropriate, as the C_D is expected to be constant in the applied range of Reynolds numbers. Note that the zero intercept in this case is not automatic, and it must be enforced as the zero force should be correlated with zero velocity. Eventually, the drag coefficients can be calculated from the gradients of the lines using the following formulation, which can be derived from (4)

$$\hat{C}_D = \frac{2\,K}{\rho\,A} \tag{7}$$

where K is the gradient of the linear relationship between $\overline{F}_D$ and $\overline{V}^2$. The coefficients found using this method are very similar to the average coefficients, however this method results in coefficients that are closer to the known values from textbooks [43–45].

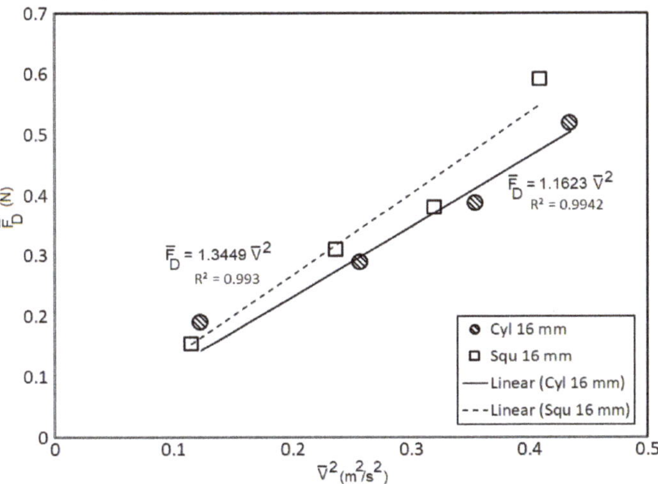

Figure 20. Linear correlation between the drag force and $\overline{V}^2$ (square of the velocity) can be used to obtain the drag coefficients. The high R^2 numbers suggest that the linear regression fits the data very well. The gradient for the square bar data (empty square markers) is slightly higher than that of the cylindrical bar (circle markers).

Presented in Table 8 is a list of the C_D values obtained in our experiments compared to those presented in textbooks. The 2D and 3D refer to the two-dimensional and three-dimensional assumptions of the immersed bodies. The $\overline{C}_D$ refers to the averaged coefficient of drag. This $\overline{C}_D$ is obtained by simply taking the average of all C_Ds calculated using different mean velocities and their corresponding drag forces. The alternative $\hat{C}_D$ is obtained by first plotting the drag force versus the square of the mean velocities ($\overline{F}_D$ vs. $\overline{V}^2$) and then computing the gradients of their linear relationship. The last column of the table shows the C_D value when the cylinder is considered as a three-dimensional (3D) body. The length-to-diameter ratio of our sample is $\frac{L}{D} \sim 7$, and the corresponding C_D value is in between 0.74 and 0.82 [44]. Compared to the 2D version of the known values, both methods result in drag coefficients within reasonable margin errors for the given experiment setup and range of applied Reynolds numbers ($5000 \leq Re \leq 10^4$). The known 2D C_D is stated to be valid for $10^3 \leq Re \leq 10^4$ (laminar regime), while the 3D C_D is valid for $Re \geq 10^4$ (turbulent regime) [44].

Table 8. List of the C_D values obtained in our experiments compared to that presented in textbooks. The 2D and 3D refer to the two-dimensional and three-dimensional assumptions of the immersed bodies. The $\overline{C}_D$ refers to the averaged coefficient of drag. The $\hat{C}_D$ is obtained using the linear regression analysis of $\overline{F}_D$ vs. $\overline{V}^2$.

Sample	Diameter/Side	$\overline{C}_D$	$\hat{C}_D$	Known C_D (2D) [44]	Known C_D (3D) [44]
Cylinder	15.94 mm	1.48 ± 0.25	1.39	1.2	0.74–0.82
Square bar	15.93 mm	1.58 ± 0.13	1.61	2.1 or 2.2	

Given the limitation of the support system and the accuracy of the sensors, the drag coefficients obtained in this experiment can be considered reasonable. The close walls presented by the observation chamber introduce effects on the vortex trail, hence the drag forces. Numerical studies at low Reynolds numbers showed that the wall increases the drag coefficient [55,56]. Experimentation at high Reynolds numbers (above 3000) on a cylinder, flat plates, and a square bar indicates the effects of walls on the pattern of the vortex-shedding [57]. The presence of the flow meter upstream of the sample, as mentioned previously, should have affected the force reading. Simultaneous flow rate reading is not necessary, as the relationship between the pump's frequency and flow speed has been known. Alternative supporting systems for the bar, for example pin support or fixed support on both ends, should be studied in future. Lastly, the absence of flow straightener needs to be considered, as it is known to affect the flow characteristics in the observation chamber.

7. Budget

Table 9 shows major items purchased for the flow tank, totaling approximately USD 5550. This is slightly higher than the USD 3500 cost of a custom-made water tunnel with a 30 cm wide and 180 cm long test section reported by Darfler and Tsai [2]. Our budget matches well with a flow loop with similar scale testing chamber reported by Northern Arizona State University [58]. However, this system can only perform laminar flow due to the small pump size. The centrifugal pump, as is expected, is the most expensive item in our project. However, its cost is still less than 22% of the total budget. The cost for the speed controller is also modest. The remaining cost covers the materials, such as stainless steel plates for the tank body, an acrylic sheet for the observation chamber, various pipes and valves for connectors, and steel tubes for the supporting frames. Gas supply for the welding takes less than USD 1000. The total cost listed here does not include the labor cost for the welding, assembly, design, transportation, wiring, and painting. All price is based on 2018 market values. An up-to-date pump price may be found on the manufacturer's website [59].

Table 9. List of major components of the flume, amount and its 2018 price. Note that the labor cost is not included here.

Item	Total Price (USD)	Percent of Total Cost (%)
3″ Self-priming centrifugal pump	1105	19.91
Programmable 3-phase AC motor speed control	723	13.03
Acrylic Sheet 0.7″ thickness	298	5.37
Stainless Steel (various thickness)	805.9	14.52
Low-carbon steel square tubes and sheets	694.5	12.51
Various hoses, pipes, and fittings	654	11.78
Various bearings and mountings	138.8	2.50
PVC pipes, fitters, and elbows	159	2.86
Welding screen	240	4.32
Argon gas supply for welding	504	9.08
Various shipping of items	228.5	4.12
Total	5551	100

8. Conclusions

The design and manufacturing process of an educational water flume has been shown to provide a learning platform for engineering students. The project is suitable for a senior design capstone project that would take two (2) semesters to complete. The design process has provided complex hands-on exercise materials for students desiring to master engineering fundamentals and skills such as computer aided designs, computer analysis of structures and fluid flows, electrical circuits, and experimentation. The open flume can be an expensive investment, but its versatility truly provides avenues for multi-disciplinary creative research. For a small-scale water flume, the required pump is governed by the height of the observation chamber and desired flow rate for the experiments. Welding of the tank body and supporting frames take the most challenging step and must be facilitated well. Professional welding and manufacturing, when budget is allowed, should be preferred to obtain a quality product, as rusting can become imminent.

Funding: The construction of the flume was funded by the Vibration Institute through its Academic Grant program for 2018. The Capstone Senior Design Project was partially funded by the Division of Engineering, State University of New York (SUNY) at New Paltz, New York.

Institutional Review Board Statement: Not Applicable.

Informed Consent Statement: Not Applicable.

Data Availability Statement: All essential data have been presented in the paper. Additional data are available on request from the corresponding author.

Acknowledgments: The authors wish to thank many alumni of the Division of Engineering who have contributed to this project: Matthew Bogue, Peter Demertzis, Jaime De La Vega, Roneil Harris, Apurba Sharma, Seth Pearl, Jessy Li, Michelle Wong, Kieran Cavanagh, Kira A. Solano, Christopher M. Rodi, Hadi Rabadi, Maritsa Rehma, and Istvan "Steve" Benyei. Many thanks for the continuing support from faculty colleagues during the project: Yi-Chung Chen, Faramarz Vaziri, Eric Myers, and Ashwin Vaidya as well as the technical lab support from Anthony Denizard.

Conflicts of Interest: The author declares no conflict of interest.

References

1. Bixler, B.; Pease, D.; Fairhurst, F. The accuracy of computational fluid dynamics analysis of the passive drag of a male swimmer. *Sports Biomech.* **2007**, *6*, 81–98. [CrossRef] [PubMed]
2. Darfler, R.C.; Tsai, W.W. Method for a Low Cost Hydrokinetic Test Platform: An Open Source Water Flume. In Proceedings of the ASEE Annual Conference and Exposition, Columbus, OH, USA, 25–28 June 2017. [CrossRef]
3. Washuta, N.J.; Howison, J.; Clark, B.L.; Imhoff IV, R.H.; Dos Reis, L. Water Tunnel Design; A Senior Capstone Project to Promote Hands-on Learning in Fluids. In Proceedings of the ASEE Annual Conference and Exposition, Salt Lake City, UT, USA, 24–27 June 2018. [CrossRef]
4. Panah, A.E.; Barakati, A. Design and Build a Water Channel for a Fluid Dynamics Lab. In Proceedings of the ASEE Annual Conference and Exposition, Columbus, OH, USA, 25–28 June 2017. [CrossRef]
5. Cohrs, M.; Ernst, W.; Vaidya, A. Potential for energy harvesting from vortex induced oscillations. *Int. J. Ecol. Dev.* **2013**, *26*, 1–9.
6. Camassa, R.; Chung, B.J.; Howard, P.; McLaughlin, R.; Vaidya, A. Vortex induced oscillations of cylinders at low and intermediate reynolds numbers. In *Advances in Mathematical Fluid Mechanics—Dedicated to Giovanni Paolo Galdi on the Occasion of His 60th Birthday*; Springer: Berlin/Heidelberg, Germany, 2010; pp. 135–145. [CrossRef]
7. Araneo, J.; Chung, B.J.; Cristaldi, M.; Pateras, J.; Vaidya, A.; Wulandana, R. Experimental control from wake induced autorotation with applications to energy harvesting. *Int. J. Green Energy* **2019**, *16*, 1400–1413. [CrossRef]
8. Wulandana, R.; Foote, D.; Vaidya, A.; Chung, B.J. Vortex-Induced Autorotation Potentials of Bladeless Turbine Models. *Int. J. Green Energy* **2021**. (accepted for publication).
9. Sun, W.; Zhao, D.; Tan, T.; Yan, Z.; Guo, P.; Luo, X. Low velocity water flow energy harvesting using vortex induced vibration and galloping. *Appl. Energy* **2019**, *251*, 113392. [CrossRef]
10. Cao, D.; Ding, X.; Guo, X.; Yao, M. Design, Simulation and Experiment for a Vortex-Induced Vibration Energy Harvester for Low-Velocity Water Flow. *Int. J. Precis. Eng. Manuf. Green Technol.* **2020**, *8*, 1–14. [CrossRef]
11. Cao, D.; Ding, X.; Guo, X.; Yao, M. Improved Flow-Induced Vibration Energy Harvester by Using Magnetic Force: An Experimental Study. *Int. J. Precis. Eng. Manuf. Green Technol.* **2020**, *8*, 879–887. [CrossRef]
12. Arionfard, H.; Nishi, Y. Experimental investigation of a drag assisted vortex-induced vibration energy converter. *J. Fluids Struct.* **2017**, *68*, 48–57. [CrossRef]
13. Fernandes, A.C.; Armandei, M. Low-head hydropower extraction based on torsional galloping. *Renew. Energy* **2014**, *69*, 447–452. [CrossRef]
14. Rostami, A.B.; Fernandes, A.C.; Bakhshandeh Rostami, A.; Fernandes, A.C.; Rostami, A.B.; Fernandes, A.C. The effect of inertia and flap on autorotation applied for hydrokinetic energy harvesting. *Appl. Energy* **2015**, *143*, 312–323. [CrossRef]
15. Fernandes, A.C.; Bakhshandeh Rostami, A. Hydrokinetic energy harvesting by an innovative vertical axis current turbine. *Renew. Energy* **2015**, *81*, 694–706. [CrossRef]
16. Rostami, A.B.; Fernandes, A.C. Mathematical model and stability analysis of fluttering and autorotation of an articulated plate into a flow. *Commun. Nonlinear Sci. Numer. Simul.* **2018**, *56*, 544–560. [CrossRef]
17. Raghavan, K.; Garcia, E.M.H.; Bernistas, M.M.; Ben-Simon, Y. The Vivace converter: Model tests at reynolds numbers around 100,000. J. *Offshore Mech. Arct. Eng.* **2009**, *131*, 1–13.
18. Chang, C.-C.J.; Kumar, R.A.; Bernitsas, M.M. VIV and galloping of single circular cylinder with surface roughness at $3.0 \times 10^4 \leq Re \leq 1.2 \times 10^5$. *Ocean Eng.* **2011**, *38*, 1713–1732. [CrossRef]
19. Bernitsas, M.M.; Raghavan, K.; Ben-Simon, Y.; Garcia, E. VIVACE (Vortex Induced Vibration Aquatic Clean Energy): A New Concept in Generation of Clean and Renewable Energy From Fluid Flow. *ASME J. Offshore Mech. Arct. Eng.* **2008**, *130*, 41101–41115. [CrossRef]

20. Shao, N.; Lian, J.; Liu, F.; Yan, X.; Li, P. Experimental investigation of flow induced motion and energy conversion for triangular prism. *Energy* **2020**, *194*, 116865. [CrossRef]
21. Yan, X.; Lian, J.; Liu, F.; Wang, X.; Shao, N. Hydrokinetic energy conversion of Flow-induced motion for triangular prism by varying magnetic flux density of generator. *Energy Convers. Manag.* **2021**, *227*, 113553. [CrossRef]
22. Barber, R.B.; Hill, C.S.; Babuska, P.F.; Wiebe, R.; Aliseda, A.; Motley, M.R. Flume-scale testing of an adaptive pitch marine hydrokinetic turbine. *Compos. Struct.* **2017**, *168*, 465–473. [CrossRef]
23. Okulov, V.L.; Naumov, I.N.; Kabardin, I.; Mikkelsen, R.; Sørensen, J.N. Experimental investigation of the wake behind a model of wind turbine in a water flume. *J. Phys. Conf. Ser.* **2014**, *555*, 012080. [CrossRef]
24. Talukdar, P.K.; Sardar, A.; Kulkarni, V.; Saha, U.K. Parametric analysis of model Savonius hydrokinetic turbines through experimental and computational investigations. *Energy Convers. Manag.* **2018**, *158*, 36–49. [CrossRef]
25. Sarma, N.K.; Biswas, A.; Misra, R.D. Experimental and computational evaluation of Savonius hydrokinetic turbine for low velocity condition with comparison to Savonius wind turbine at the same input power. *Energy Convers. Manag.* **2014**, *83*, 88–98. [CrossRef]
26. Kailash, G.; Eldho, T.I.; Prabhu, S.V. Performance study of modified savonius water turbine with two deflector plates. *Int. J. Rotating Mach.* **2012**, *2012*. [CrossRef]
27. Patel, V.; Eldho, T.I.; Prabhu, S.V. Performance enhancement of a Darrieus hydrokinetic turbine with the blocking of a specific flow region for optimum use of hydropower. *Renew. Energy* **2019**, *135*, 1144–1156. [CrossRef]
28. Kyozuka, Y.; Akira, H.; Duan, D.; Urakata, Y. An Experimental Study On the Darrieus-Savonius Turbine For the Tidal Current Power Generation. In Proceedings of the Nineteenth International Offshore and Polar Engineering Conference, Osaka, Japan, 21–26 June 2009.
29. Jahangir Alam, M.; Iqbal, M.T. Design and development of hybrid vertical axis turbine. In Proceedings of the Canadian Conference on Electrical and Computer Engineering, St. John's, NL, Canada, 3–6 May 2009; pp. 1178–1183. [CrossRef]
30. Volpe, M.A.; Beninati, M.L.; Riley, D.R.; Krane, M.H.; Fontaine, A.A. Development of measurement methods for testing of hydrokinetic devices to evaluate the environmental effect on local substrate. In Proceedings of the OCEANS'11-MTS/IEEE Kona, Waikoloa, HI, USA, 19–22 September 2011. [CrossRef]
31. Zhou, T.; Endreny, T.A. Reshaping of the hyporheic zone beneath river restoration structures: Flume and hydrodynamic experiments. *Water Resour. Res.* **2013**, *49*, 5009–5020. [CrossRef]
32. Duan, G. Upgrading an Experimental Flume for Engineering Research Education. 2009. Available online: https://apps.dtic.mil/sti/citations/ADA516451 (accessed on 1 December 2020).
33. Ubing, C.; Ettema, R.; Thornton, C.I. Flume experiments on baffle-posts for retarding open channel flow. *J. Hydraul. Res.* **2017**, *55*, 430–437. [CrossRef]
34. Bong, C.H.J.; Liow, C.V. Hydraulic characteristics of flow through angled baffle-plates in an open channel. *Int. J. River Basin Manag.* **2020**, *18*, 377–382. [CrossRef]
35. Roussinova, V.; Balachandar, R. Open channel flow past a train of rib roughness. *J. Turbul.* **2011**, *12*, 1–17. [CrossRef]
36. Yiping, L.; Anim, D.O.; Wang, Y.; Tang, C.; Du, W.; Lixiao, N.; Yu, Z.; Acharya, K.; Chen, L. Laboratory simulations of wave attenuation by an emergent vegetation of artificial Phragmites australis: An experimental study of an open-channel wave flume. *J. Environ. Eng. Landsc. Manag.* **2015**, *23*, 251–266. [CrossRef]
37. Camassa, R.; Chung, B.J.; Gipson, G.; McLaughlin, R.; Vaidya, A. Vortex Induced Oscillations of Cylinders. 2008. Available online: https://ecommons.cornell.edu/handle/1813/11484 (accessed on 1 November 2020).
38. Gharahjeh, S.; Aydin, I. Application of video imagery techniques for low cost measurement of water surface velocity in open channels. *Flow Meas. Instrum.* **2016**, *51*, 79–94. [CrossRef]
39. Crimaldi, J.P.; Knight, D.W. A Laser-Based Flow Visualization System for Fluid Mechanics Instruction. In Proceedings of the 2005 American Society for Engineering Education Annual Conference & Exposition, Session 1526, Portland, OR, USA, 12–15 June 2005.
40. Korkischko, I.; Meneghini, J.R. Experimental investigation of flow-induced vibration on isolated and tandem circular cylinders fitted with strakes. *J. Fluids Struct.* **2010**, *26*, 611–625. [CrossRef]
41. Chung, B.; Cohrs, M.; Ernst, W.; Galdi, G.P.; Vaidya, A. Wake—Cylinder interactions of a hinged cylinder at low and intermediate Reynolds numbers. *Arch. Appl. Mech.* **2015**, *86*, 627–641. [CrossRef]
42. Cengel, Y.A.; Cimbala, J.M. *Fluid Mechanics Fundamental and Applications*, 1st ed.; McGraw Hill: New York, NY, USA, 2006.
43. Fox, R.W.; Pritchard, P.J.; McDonald, A.T. *Introduction to Fluid Mechanics*, 6th ed.; John Wiley & Sons, Inc.: Hoboken, NJ, USA, 2004.
44. White, F.M. *Fluid Mechanics*, 7th ed.; McGraw Hill: New York, NY, USA, 2011.
45. Janna, W.S. *Introduction to Fluid Mechanics*, 4th ed.; CRC Press: Boca Raton, FL, USA, 2009.
46. McMaster-Carr Website. Available online: Mcmaster.com (accessed on 1 April 2018).
47. AMT. 3″ Self-Priming Centrifugal Pumps. Available online: http://amtpumps.com/site/wp-content/uploads/2017/01/SPE-15-16.pdf (accessed on 1 June 2018).
48. Cimbala, J. The Role of CFD in Undergraduate Fluid Mechanics Education. In Proceedings of the 59th Annual Meeting of the APS (American Physical Society) Division of Fluid Dynamics, Tampa Bay, FL, USA, 19–21 November 2006; pp. 19–21.
49. Lowe, S.A. Omission of critical Reynolds number for open channel flows in many textbooks. *J. Prof. Issues Eng. Educ.* **2003**, *129*, 58–59. [CrossRef]

50. Frei, W. Which Turbulence Model Should I Choose for My CFD Application? | COMSOL Blog. 2017. Available online: https://www.comsol.com/blogs/which-turbulence-model-should-choose-cfd-application/ (accessed on 25 May 2021).
51. Subramanian, A.; Goudarzi, N. CFD Analysis of a Water Flume Design for Testing Marine and Hydrokinetics Energy Converters. In *ASME 2018 Power Conference*; ASME International: Lake Buena Vista, FL, USA, 2018. [CrossRef]
52. Pullinger, M.G.; Sargison, J.E. Using CFD to improve the design of a circulating water channel. In Proceedings of the 16th Australasian Fluid Mechanics Conference, Gold Coast, Australia, 3–7 December 2007; pp. 94–98.
53. Heyrani, M.; Mohammadian, A.; Nistor, I.; Dursun, O.F. Numerical modeling of venturi flume. *Hydrology* **2021**, *8*, 27. [CrossRef]
54. Tang, H.; Tian, Z.; Yan, J.; Yuan, S. Determining drag coefficients and their application in modelling of turbulent flow with submerged vegetation. *Adv. Water Resour.* **2014**, *69*, 134–145. [CrossRef]
55. Singha, S.; Sinhamahapatra, K.P. Flow past a circular cylinder between parallel walls at low Reynolds numbers. *Ocean Eng.* **2010**, *37*, 757–769. [CrossRef]
56. Chakraborty, J.; Verma, N.; Chhabra, R.P. Wall effects in flow past a circular cylinder in a plane channel: A numerical study. *Chem. Eng. Process. Process Intensif.* **2004**, *43*, 1529–1537. [CrossRef]
57. Wang, X.; Chen, J.; Zhou, B.; Li, Y.; Xiang, Q. Experimental investigation of flow past a confined bluff body: Effects of body shape, blockage ratio and Reynolds number. *Ocean Eng.* **2021**, *220*, 108412. [CrossRef]
58. Mesharri, A.; Flah, A.; Jace, B.; Jessica, L.; Jacob, G.; Matt, S. Educational Water Tunnel. 2016. Available online: https://ceias.nau.edu/capstone/projects/ME/2016/EducationalWaterTunnel/documents.html (accessed on 31 December 2020).
59. AMT. 3″ Self Priming Centrifugal. Available online: https://amtpumps.com/site/product/3-self-priming-centrifugal/ (accessed on 30 June 2020).

 *fluids*

Review

An Introduction of Droplet Impact Dynamics to Engineering Students

Sara Moghtadernejad [1,*], **Christian Lee** [1] **and Mehdi Jadidi** [2]

[1] Department of Chemical Engineering, California State University, Long Beach, CA 90840, USA; Christian.Lee01@student.csulb.edu

[2] Department of Mechanical and Industrial Engineering, Concordia University, Montreal, QC H3G 1M8, Canada; m_jadi@encs.concordia.ca

* Correspondence: sara.moghtadernejad@csulb.edu

Received: 1 June 2020; Accepted: 1 July 2020; Published: 2 July 2020

Abstract: An intensive training course has been developed and implemented at the California State University Long Beach based on 8 years of experience in the multiphase flow area with the specific focus on droplet–solid interactions. Due to the rapid development of droplet-based equipment and industrial techniques, numerous industries are concerned with understanding the behavior of droplet dynamics and the characteristics that govern them. The presence and ensuing characteristics of the droplet regimes (spreading, receding, rebounding, and splashing) are heavily dependent on droplet and surface conditions. The effect of surface temperature, surface wettability, impact velocity, droplet shape and volume on droplet impact dynamics, and heat transfer are discussed in this training paper. Droplet impacts on moving solid surfaces and the effects of normal and tangential velocities on droplet dynamics are other topics that are discussed here. Despite the vast amount of studies into the dynamics of droplet impact, there is still much more to be investigated as research has expanded into a myriad of different conditions. However, the current paper is intended as a practical training document and a source of basic information, therefore, the scope is kept sufficiently broad to be of interest to readers from different engineering disciplines.

Keywords: droplet impact; undergraduate education; applications of fluids

---

1. Introduction

This paper summarizes an intensive course on the subject of multiphase flow with the focus on both the theoretical and practical knowledge on droplet–solid interactions. More specifically, the goal of this training manuscript is to provide students with in-depth practical knowledge of droplet dynamics due to its numerous industrial and scientific applications. The dynamics of droplet impacts on solid surfaces has been studied for years starting with the work of Worthington [1,2]. To this day, droplet impact dynamics are widely studied as further improvements in understanding of droplet behavior due to the variety of their industrial applications. In particular, droplet impacts are imperative to industrial processes employing spray coating and painting, spray cooling, inkjet printing, combustion engines, and anti-icing characteristics of critical industrial components such as aircraft surface, powder lines, and wind turbines [3–7]. A brief discussion for some of these usages is presented below.

One of the prominent applications of droplet studies is aircraft icing, which refers to creation of ice on the surface of flying objects. This usually happens at the presence of super-cooled water droplets and below icing temperature. Those droplets are formed as droplets with various size merge in clouds, or when the falling snow melts as it passes through a warmer layer (weather inversion). In either case, they are very unstable, and any disturbance will cause ice formation. During a flight, a fraction of super-cooled droplets tracks through the airflow and impact the aircraft surface. In most cases, the low energy droplets freeze upon impact, whereas the ones with higher energy flow along the surface until their energy is depleted and then freeze. Consequently, the resulting ice changes the surface structure of the aircraft and this reduction in the aerodynamic of the wing increases the fuel consumption [8–13]. As such, a detailed understanding of this process is much needed to engineer more efficient wings.

Other application of droplets is in direct injection combustion engines, where the behavior of fuel droplet impingement on the piston and surrounding cylinder walls is heavily studied [14–18]. The formation of the wall-fuel film and the improvement in the efficacy of combustion engine promote a wide variety of research into droplet dynamics [14–18]. The properties of superhydrophobic surfaces are of major interest in aerospace and power industries for their anti-icing characteristics, to prevent ice formation on wind turbine blades, power-lines, or aircraft wings [11,19–21]. Methods improving the repulsion of water droplets off surfaces have been heavily studied. Such methods involve variable surface inclination and application of superhydrophobic coatings [4,22–26].

Another interesting application of droplet dynamics is in the thermal spray process, where molten or semi-molten metal and ceramic particles are deposited on a substrate to generate various types of coatings such as thermal barrier, wear resistance, and corrosion resistance ones [27–32]. During the thermal spray process first, a heated gas is created via chemical combustion or electrical energy in a torch. Then the gas is used to melt the coating powder or wire into droplets and accelerate them toward a substrate, where the particles generate a splat. Finally, a coating is formed via multiple layers of splats. In that regards, most research studies the effect of droplet's inflight behaviors such as velocity, temperature, and trajectory, and their interaction with the substrate's temperature and roughness on the process of droplet solidification and coating structure [27–31,33–38].

Yet another importance of fundamental study in the droplet impact field is its application in spray cooling, where an array of small droplets is applied to a heated surface as a cooling mechanism to enhance its heat transfer [39–44]. In this operation, the cooling effectiveness is strongly influenced by fluid properties and droplet's size and velocity. Amongst the numerous usages here are dermatological operation, fire protection, and cooling of hot surfaces like hot strip mill and high-performance electronic devices. For example, cryogenic spray cooling is selectively directed to pre-cool human skin in laser treatments and hair removal procedures. Similarly, in steel strip casting, a jet gas with water droplets is guided to cool a high temperature (up to 1800 K) steel surface that shapes the final microstructure optimization. All in all, new developments in spray cooling technology demand improving the heat transfer rate, while maintaining uniform heat removal, and preventing temperature overshoot. They also require uniform operating temperatures maintenance, the removal of high heat flux, or the adaptability to changes in heat flux [39,42,45].

Furthermore, droplet dynamics are widely used in ink-jet printing as it involves the generation and deposition of small droplets, usually containing colorants, onto a substrate in certain patterns. Main challenges in the process are when droplets bounce back or spread unevenly on the paper. The development in understanding droplet dynamics is thus motivated by the vast applications of ink-jet printing, where improvements in droplet positioning, volume, and directionality are the main concerns [46–48]. Note that, the input substance (ink) can be a tiny liquid of "smart" material. As such today's applications of ink-jet technologies include printing solar cells, medical sensors, and electronic circuits. Particularly, ink-jet printing techniques have rapid developments in medical industry, where the manipulation of small amounts of liquid is essential in cell culture growth and pharmaceutical drug production [49–51].

Lastly, there has been rapid developments in microfluidic technologies, where liquid drops are handled on nano or micro scales. Microfluidic technologies have heavily impacted the biomedical, environmental, food, and chemical industries [52–54]. Other notable industrial applications of droplet impacts include quenching of aluminum alloys and steel, fire suppression, incinerators, soil erosion, and crop spraying [3,55–58].

Key Dimensionless Numbers in Droplet Impact Dynamics

To facilitate the characterization of droplet dynamics several dimensionless numbers have been introduced in the literature. The main dimensionless numbers are given in Table 1 [3,59,60], where ρ_l denotes the liquid density, d_0 is the droplet diameter, u_0 is the droplet impact velocity, σ is the surface tension, μ_l is the liquid viscosity, g is the gravity, and t is time. Furthermore, in this table, d_{cl}, c_l, k_l, T_0, h_{lv}, Q, and M_{evap} denote the diameter of the wetted region, liquid specific heat, liquid thermal conductivity, initial temperature, enthalpy of vaporization, total heat that is transferred to the droplet, and the evaporated mass at a given time, respectively. Subscripts l and s also stand for liquid and solid, respectively. It is worth mentioning that Table 1 only covers the main dimensionless numbers, while additional dimensionless numbers can be defined for the contact angle of the droplet on the surface, the boundary layer thickness, vapor properties, etc.

Table 1. Key dimensionless numbers to study droplet impact dynamics.

Dimensionless Number	Formulation
Weber number (ratio of inertial to surface tension forces)	$We = \dfrac{\rho_l d_0 u_0^2}{\sigma}$
Reynolds number (ratio of inertial to viscous forces)	$Re = \dfrac{\rho_l d_0 u_0}{\mu_l}$
Ohnesorge number	$Oh = We^{0.5} Re^{-1}$
Bond number (ratio of gravitational to surface tension forces)	$Bo = \dfrac{\rho_l g d_0^2}{4\sigma}$
Dimensionless time	$\tau = \dfrac{u_0 t}{d_0}$
Spreading ratio	$S = \dfrac{d_{cl}}{d_0}$
Prandtl number	$Pr = \dfrac{\mu_l c_l}{k_l}$
Stefan number	$St = \dfrac{c_l \Delta T_0}{h_{lv}}$
Ratio of the thermal effusivity of the liquid and the solid	$R_{eff} = \dfrac{\sqrt{(\rho c k)_l}}{\sqrt{(\rho c k)_s}}$
Ratio of specific heat capacities	$R_c = \dfrac{(\rho c)_l}{(\rho c)_s}$
Ratio of the total heat Q transferred to the droplet to the maximal possible heat transfer	$E^* = \dfrac{6Q}{\pi \rho_l d_0^3 h_{lv}}$
Evaporation efficiency	$E^*_{evap} = \dfrac{6M_{evap}}{\pi \rho_l d_0^3}$

2. Droplet Impact on Static Solid Surfaces

Experimental investigations suggest the presence of six possible regimes for droplet impact on dry surfaces, including deposition, receding breakup, rebound, and splashing (see Figure 1). The characteristics of these regimes after droplet impact has been deeply investigated [3,57,59,61–64]. The presence, or lack of each regime is dependent upon multiple conditions of the impacting droplet and characteristics of the surface. Such properties include impact velocity, droplet size, liquid surface tension and viscosity, surface temperature, wettability [3,4], etc.

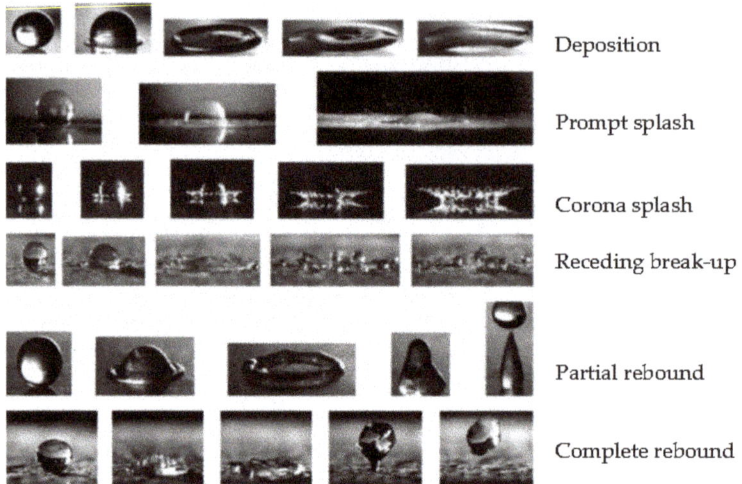

Figure 1. Droplet impact regimes on dry surfaces (reproduced with permission from [65]).

The expansion of the droplet after its impact on the surface is known as the spreading regime. This phenomenon is observed at relatively low Weber number values and low surface temperatures [3,63,66]. One of the most important parameters for characterizing the spreading regime during the droplet impact is the maximum spreading diameter of the droplet. The maximum spreading diameter of the droplet is imperative when heat transfer is a concern as the maximum spreading defines how much surface area is available for heat transfer [66]. Several characteristics of both the droplet and surface can affect the maximum spreading, such as droplet impact velocity and surface wettability [67–70]. Initial droplet velocity heavily impacts its spreading. As the droplet impacts on the surface, the kinetic energy in the vertical direction is transferred to the radial direction, promoting the spreading of the droplet. Lann et al. [69] described in their work that droplets spreading behavior is overseen by the conversion of kinetic energy into surface energy or dissipated heat. Different models have been developed to predict the maximum spreading ratio, $S_{max} = d_{cl,max}/d_0$, as a function of the impact parameters. Some of the most commonly used formulas in the literature are as follows:

Jones [71]

$$S_{max} = \sqrt{\frac{4}{3}Re^{1/4}} \tag{1}$$

Chandra and Avedisian [72]

$$\frac{3We}{2Re}S_{max}^4 + (1-cos\theta)S_{max}^2 - \left(\frac{We}{3}+4\right) = 0 \tag{2}$$

Asai et al. [73]

$$S_{max} = 1 + 0.48We^{0.5} \times \exp\left[-1.48\ We^{0.22}\ Re^{-0.21}\right] \tag{3}$$

Pasandideh-Fard et al. [74]

$$S_{max} = \sqrt{\frac{We+12}{3(1-cos\theta_A) + 4\left(We/\sqrt{Re}\right)}} \tag{4}$$

Mao et al. [75]

$$\left[\frac{1}{4}(1-cos\theta) + 0.2\frac{We^{0.83}}{Re^{0.33}}\right]S_{max}^3 - \left(\frac{We}{12}+1\right)S_{max} + \frac{2}{3} = 0 \tag{5}$$

Ukiwe and Kwok [76]

$$(We + 12)\, S_{max} = 8 + S_{max}^3 \times \left[3(1 - cos\theta_A) + 4\frac{We}{\sqrt{Re}}\right] \tag{6}$$

where θ_A stands for the advancing contact angle.

The receding regime involves the contraction of the droplet once it has reached the maximum spreading. Droplet receding is due to the liquid surface tension, which pulls the droplet together [3,66]. The receding time depends on several factors including the surface shape and the liquid surface tension [67,77]. The suppression of this regime has also been observed under various conditions. For example during the impact of liquid fuels such as decane, ethanol, and tetradecane due to their low surface tension, the contraction phase is suppressed [77]. This regime has also been shown to be heavily affected by surface temperature. Receding can be hindered or even vanished by sufficiently low surface temperatures. For instance, during the water droplet impact on ice films, once the droplet reaches maximum spreading, the contact interface between the droplet and the ice film freezes, preventing the formation of a receding regime [78–82].

After receding, the droplet may show rebounding depending on various parameters such as surface wettability, temperature, and kinetic energy. In the rebound regime, if the droplet has a high contact angle and sufficient kinetic energy, it may bounce off the surface. The rebound regime has also been shown to be absent under various conditions. For instance, upon impact on a super-cooled superhydrophobic surface, water droplet freezes and adheres to the surface as it spreads, therefore, no rebounding occurs [83].

The appearance of the splashing regime is also dependent on various conditions of the droplet and surface. During impact with high enough energy, the droplet hits the surface and disintegrates into secondary droplets. Splashing can be induced by changing the orientation of the surface, such as having spherical or inclined surfaces, or impact on a moving liquid film [84–86]. There are several types of splash such as corona, prompt, and fingering within the splashing regime [77,87–89]. In corona splashes (see Figure 1) the outer rim of the lamella lifts off the surface forming a crown. From the crown, the lamella breaks apart forming secondary droplets. In prompt splashes, the droplet disintegrates, and secondary droplets form immediately after impact. Finally, in fingering splashes, protrusions extend from the droplet, eventually disintegrating and forming secondary droplets [77,87–89]. Various boiling regimes have also been experimentally studied at super-heated temperatures. This causes different splash phenomena, such as boiling-induced breakup or boiling-induced splashing, due to nucleate boiling in the contact interface [66].

As mentioned earlier in this manuscript, the regimes and phenomena observed during droplet impact significantly depend on a variety of droplet/surface conditions. Here we aim to describe the key conditions affecting the impact dynamics of droplet on solid substrates. As such the article is divided into two sections describing the droplet impacts on (1) stationary solid surfaces and (2) moving solid surfaces. Noting that due to the vast number of characteristics defining the behavior of droplets only the major ones will be discussed in this article; namely impact velocity, droplet shape and size, surface wettability, and temperature.

2.1. Effect of Surface Temperature

Surface temperature is an important characteristic affecting the dynamics of impacting droplets, particularly when heat transfer is a concern. In the case of a hot, dry solid surface, the droplet impact outcomes are classified into several regimes: evaporation, nucleate boiling, foaming, transitional boiling, and film boiling. In the nucleate boiling regime, which occurs at relatively high surface temperatures, the droplet is in direct contact with the surface and vapor bubbles are formed at various isolated nucleation sites. These bubbles rise and the droplet will eventually boil off (see Figure 2b) [67]. In the foaming regime, which is a subcategory of nucleate boiling, the entire drop starts to foam. The vapor bubbles grow much larger in this regime while no separation from the liquid–gas interface and no coalescence are detected. In the transition boiling, due to high wall temperature, the generation rate of the vapor bubble increases quickly. Owing to bubbles coalescence, a vapor layer is formed over some portions of the area between the drop and the surface, while the rest of the drop wets the surface. In this regime, liquid layers frequently collapse, therefore, this regime is very unstable and secondary droplets are also generated [90]. Further increasing the temperature changes the regime to film boiling. In this regime, a vapor layer forms preventing the complete contact between the liquid and the surface [67]. In the film boiling regime, the droplet levitates on a vapor layer, as shown in Figure 3. This phenomenon is heavily studied and called the Leidenfrost effect [91]. The Leidenfrost effect also promotes rebounding of the droplet without disintegration into secondary droplets.

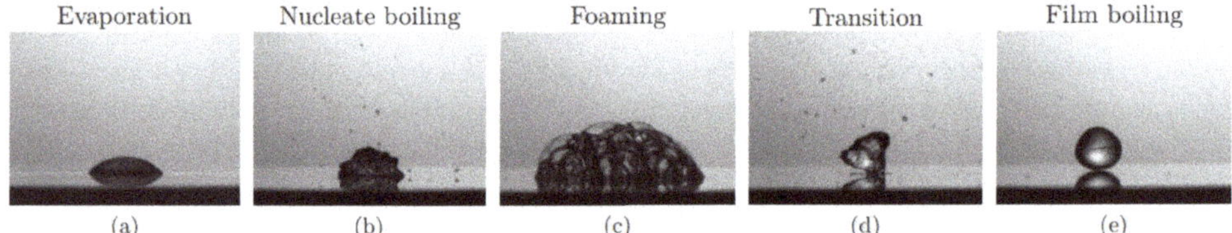

Figure 2. Outcomes of droplet collision with a hot, dry solid surface: (**a**) evaporation, nucleate (**b**) boiling, (**c**) foaming, (**d**) transition, and (**e**) film boiling (reproduced with permission from [90]).

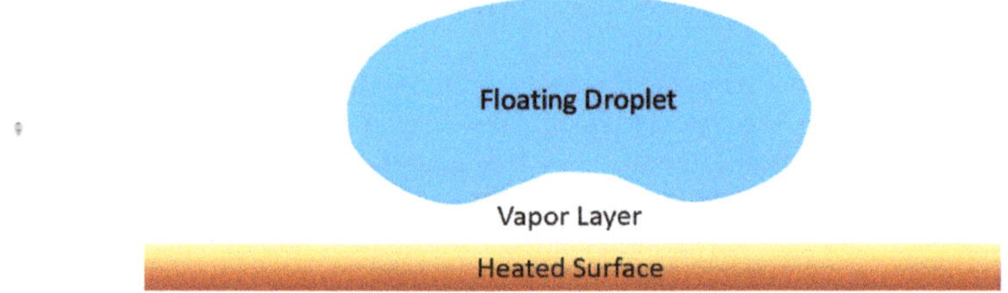

Figure 3. The Leidenfrost phenomenon forms vapor layer preventing droplet from contacting the surface.

One important characteristic of heat transfer in droplet dynamics is the evaporation time. When the surface temperature is within the nucleate boiling regime, it was observed that as surface temperature increases, the evaporation time decreases due to an increase heat transfer. During the transitional regime, the evaporation time is variable due to uneven contact between the liquid and surface. However, in the film boiling regime, the evaporation time sharply increases due to the Leidenfrost effect. When the droplet enters the film boiling regime, it is no longer in contact with the surface as the vapor layer separates the two. This vapor layer acts as an insulating layer, slowing down evaporation [67]. Further increasing the temperature past the Leidenfrost effect will decrease the evaporation time [92].

Liquid surface tension has significant influence on the dynamic Leidenfrost temperature. The dynamic Leidenfrost temperature is the minimum surface temperature at which the impacting droplet bounces without splashing, in other words the minimum temperature to induce the Leidenfrost effect [93]. Chen et al. [93] studied the effects of surfactants on the dynamic Leidenfrost temperature. In these sets of experiments alcohol surfactants, octanol and ethyl-hexanol, were added to water droplets causing a reduction in surface tension of the droplet. As a result, the maximum spreading diameter increased during the spreading regime. Additionally, it was observed that the time to reach maximum spreading was decreased. Chen et al. [93] found that by adding surfactants the dynamic Leidenfrost temperature increased. The addition of surfactants promotes splashing and prevents the formation of vapor layer. The reduced surface tension promotes a thinner lamella during spreading, this makes it easier for vapor bubbles to burst out from the boiling film and break the vapor layer, preventing the Leidenfrost state [93]. The reduced surface tension also makes it difficult for vapor bubbles to coalesce and form a vapor layer as a result of the lower bubble departure diameter. The dynamic Leidenfrost temperature can also be raised higher by increasing the concentration of surfactants in the droplet due to increased reduction in surface tension. Surface wettability is another parameter that affects the Leidenfrost point in a way that more hydrophilic surface leads to a higher Leidenfrost temperature [67,90].

Cooled and super-cooled surface temperatures also have an intense effect on droplet dynamics. At sufficiently cool temperatures, droplets begin to nucleate and freeze. The sessile droplet undergoes a complex solidification process, typically split into five stages [83,94,95], which is similar to the freezing of suspended water droplets. First, the droplet is cooled from its initial temperature to temperatures below the equilibrium freezing temperature during a cooling stage. Second, the droplet experiences a nucleation stage, where ice crystal nucleation occurs. Third, rapid crystal forms from the nucleation points driven by supercooling during the recalescence stage until it reaches the equilibrium temperature. Fourth, crystal growth is driven by heat transfer until the droplet is completely frozen during the freezing stage. Lastly, during the solid cooling stage, the temperature of the solidified droplet decreases due to the continuous cooling of the cold plate [83,94,95]. At this stage the ensuing droplet forms a peculiar shape with a pointed tip as shown in the last sequence of Figure 4.

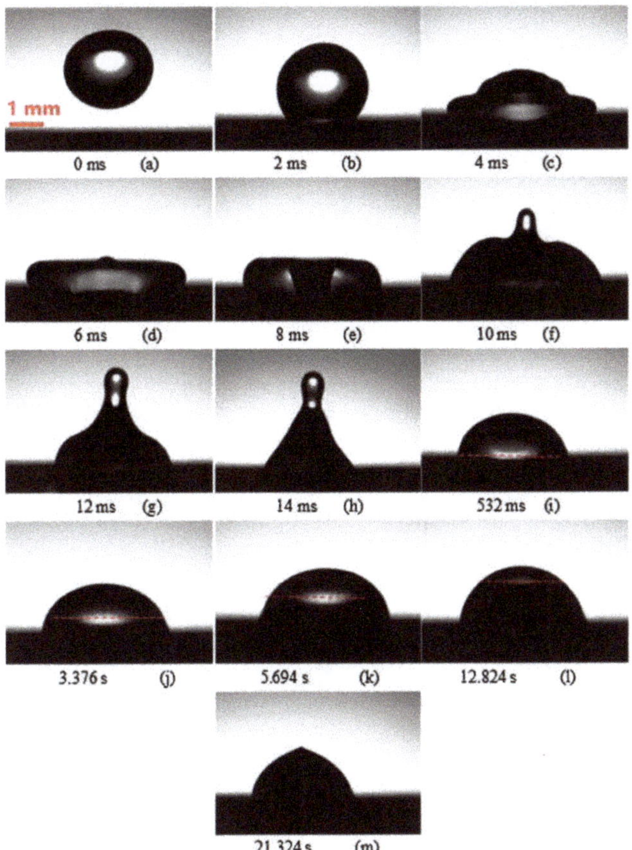

Figure 4. Sequences of water droplet impact on a cold hydrophobic surface, in a way (**a–m**): the starting point of the process to the end of the phenomenon. Droplet diameter = 2.82 mm, impact velocity = 0.77 m/s, and surface temperature = −10 °C (reproduced with permission from [68]).

The deformation of an impacting water droplet throughout the freezing process on a hydrophobic horizontal surface is shown in Figure 4. Experimental studies have shown that sub-cooled temperature has no significant effect on the spreading of the droplet after impact [80,96]. However, it shows significant suppression of the receding speed and height when the period of receding is sufficiently long [96,97].

2.2. Effect of Surface Wettability

Surface wettability is the ability of surface to be wetted by a liquid and is mainly determined by surface roughness and chemistry. The Wenzel [98] and Cassie-Baxter [99] states describe the droplet wetting regime on the surface (see Figure 5). In the Wenzel state, water penetrates into the surface structures and conforms to the surface while in the Cassie-Baxter state, the water droplet remains above the surface structures, maintaining an almost spherical shape [4].

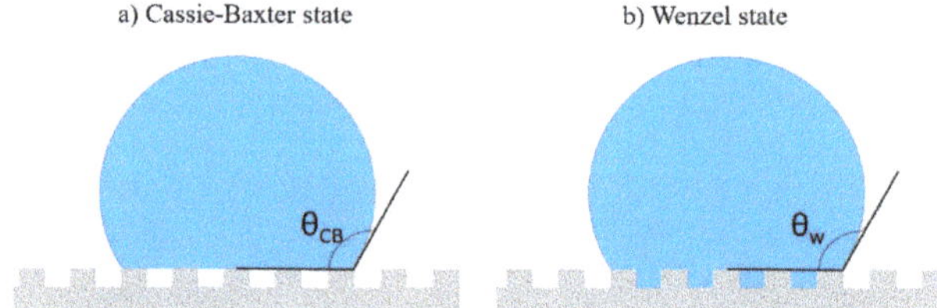

Figure 5. Wetting regimes of droplet on a solid surface: (**a**) Cassie-Baxter state and (**b**) Wenzel state (reproduced with permission from [4]).

Static contact angle, θ, which is defined as the angle between the droplet and the surface at the contact line (see Figure 6) is a characteristic used to define the degree of surface wettability. The advancing contact angle, θ_A, is the contact angle for a droplet with an advancing contact line (e.g., for a growing droplet), while the receding contact angle, θ_R, is the contact angle for a receding contact line (e.g., a shrinking droplet). Contact angle hysteresis is defined as $\theta_A - \theta_R$ and is an indication of the droplet mobility on the surface, in a way that the lower is the hysteresis the easier the droplet moves on the surface (higher mobility). In general, in contact with water, a hydrophilic surface displays a contact angle of less than 90°, while a hydrophobic surface shows a contact angle of more than 90°. On a superhydrophobic surface, contact angle is more than 150°, and the contact angle hysteresis should be less than 10° [4,68].

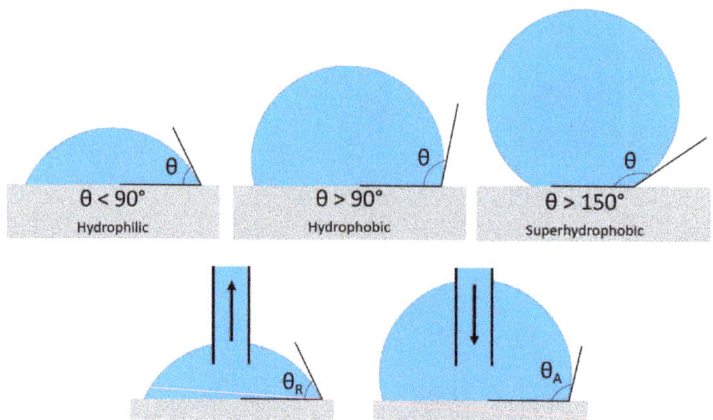

Figure 6. Schematic of static, θ, receding, θ_R, and advancing, θ_A, contact angles [100].

The wettability of a surface considerably affects the droplet dynamics after impact. It is shown that the wetted area during the spreading regime decreases with lower surface wettability [68]. It is also found that as wettability decreased a longer time was required for a droplet to reach equilibrium. This is due to oscillations in the droplet as more kinetic energy remains after the spreading regime. As spreading is suppressed, less energy is dissipated [68]. Tang et al. [77] studied the droplet dynamics of various liquid on surfaces with variable roughness. It was shown that as surface roughness increases, the droplet has a slower spreading time and smaller maximum spreading diameter. It was also observed that increase of surface roughness, promotes droplet splashing. Moreover, it was demonstrated that surface roughness has a prominent impact on the promotion of splashing in liquids with a smaller Ohnesorge number.

Surface wettability has a significant impact on the heat transfer rate in droplet dynamics. Pan et al. [68] studied droplet impacts on cold surfaces and observed how wettability affects the freezing process. As surface wettability decreases the total icing time of the droplet increases. This is because of a lower heat transfer rate due to a smaller heat transfer area. The lower surface contact area is due to the high interfacial tension in low wettability surfaces, which prevents droplet spreading. As expected, on hydrophilic surfaces the total freezing time is shorter comparing to the hydrophobic surfaces. On hydrophilic surfaces spreading is promoted due to more heat transfer area. Overall, more contact surface area leads to a higher heat transfer rate and in turn a shorter freezing time [68].

Surface wettability also has an impact on the boiling process during droplet dynamics. In general, as wettability increases the rate of phase-change heat transfer enhances [92,101]. As surface becomes more hydrophilic, smaller and faster bubble growth occurs [102,103]. Kim et al. [92] studied the effects of surface wettability on droplet rebounding on hot surfaces above the Leidenfrost temperature. In their experiments four surfaces were prepared, a smooth hydrophilic, a smooth hydrophobic, a hydrophilic, and a hydrophobic surface both with nanoscale structures. For both smooth and nanoscale hydrophobic surfaces, rebounding of the droplet was achieved, but at varying surface temperatures. For both hydrophilic surfaces, it was found that a higher surface temperature was needed to induce rebounding as more energy was required to disperse the droplet from the surface. On hydrophilic surfaces, there is more contact area and more surface tension, thus more work is needed to break the droplet adherence to the surface. Finally, on the nanoscale hydrophilic surface, splashing was induced caused by the capillary effects and cavities of the surface [92].

2.3. Effect of Impact Velocity

One of the most important characteristics affecting the dynamics of droplet impact is the droplet impact velocity. Specifically, impact velocity has a profound effect on the droplet spreading regime. Clearly, when the impact velocity increases, more kinetic energy is given to the droplet. The momentum in the vertical direction is then transferred in the radial direction upon impact [3,67,68]. As higher impact velocities a higher degree of spreading is promoted, leading to more surface area for heat transfer [62,66–68,93,104]. Rajesh et al. [67] showed that on superheated concave and convex surfaces, an increase in impact velocity translated to an increase in both maximum droplet spreading and contact time. This increase in maximum spreading diameter is due to a rise in impact kinetic energy allowing the droplet to spread more. Similarly, the increase in maximum droplet spreading, due to higher impact velocity, affects the freezing time of the droplet on cold surfaces. With higher impact velocities, a trend of shorter icing time and faster freezing is observed (faster spreading and higher heat transfer dispersal occurs upon impact [68]). Furthermore, as mentioned above, if the droplet has enough energy, it rebounds after retraction. Increase of the impact velocity has shown to promote rebounding and further increase of the impact velocity can induce splashing [3,59]. Chen et al. [93] studied the effect of the initial droplet impact momentum on the Leidenfrost point and detected that at high impact velocity, a higher surface temperature is needed to induce the Leidenfrost effect. The increased droplet impact momentum leads to higher kinetic energy in the spreading droplet, this causes a thinner liquid disk, hence nucleate bubbles are easier to burst out from the disk, which breaks the Leidenfrost state [93].

2.4. Effects of Droplet Shape and Volume

The shape of the droplet can impact its dynamics and the heat transfer rate. By increasing the ellipticity and asymmetry of the droplet, the spreading and retraction regimes are significantly changed. Generally, asymmetrical droplets require a higher impact velocity for rebounding [105–108]. Yun [109] studied the effects of asymmetry and ellipticity on droplet dynamics, specifically on its rebound regime. Asymmetric droplets, in the shape of an egg, showed uneven spreading and retraction in both the x and y axis (see Figure 7). To investigate the effect of asymmetry and ellipticity on the rebound regime, the height of the droplet at the center of mass was measured. It was found that with increasing ellipticity the height of the droplet decreased [109]. Additionally, with decreasing asymmetry the droplet showed a reduced rebound height. This suggests that suppression of droplet rebound can be induced by variances in its ellipticity and asymmetry. In other words, the degree of droplet deposition can be promoted with variances in the droplet's shape. The suppression of droplet bouncing is due to a break in the horizontal momentum of the droplet at low asymmetry and high ellipticity [109].

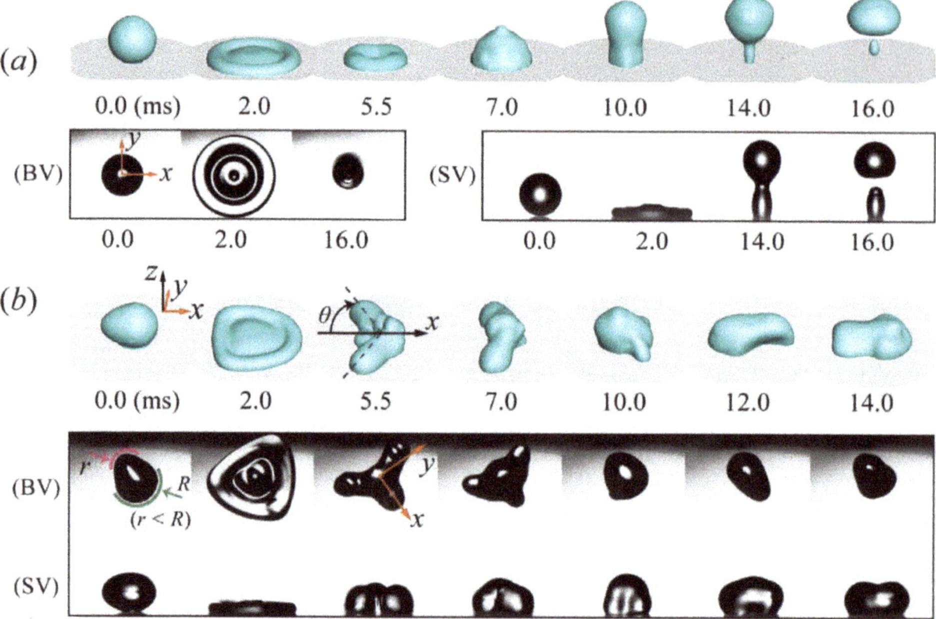

Figure 7. Simulation and experimental snapshots of the impact dynamics of (**a**) spherical and (**b**) asymmetric droplets on a hydrophobic surface (reproduced with permission from [109]); the equivalent droplet diameter is 2 mm and $We = 17$. SV and BV stand for the side and bottom-views in the experiment, respectively.

The size of the impacting droplet can also affect multiple characteristics of droplet dynamics. Pan et al. [68] studied how the initial droplet diameter affected its maximum spreading diameter and heat transfer rate. By increasing the initial droplet size, water droplet spreads more and showcases more oscillation during the spreading regime owing to an increase in initial kinetic energy. Additionally, upon impact on a cold surface, larger droplets showed increase in freezing time. Despite an increase in surface area during the spreading regime, larger droplets needed a longer time to freeze due to an increase in mass [68]. This experiment demonstrated the predominance of droplet volume over the effects of droplet dynamics on freezing time due to longer time spent during the heat transfer process compared to the droplet dynamic process. It is worth mentioning that an increase in droplet volume can induce splashing due to an increase in the Weber number.

2.5. Effect of Relative Humidity

Relative humidity has considerable influence on droplet dynamics. Bobinski et al. [110] qualitatively observed the effects of relative humidity and surface temperature on droplet impact in icing conditions. In their studies, relative humidity was kept constant and equal to 75%, while the surface temperature was reduced to −10 °C. It was found that droplet freezing is triggered during the spreading phase and it freezes when the maximum spreading is reached, even on the superhydrophobic surface. By decreasing the relative humidity to 9%, it was shown that the impact process on the superhydrophobic surface is similar to the case of room temperature tests and the droplet does not freeze. Experimental study was also performed by Jadidi et al. [111] to understand the effects of relative humidity, surface temperature, and wettability, and the frosting mechanism on water droplet impact dynamics. In their study superhydrophobic, aluminum, and glass surfaces with three different surface temperatures (i.e., 20, 2, and −2 °C) were used. Furthermore, three different relative humidities (i.e., 10%, 20%, and 30%) were applied while the droplet Weber and Reynolds numbers, and the air temperature were fixed. It was revealed that the ratio of the surface temperature to the dew point temperature, which depends on relative humidity and air temperature, has a significant impact on droplet spreading, recoil, and contact angle. In general, when the mentioned ratio is less than one and decreases (it can be done by increasing the relative humidity or decreasing the surface temperature), condensation and frost formation become important, droplet spreading diameter increases significantly, and the equilibrium contact angle decreases.

2.6. Effect of Dimensionless Numbers

The effect of dimensionless numbers on the spreading ratio, S, during droplet deposition on a dry surface is shown in Figure 8 [60]. To obtain these results, one dimensionless number was individually changed while other dimensionless numbers as well as contact angles were kept constant. As shown in Figure 8a, the maximal spreading ratio increases with Reynolds number. On the other hand, the minimal spreading ratio at the end of the receding phase slightly decreases as the Reynolds number increases. Furthermore, the Reynolds number does not significantly affect the duration of spreading and receding phases in terms of dimensionless time. In contrast, as can be seen in Figure 8b, the Weber number does not affect the spreading phase. However, by increasing the Weber number, receding is slowed down. Increasing the Bond number, results in a slight increase of the spreading ratio in all impact stages (see Figure 8c). In short, the competition of inertial and viscous forces has significant influence on the spreading phase, while the surface tension force mainly affects the receding phase [60].

Herbert et al. [60] also showed that by increasing the Reynolds number, the total heat transfer from the substrate to the droplet significantly decreases. It was explained that although the heat transfer area is larger during the spreading and the beginning of the receding stages for large values of Reynolds, the heat flow is noticeably lower in these phases. In addition, it was shown that increasing the Weber number causes the dimensionless heat flow during the spreading and the receding phases to decrease and enhance, respectively. Moreover, increasing the Prandtl number results in significant reduction of heat transfer due to a suppressed convective heat transfer at the solid–liquid interface. It was also revealed that a higher value of Bond number causes the heat transfer rate to increase throughout the entire process since the wetted area is enlarged, as shown in Figure 8c.

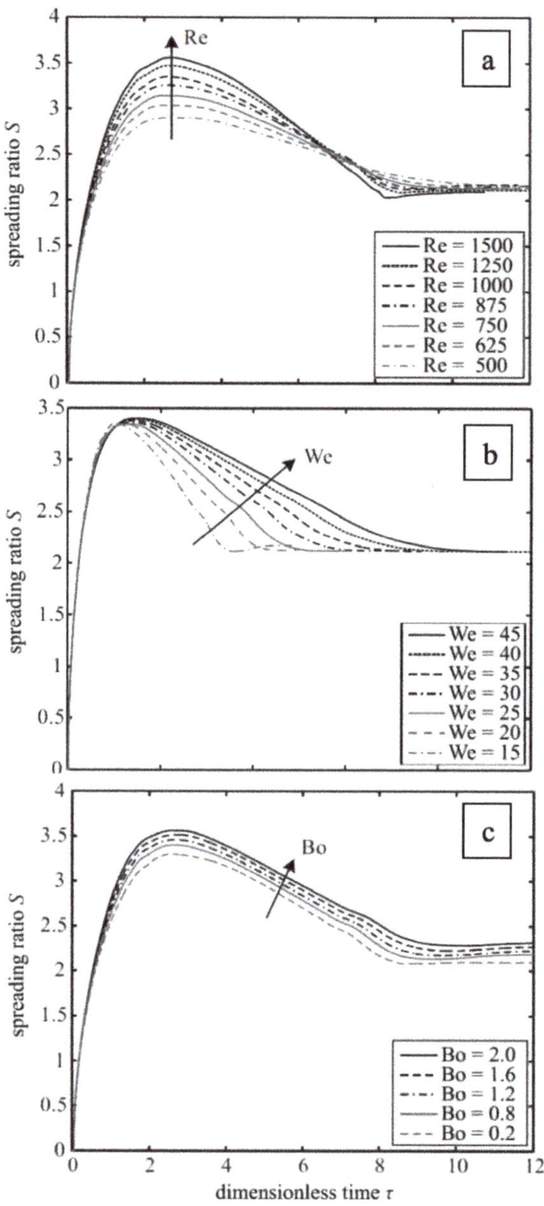

Figure 8. Effect of dimensionless numbers on the droplet spreading ratio; (**a**) effect of the Reynolds number (*We* = 15, *Bo* = 0.5, *Pr* = 9.54), (**b**) effect of the Weber number (*Re* = 1000, *Bo* = 0.5, *Pr* = 9.54), and (**c**) effect of the Bond number (*We* = 15, *Re* = 1000, *Pr* = 9.54; reproduced with permission from [60]).

3. Droplet Impact on Moving Surfaces

Droplet impact onto moving surfaces shows vast differentiations in their dynamics. The degree of droplet deformation and induction of splashing are heavily dependent on droplet impact velocity and surface tangential velocity. Povarov et al. [112] observed the presence of the air layer causing the droplet to lift off at high tangential velocities. At low surface velocity, the droplet is deposited onto the surface with no lift off. The bottom layer of the droplet adheres to the surface and spreads in a tear-like shape. At higher velocities, there is partial lift off of the droplet. The bottom layer of the droplet adheres to the surface and moves with it while the remainder of the droplet stays above the impact point. As the bottom layer is dragged away, sufficiently high tangential velocities cause the formation of an air layer. This triggers the rear edge of the droplet to splash and partially rebound off the surface. Total rebounding occurs at even higher tangential velocities as the droplet becomes significantly deformed [112]. Figure 9 illustrates droplet deformation on a moving surface. Mundo et al. [113] studied the splashing threshold of a droplet during impact on a moving substrate. Based on their observations, for high impact velocities, as the droplet impacts on the surface, a liquid film forms and spreads. The liquid film then deforms with the moving surface forming corona around the droplet. As the droplet continues to impact the surface, the fluid is fed into the corona film allowing it to grow. Whereas when less fluid is fed into the corona film, it becomes thinner and eventually disintegrates into secondary droplets. At lower impact velocities, the droplet impacts and spreads on the surface since there is not enough normal momentum to allow the corona formation and the kinetic energy is lost during the deformation process [86,113]. Mundo et al. also investigated the effects of surface roughness on droplet impacts on moving surfaces. They figured out that at increased surface roughness the droplet shows irregular deformation. Here the formation of the corona film and sequential instabilities were not observed due to vigorous splashing. Increased surface roughness also showed an increase in the formation of secondary droplets. On the other hand, decreased surface roughness promotes deposition of the droplet and splashing can be suppressed [113].

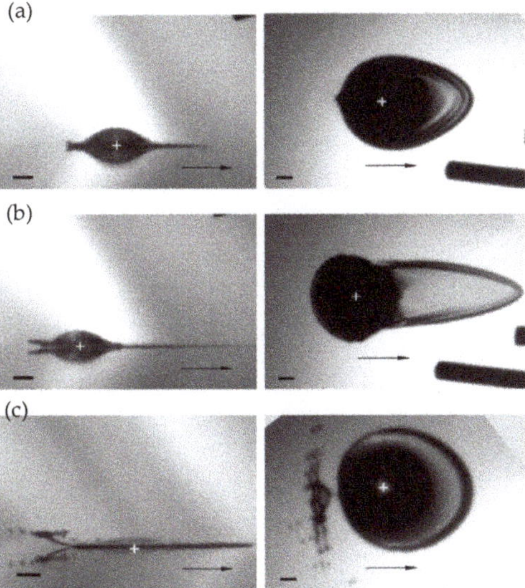

Figure 9. Side (**Left**) and top (**Right**) view of droplet deformation on a moving surface showing (**a**) deposition, (**b**) partial lift off, and (**c**) partial lift off with side-splash; scale bar is 1mm (reproduced with permission from [86]).

Zen et al. [86] studied how impacting velocity affected the splashing on moving surfaces. They described two splash phenomena, splash-around and side-splash in the rear edge. The threshold between splash-around and side-splash is dependent on the impact and surface velocities. In general, increasing the impact velocity induces side-splash and further increases in velocity lead to splash-around. It was also observed that increase of the impact velocity promotes the formation of smaller secondary droplets. Increasing the tangential velocity lead to promotion of side-splashing in the rear edge and lower tangential velocities promotes splash-around in all edges. Yao and Cai [114] demonstrated the predominance of tangential velocity on splashing threshold. They observed that as the surface velocity increased, the critical impact velocity to induce splashing significantly decreased.

4. Conclusions

The importance of droplet impact dynamics in a vast number of industrial applications such as spray coating and painting, spray cooling, ink-jet printing, combustion engines, etc., drives numerous studies into this area. This paper summarized the intensive training course aimed at discussing some of the main governing factors on droplet impact dynamics and heat transfer on both static and moving solid surfaces. In details on solid static surfaces, the effects of surface temperature, wettability, impact velocity, dimensionless numbers, and droplet shape and volume were discussed. Surface temperature played a major role on droplet dynamics, particularly at super-heated or super-cooled temperatures. At ample temperatures, the Leidenfrost effect or the freezing process can be induced, wildly changing the droplet dynamics. Surface wettability also greatly affected the droplet dynamics as spreading, retraction, and rebounding can be promoted or suppressed depending on impact conditions. Impact velocity of the droplet heavily affects spreading and splashing regimes owing to transfer of kinetic energy after impact.

The effects of tangential and impact velocity on the impact dynamics on moving surfaces were also discussed in this article. Droplet impact dynamics are heavily dependent on the impact and surface velocity. Splashing is easily promoted on moving surfaces due to deformations in droplet caused by the moving surface. In general, having training on droplet impact subject for engineering students is essential due to the real-world applications of this topic.

Author Contributions: Conceptualization, supervision, writing-review and editing, S.M.; writing-original draft preparation, C.L. and M.J. All authors have read and agreed to the published version of the manuscript.

Funding: This research received no external funding.

Acknowledgments: The authors gratefully acknowledge the support of the "Honors Program" at California State University Long Beach.

Conflicts of Interest: The authors declare no conflict of interest.

References

1. Worthington, A.M. On the forms assumed by rops of liquids falling vertically on a horizontal plate. *Proc. R. Soc. Lond.* **1876**, *25*, 261–272.
2. Worthington, A.M. *A Study of Splashes*; Longmans, Green, and Company: London, UK, 1908. Available online: https://archive.org/details/studyofsplashes00wortrich/mode/2up (accessed on 1 July 2020).
3. Yarin, A.L. Drop Impact Dynamics: Splashing, Spreading, Receding, Bouncing . . . . *Annu. Rev. Fluid Mech.* **2006**, *38*, 159–192. [CrossRef]
4. Khojasteh, D.; Kazerooni, M.; Salarian, S.; Kamali, R. Droplet impact on superhydrophobic surfaces: A review of recent developments. *J. Ind. Eng. Chem.* **2016**, *42*, 1–14. [CrossRef]
5. Bolleddula, D.A.; Berchielli, A.; Aliseda, A. Impact of a heterogeneous liquid droplet on a dry surface: Application to the pharmaceutical industry. *Adv. Colloid Interface Sci.* **2010**, *159*, 144–159. [CrossRef]
6. Moghtadernejad, S.; Jadidi, M.; Esmail, N.; Dolatabadi, A. Shear-driven droplet coalescence and rivulet formation. *Proc. Inst. Mech. Eng. Part. C J. Mech. Eng. Sci.* **2016**. [CrossRef]
7. Moghtadernejad, S. *Dynamics of Droplet Shedding and Coalescence under the Effect of Shear Flow*; Concordia University: Montréal, QC, Canada, 2014.
8. Bragg, M.B. Aicraft aerodynamic effects due to large droplet ice accretions. In Proceedings of the 34th Aerospace Sciences Meeting and Exhibit, Reno, NV, USA, 15–18 January 1996. [CrossRef]
9. Bragg, M.; Basar, T.; Perkins, W.; Selig, M.; Voulgaris, P.; Melody, J.; Sarter, N. Smart icing systems for aircraft icing safety. In Proceedings of the 40th AIAA Aerospace Sciences Meeting and Exhibit, Reno, NV, USA, 14–17 January 2002. [CrossRef]
10. Cao, Y.; Tan, W.; Wu, Z. Aircraft icing: An ongoing threat to aviation safety. *Aerosp. Sci. Technol.* **2018**. [CrossRef]

11. Moghtadernejad, S.; Jadidi, M.; Tembely, M.; Esmail, N.; Dolatabadi, A. Concurrent droplet coalescence and solidification on surfaces with various wettabilities. *J. Fluids Eng. Trans. ASME* **2015**. [CrossRef]

12. Myong, R.S. Droplet impingement in aircraft icing. In *Progress in Aircraft Icing and Aircraft Erosion Research*; Springs of Dreams Corporation: Orange County, CA, USA, 2017.

13. Potapczuk, M.G. Aircraft icing research at NASA Glenn research center. *J. Aerosp. Eng.* **2013**. [CrossRef]

14. Liu, Y.C.; Farouk, T.; Savas, A.J.; Dryer, F.L.; Avedisian, C.T. On the spherically symmetrical combustion of methyl decanoate droplets and comparisons with detailed numerical modeling. *Combust. Flame* **2013**, *160*, 641–655. [CrossRef]

15. Cen, C.; Wu, H.; Lee, C.f.; Liu, F.; Li, Y. Experimental investigation on the characteristic of jet break-up for butanol droplet impacting onto a heated surface in the film boiling regime. *Int. J. Heat Mass Transf.* **2018**, *123*, 129–136. [CrossRef]

16. Zama, Y.; Odawara, Y.; Furuhata, T. Experimental investigation on velocity inside a diesel spray after impingement on a wall. *Fuel* **2017**, *203*, 757–763. [CrossRef]

17. Pei, Y.; Qin, J.; Li, X.; Zhang, D.; Wang, K.; Liu, Y. Experimental investigation on free and impingement spray fueled with methanol, ethanol, isooctane, TRF and gasoline. *Fuel* **2017**, *208*, 174–183. [CrossRef]

18. Labergue, A.; Gradeck, M.; Lemoine, F. Experimental investigation of spray impingement hydrodynamic on a hot surface at high flow rates using phase Doppler analysis and infrared thermography. *Int. J. Heat Mass Transf.* **2016**, *100*, 65–78. [CrossRef]

19. Moghtadernejad, S.; Tembely, M.; Jadidi, M.; Esmail, N.; Dolatabadi, A. Shear driven droplet shedding and coalescence on a superhydrophobic surface. *Phys. Fluids* **2015**. [CrossRef]

20. Moghtadernejad, S.; Mohammadi, M.; Jadidi, M.; Tembely, M.; Dolatabadi, A. Shear Driven Droplet Shedding on Surfaces with Various Wettabilities. *SAE Int. J. Aerosp.* **2013**. [CrossRef]

21. Moghtadernejad, S.; Jadidi, M.; Ahmmed, K.M.T.; Lee, C.; Dolatabadi, A.; Kietzig, A.M. Experimental study of droplet shedding on laser-patterned substrates. *Phys. Fluids* **2019**. [CrossRef]

22. Knausgård, K. Superhydrophobic Anti-Ice Nanocoatings. Master's Thesis, Norwegian University of Science and Technology, Trondheim, Norway, 2012.

23. Antonini, C.; Innocenti, M.; Horn, T.; Marengo, M.; Amirfazli, A. Understanding the effect of superhydrophobic coatings on energy reduction in anti-icing systems. *Cold Reg. Sci. Technol.* **2011**. [CrossRef]

24. Sarkar, D.K.; Farzaneh, M. Superhydrophobic coatings with reduced ice adhesion. *J. Adhes. Sci. Technol.* **2009**. [CrossRef]

25. Cao, L.; Jones, A.K.; Sikka, V.K.; Wu, J.; Gao, D. Anti-Icing superhydrophobic coatings. *Langmuir* **2009**, *25*, 12444–12448. [CrossRef]

26. Moghtadernejad, S.; Jadidi, M.; Dolatabadi, A.; Esmail, N. SPH Simulation of Rivulet Dynamics on Surfaces with Various Wettabilities. *SAE Int. J. Aerosp.* **2015**. [CrossRef]

27. Chandra, S.; Fauchais, P. Formation of solid splats during thermal spray deposition. *J. Therm. Spray Technol.* **2009**. [CrossRef]

28. Pasandideh-Fard, M.; Pershin, V.; Chandra, S.; Mostaghimi, J. Splat shapes in a thermal spray coating process: Simulations and experiments. *J. Therm. Spray Technol.* **2002**. [CrossRef]

29. Jadidi, M.; Moghtadernejad, S.; Dolatabadi, A. A comprehensive review on fluid dynamics and transport of suspension/liquid droplets and particles in High-Velocity Oxygen-Fuel (HVOF) thermal spray. *Coatings* **2015**, *5*, 576–645. [CrossRef]

30. Jadidi, M.; Vardelle, A.; Dolatabadi, A.; Moreau, C. Heat transfer in suspension plasma spraying. In *Handbook of Thermal Science and Engineering*; Springer International Publishing: Cham, Switzerland, 2018.

31. Fauchais, P.L.; Heberlein, J.V.R.; Boulos, M.I. *Thermal Spray Fundamentals: From Powder to Part*; Springer: Berlin/Heidelberg, Germany, 2014.

32. Jadidi, M.; Moghtadernejad, S.; Dolatabadi, A. Penetration and breakup of liquid jet in transverse free air jet with application in suspension-solution thermal sprays. *Mater. Des.* **2016**. [CrossRef]

33. Jabbari, F.; Jadidi, M.; Wuthrich, R.; Dolatabadi, A. A numerical study of suspension injection in plasma-spraying process. *J. Therm. Spray Technol.* **2014**. [CrossRef]

34. Jadidi, M.; Yeganeh, A.Z.; Dolatabadi, A. Numerical Study of Suspension HVOF Spray and Particle Behavior Near Flat and Cylindrical Substrates. *J. Therm. Spray Technol.* **2018**. [CrossRef]

35. Parizi, H.B.; Rosenzweig, L.; Mostaghimi, J.; Chandra, S.; Coyle, T.; Salimi, H.; Pershin, L.; McDonald, A.; Moreau, C. Numerical simulation of droplet impact on patterned surfaces. *J. Therm. Spray Technol.* **2007**. [CrossRef]

36. Jadidi, M.; Moghtadernejad, S.; Dolatabadi, A. Numerical Modeling of Suspension HVOF Spray. *J. Therm. Spray Technol.* **2016**. [CrossRef]

37. Pasandideh-Fard, M.; Chandra, S.; Mostaghimi, J. A three-dimensional model of droplet impact and solidification. *Int. J. Heat Mass Transf.* **2002**. [CrossRef]

38. Jadidi, M.; Mousavi, M.; Moghtadernejad, S.; Dolatabadi, A. A Three-Dimensional Analysis of the Suspension Plasma Spray Impinging on a Flat Substrate. *J. Therm. Spray Technol.* **2014**. [CrossRef]

39. Yan, Z.; Zhao, R.; Duan, F.; Wong, T.N.; Toh, K.C.; Choo, K.F.; Chan, P.K.; Chua, Y.S. Spray Cooling. In *Two Phase Flow, Phase Change and Numerical Modeling*; Ahsan, A., Ed.; IntechOpen: London, UK, 2011. Available online: https://www.intechopen.com/books/two-phase-flow-phase-change-and-numerical-modeling/spray-cooling (accessed on 1 July 2020). [CrossRef]

40. Horacek, B.; Kiger, K.T.; Kim, J. Single nozzle spray cooling heat transfer mechanisms. *Int. J. Heat Mass Transf.* **2005**. [CrossRef]

41. Jia, W.; Qiu, H.H. Experimental investigation of droplet dynamics and heat transfer in spray cooling. *Exp. Therm. Fluid Sci.* **2003**. [CrossRef]

42. Kim, J. Spray cooling heat transfer: The state of the art. *Int. J. Heat Fluid Flow* **2007**. [CrossRef]

43. Sodtke, C.; Stephan, P. Spray cooling on micro structured surfaces. *Int. J. Heat Mass Transf.* **2007**. [CrossRef]

44. Breitenbach, J.; Roisman, I.V.; Tropea, C. From drop impact physics to spray cooling models: A critical review. *Exp. Fluids* **2018**. [CrossRef]

45. Incropera, F.P.; DeWitt, D.P.; Bergman, T.L.; Lavine, A.S. *Fundamentals of Heat and Mass Transfer*, 6th ed.; Wiley: Hoboken, NJ, USA, 2007.

46. Wijshoff, H. Drop dynamics in the inkjet printing process. *Curr. Opin. Colloid Interface Sci.* **2018**. [CrossRef]

47. Gan, H.Y.; Shan, X.; Eriksson, T.; Lok, B.K.; Lam, Y.C. Reduction of droplet volume by controlling actuating waveforms in inkjet printing for micro-pattern formation. *J. Micromech. Microeng.* **2009**. [CrossRef]

48. Castrejón-Pita, J.R.; Martin, G.D.; Hoath, S.D.; Hutchings, I.M. A simple large-scale droplet generator for studies of inkjet printing. *Rev. Sci. Instrum.* **2008**. [CrossRef]

49. Daly, R.; Harrington, T.S.; Martin, G.D.; Hutchings, I.M. Inkjet printing for pharmaceutics—A review of research and manufacturing. *Int. J. Pharm.* **2015**. [CrossRef]

50. Yusof, A.; Keegan, H.; Spillane, C.D.; Sheils, O.M.; Martin, C.M.; O'Leary, J.J.; Zengerle, R.; Koltay, P. Inkjet-like printing of single-cells. *Lab. Chip* **2011**. [CrossRef]

51. Williams, C. Ink-jet printers go beyond paper. *Phys. World* **2006**. [CrossRef]

52. Fan, Y.Q.; Wang, H.L.; Gao, K.X.; Liu, J.J.; Chai, D.P.; Zhang, Y.J. Applications of Modular Microfluidics Technology. *Chin. J. Anal. Chem.* **2018**, *46*, 1863–1871. [CrossRef]

53. Teh, S.Y.; Lin, R.; Hung, L.H.; Lee, A.P. Droplet microfluidics. *Lab. Chip* **2008**. [CrossRef]

54. Lee, C.Y.; Chang, C.L.; Wang, Y.N.; Fu, L.M. Microfluidic mixing: A review. *Int. J. Mol. Sci.* **2011**, *12*, 3263–3287. [CrossRef] [PubMed]

55. Grant, G.; Brenton, J.; Drysdale, D. Fire suppression by water sprays. *Prog. Energy Combust. Sci.* **2000**. [CrossRef]

56. Moon, J.H.; Kim, D.Y.; Lee, S.H. Spreading and receding characteristics of a non-Newtonian droplet impinging on a heated surface. *Exp. Therm. Fluid Sci.* **2014**. [CrossRef]

57. Rein, M. Phenomena of liquid drop impact on solid and liquid surfaces. *Fluid Dyn. Res.* **1993**. [CrossRef]

58. Moon, J.H.; Lee, J.B.; Lee, S.H. Dynamic Behavior of Non-Newtonian Droplets Impinging on Solid Surfaces. *Mater. Trans.* **2013**. [CrossRef]

59. Josserand, C.; Thoroddsen, S.T. Drop Impact on a Solid Surface. *Annu. Rev. Fluid Mech.* **2016**. [CrossRef]

60. Herbert, S.; Gambaryan-Roisman, T.; Stephan, P. Influence of the governing dimensionless parameters on heat transfer during single drop impingement onto a hot wall. *Colloids Surf. A Physicochem. Eng. Asp.* **2013**. [CrossRef]

61. Marengo, M.; Antonini, C.; Roisman, I.V.; Tropea, C. Drop collisions with simple and complex surfaces. *Curr. Opin. Colloid Interface Sci.* **2011**. [CrossRef]

62. Wang, H.; Zhu, X.; Chen, R.; Liao, Q.; Ding, B. How supercooled superhydrophobic surfaces affect dynamic behaviors of impacting water droplets? *Int. J. Heat Mass Transf.* **2018**, *124*, 1025–1032.

63. Bange, P.G.; Patil, N.D.; Bhardwaj, R. Impact Dynamics of a Droplet on a Heated Surface. In Proceedings of the 5th International Conference of Fluid Flow, Heat and Mass Transfer (FFHMT'18), Niagara Falls, ON, Canada, 7–9 June 2018; Volume 190, pp. 232–247. [CrossRef]

64. Guo, Y.; Shen, S.; Yang, Y.; Liang, G.; Zhen, N. Rebound and spreading during a drop impact on wetted cylinders. *Exp. Therm. Fluid Sci.* **2013**, *52*, 97–103. [CrossRef]

65. Rioboo, R.; Tropea, C.; Marengo, M. Outcomes from a drop impact on solid surfaces. *At. Sprays* **2001**. [CrossRef]

66. Zhao, P.; Hargrave, G.K.; Versteeg, H.K.; Garner, C.P.; Reid, B.A.; Long, E.J.; Zhao, H. The dynamics of droplet impact on a heated porous surfac. *Chem. Eng. Sci.* **2018**. [CrossRef]

67. Rajesh, R.S.; Naveen, P.T.; Krishnakumar, K.; Ranjith, S.K. Dynamics of single droplet impact on cylindrically-curved superheated surfaces. *Exp. Therm. Fluid Sci.* **2019**, *101*, 251–262. [CrossRef]

68. Pan, Y.; Shi, K.; Duan, X.; Naterer, G.F. Experimental investigation of water droplet impact and freezing on micropatterned stainless steel surfaces with varying wettabilities. *Int. J. Heat Mass Transf.* **2019**, *129*, 953–964. [CrossRef]

69. Laan, N.; de Bruin, K.G.; Bartolo, D.; Josserand, C.; Bonn, D. Maximum diameter of impacting liquid droplets. *Phys. Rev. Appl.* **2014**, *2*, 044018. [CrossRef]

70. Lee, J.B.; Laan, N.; de Bruin, K.G.; Skantzaris, G. Universal rescaling of drop impact on smooth and rough surfaces. *J. Fluid Mech.* **2016**, *786*, R4. [CrossRef]

71. Jones, H. Cooling, freezing and substrate impact of droplets formed by rotary atomization. *J. Phys. D Appl. Phys.* **1971**. [CrossRef]

72. Chandra, S.; Avedisian, C.T. On the collision of a droplet with a solid surface. *Proc. R. Soc. A Math. Phys. Eng. Sci.* **1991**. [CrossRef]

73. Asai, A.; Shioya, M.; Hirasawa, S.; Okazaki, T. Impact of an ink drop on paper. *J. Imaging Sci. Technol.* **1993**, *37*, 205–207.

74. Pasandideh-Fard, M.; Qiao, Y.M.; Chandra, S.; Mostaghimi, J. Capillary effects during droplet impact on a solid surface. *Phys. Fluids* **1996**. [CrossRef]

75. Mao, T.; Kuhn, D.C.S.; Tran, H. Spread and Rebound of Liquid Droplets upon Impact on Flat Surfaces. *AIChE J.* **1997**. [CrossRef]

76. Ukiwe, C.; Kwok, D.Y. On the maximum spreading diameter of impacting droplets on well-prepared solid surfaces. *Langmuir* **2005**. [CrossRef] [PubMed]

77. Tang, C.; Qin, M.; Weng, X.; Zhang, X.; Zhang, P.; Li, J.; Huang, Z.; Qin, M.; Weng, X.; Zhang, X.; et al. Dynamics of droplet impact on solid surface with different roughness. *Int. J. Multiph. Flow* **2017**, *96*, 56–69. [CrossRef]

78. Jin, Z.; Zhang, H.; Yang, Z. International Journal of Heat and Mass Transfer The impact and freezing processes of a water droplet on different cold cylindrical surfaces. *Int. J. Heat Mass Transf.* **2017**, *113*, 318–323. [CrossRef]

79. Jin, Z.; Wang, Z.; Sui, D.; Yang, Z. The impact and freezing processes of a water droplet on different inclined cold surfaces. *Int. J. Heat Mass Transf.* **2016**, *97*, 211–223.

80. Zhang, H.; Jin, Z.; Jiao, M.; Yang, Z. Experimental investigation of the impact and freezing processes of a water droplet on different cold concave surfaces. *Int. J. Therm. Sci.* **2018**, *132*, 498–508. [CrossRef]

81. Jin, Z.; Zhang, H.; Yang, Z. Experimental investigation of the impact and freezing processes of a water droplet on an ice surface. *Int. J. Heat Mass Transf.* **2017**. [CrossRef]

82. Ju, J.; Jin, Z.; Zhang, H.; Yang, Z.; Zhang, J. The impact and freezing processes of a water droplet on different cold spherical surfaces. *Exp. Therm. Fluid Sci.* **2018**. [CrossRef]

83. Zhang, R.; Hao, P.; Zhang, X.; He, F. Supercooled water droplet impact on superhydrophobic surfaces with various roughness and temperature. *Int. J. Heat Mass Transf.* **2018**, *122*, 395–402. [CrossRef]

84. Liang, G.; Guo, Y.; Mu, X.; Shen, S. Experimental investigation of a drop impacting on wetted spheres. *Exp. Therm. Fluid Sci.* **2014**, *55*, 150–157. [CrossRef]

85. Burzynski, D.A.; Bansmer, S.E. Droplet splashing on thin moving films at high Weber numbers. *Int. J. Multiph. Flow* **2018**, *101*, 202–211. [CrossRef]

86. Zen, T.S.; Chou, F.C.; Ma, J.L. Ethanol drop impact on an inclined moving surface. *Int. Commun. Heat Mass Transf.* **2010**, *37*, 1025–1030. [CrossRef]

87. Xu, L.; Zhang, W.W.; Nagel, S.R. Drop splashing on a dry smooth surface. *Phys. Rev. Lett.* **2005**. [CrossRef]

88. Allen, R.F. The role of surface tension in splashing. *J. Colloid Interface Sci.* **1975**. [CrossRef]

89. Thoroddsen, S.T.; Sakakibara, J. Evolution of the fingering pattern of an impacting drop. *Phys. Fluids* **1998**. [CrossRef]

90. Breitenbach, J.; Roisman, I.V.; Tropea, C. Drop collision with a hot, dry solid substrate: Heat transfer during nucleate boiling. *Phys. Rev. Fluids* **2017**. [CrossRef]

91. Leidenfrost, J.G. On the fixation of water in diverse fire. *Int. J. Heat Mass Transf.* **1966**. [CrossRef]

92. Kim, S.H.; Jiang, Y.; Kim, H. Droplet impact and LFP on wettability and nanostructured surface. *Exp. Therm. Fluid Sci.* **2018**, *99*, 85–93. [CrossRef]

93. Chen, H.; Cheng, W.l.; Peng, Y.h.; Jiang, L.j. Dynamic Leidenfrost temperature increase of impacting droplets containing high-alcohol surfactant. *Int. J. Heat Mass Transf.* **2018**, *118*, 1160–1168. [CrossRef]

94. Chaudhary, G.; Li, R. Freezing of water droplets on solid surfaces: An experimental and numerical study. *Exp. Therm. Fluid Sci.* **2014**. [CrossRef]

95. Jung, S.; Tiwari, M.K.; Doan, N.V.; Poulikakos, D. Mechanism of supercooled droplet freezing on surfaces. *Nat. Commun.* **2012**. [CrossRef] [PubMed]

96. Li, H.; Roisman, I.V.; Tropea, C. Influence of solidification on the impact of supercooled water drops onto cold surfaces. *Exp. Fluids* **2015**. [CrossRef]

97. Mishchenko, L.; Hatton, B.; Bahadur, V.; Taylor, J.A.; Krupenkin, T.; Aizenberg, J. Design of ice-free nanostructured surfaces based on repulsion of impacting water droplets. *ACS Nano* **2010**. [CrossRef] [PubMed]

98. Wenzel, R.N. Resistance of solid surfaces to wetting by water. *Ind. Eng. Chem.* **1936**. [CrossRef]

99. Cassie, A.B.D.; Baxter, S. Wettability of porous surfaces. *Trans. Faraday Soc.* **1944**. [CrossRef]

100. Gundersen, H.; Leinaas, H.P.; Thaulow, C. Surface structure and wetting characteristics of Collembola cuticles. *PLoS ONE* **2014**, *9*, e86783. [CrossRef]

101. Vakarelski, I.U.; Patankar, N.A.; Marston, J.O.; Chan, D.Y.C.; Thoroddsen, S.T. Stabilization of Leidenfrost vapour layer by textured superhydrophobic surfaces. *Nature* **2012**. [CrossRef]

102. Phan, H.T.; Caney, N.; Marty, P.; Colasson, S.; Gavillet, J. Surface wettability control by nanocoating: The effects on pool boiling heat transfer and nucleation mechanism. *Int. J. Heat Mass Transf.* **2009**. [CrossRef]

103. Kandlikar, S.G. A Theoretical Model to Predict Pool Boiling CHF Incorporating Effects of Contact Angle and Orientation. *J. Heat Transf.* **2001**. [CrossRef]

104. Hamdan, K.S.; Kim, D.E.; Moon, S.K. Droplets behavior impacting on a hot surface above the Leidenfrost temperature. *Ann. Nucl. Energy* **2015**, *80*, 338–347. [CrossRef]

105. Yun, S.; Hong, J.; Kang, K.H. Suppressing drop rebound by electrically driven shape distortion. *Phys. Rev. E Stat. Nonlinear Soft Matter Phys.* **2013**. [CrossRef]

106. Yun, S.; Lim, G. Ellipsoidal drop impact on a solid surface for rebound suppression. *J. Fluid Mech.* **2014**. [CrossRef]

107. Yun, S.; Lim, G. Control of a bouncing magnitude on a heated substrate via ellipsoidal drop shape. *Appl. Phys. Lett.* **2014**. [CrossRef]

108. Yun, S. Bouncing of an ellipsoidal drop on a superhydrophobic surface. *Sci. Rep.* **2017**. [CrossRef]

109. Yun, S. Impact dynamics of egg-shaped drops on a solid surface for suppression of the bounce magnitude. *Int. J. Heat Mass Transf.* **2018**, *127*, 172–178. [CrossRef]

110. Bobinski, T.; Sobieraj, G.; Gumowski, K.; Rokicki, J.; Psarski, M.; Marczak, J.; Celichowski, G. Droplet impact in icing conditions—The influence of ambient air humidity. *Arch. Mech.* **2014**. [CrossRef]

111. Jadidi, M.; Trepanier, J.Y.; Farzad, M.A.; Dolatabadi, A. Effects of ambient air relative humidity and surface temperature on water droplet spreading dynamics. *FEDSM* **2018**. [CrossRef]

112. Povarov, O.A.; Nazarov, O.I.; Ignat'evskaya, L.A.; Nikol'skii, A.I. Interaction of drops with boundary layer on rotating surface. *J. Eng. Phys.* **1976**. [CrossRef]

113. Mundo, C.; Sommerfeld, M.; Tropea, C. Droplet-wall collisions: Experimental studies of the deformation and breakup process. *Int. J. Multiph. Flow* **1995**. [CrossRef]

114. Yao, S.C.; Cai, K.Y. The dynamics and leidenfrost temperature of drops impacting on a hot surface at small angles. *Exp. Therm. Fluid Sci.* **1988**. [CrossRef]

 *fluids*

MDPI

Article

A CFD Tutorial in Julia: Introduction to Laminar Boundary-Layer Theory

Furkan Oz * and Kursat Kara

School of Mechanical and Aerospace Engineering, Oklahoma State University, Stillwater, OK 74078, USA; kursat.kara@okstate.edu
* Correspondence: foz@okstate.edu

Abstract: Numerical simulations of laminar boundary-layer equations are used to investigate the origins of skin-friction drag, flow separation, and aerodynamic heating concepts in advanced undergraduate- and graduate-level fluid dynamics/aerodynamics courses. A boundary-layer is a thin layer of fluid near a solid surface, and viscous effects dominate it. Students must understand the modeling of flow physics and implement numerical methods to conduct successful simulations. Writing computer codes to solve equations numerically is a critical part of the simulation process. Julia is a new programming language that is designed to combine performance and productivity. It is dynamic and fast. However, it is crucial to understand the capabilities of a new programming language before attempting to use it in a new project. In this paper, fundamental flow problems such as Blasius, Hiemenz, Homann, and Falkner-Skan flow equations are derived from scratch and numerically solved using the Julia language. We used the finite difference scheme to discretize the governing equations, employed the Thomas algorithm to solve the resulting linear system, and compared the results with the published data. In addition, we released the Julia codes in GitHub to shorten the learning curve for new users and discussed the advantages of Julia over other programming languages. We found that the Julia language has significant advantages in productivity over other coding languages. Interested readers may access the Julia codes on our GitHub page.

Keywords: CFD; Julia; Blasius; Hiemenz; Homann; Falkner–Skan; boundary-layer

Citation: Oz, F.; Kara, K. A CFD Tutorial in Julia: Introduction to Laminar Boundary-Layer Theory. *Fluids* **2021**, *6*, 207. https://doi.org/10.3390/fluids6060207

Academic Editor: Ashwin Vaidya

Received: 13 May 2021
Accepted: 28 May 2021
Published: 3 June 2021

Publisher's Note: MDPI stays neutral with regard to jurisdictional claims in published maps and institutional affiliations.

1. Introduction

Computational fluid dynamics (CFD) simulation is one of the vital steps of the design of a product that includes fluid motion. Since fluid dynamics are extremely complex and the motion equations have nonlinear terms, usage of numerical approaches is inevitable to simulate or predict fluid motion. Predictions can also be done with experiments. However, it is usually costlier than a regular CFD simulation. On the other hand, fundamental knowledge about the flow, that is planned to simulate, is necessary because of the required numerical approach decision. Learning the canonical flows well is extremely important in this point because most of the complex flow consists of a combination of a couple of canonical flows. An airfoil CFD simulation can consist of boundary-layer flow over the smooth part of the airfoil, mixing layer flow where the tail ends, and blunt body flow in the wake region. One airfoil simulation consists of three different canonical flows. A full understanding of the canonical flows is extremely important to model an accurate airfoil simulation.

Prandtl [1] stated that some of the terms in the Navier–Stokes equations can be neglected for the boundary-layer flows. As a result of this assumption, well-known boundary-layer equations arose. These approaches are still valid after a century and the resultant system of equations inspired lots of researchers in their studies. One of them was Blasius who is the Ph.D. student of Prandtl. Blasius [2] worked on the same problem as Prandtl did. However, he aimed to overcome the enigma of turbulence by considering the phenomenon of boundary-layer flow explained by Hager [3]. He further

97

simplified the boundary-layer equations for a flat plate. He assumed that the flow is parallel. In other words, the velocity component in the parallel direction is not zero and the velocity in the transverse direction is zero. Moreover, he represented the resultant system of equations with a third-order ordinary differential equation, which is known as the Blasius similarity solution.

The Falkner–Skan similarity solution is another laminar similarity solution. Cebeci [4] explains the Falkner–Skan equations named after V. M. Falkner and Sylvia W. Skan. Falkner and Skan generalized the Blasius similarity solution for non-parallel flows, such as wedge flows and corner flows. The resultant equation can be used to predict the boundary-layer thickness for wedge flow. It has to be noted that these assumptions are available in the laminar region. Once the turbulence occurs, both of these similarity solutions will be inaccurate. In the incompressible region, the Falkner–Skan equation can be used without additional equation; however, after the compressibility limit, the temperature effect must be introduced to the system as an additional equation. If the Falkner–Skan equation is solved with the energy equation, the boundary-layer profile can be obtained in the compressible region.

One of the other Ph.D. students of Prandtl is Karl Hiemenz who worked on Hiemenz flow which is a type of stagnation point flow. Hiemenz [5] formulated and calculated the stagnation point problem as explained in the Schlichting [6]. The problem was a special case of the Falkner–Skan similarity solution. Howarth [7] also worked on the same problem and concluded similar results. Another Ph.D. student of Prandtl, Fritz Homann [8] worked on the same problem for axisymmetric bodies as Schlichting [6] explains. Homann's similarity formulation was for a sphere while Hiemenz's similarity formulation was for a cylinder. Both similarity solutions are being widely used for stagnation flows.

The aforementioned papers are the origins of the boundary-layer theory. As it is mentioned, the boundary-layer theory is the origin of many problems, such as the laminar to turbulent boundary-layer transition and flow separation. The main concern of these researches is to increase the performance of the vehicle because the transition increases the heat transfer and the vehicle requires a better thermal protection system at high speeds due to increased heat transfer rate. On the other hand, flow separation may lead to a lift force loss on the wing as a result, the performance of aircraft decreases. Since the focus of the present paper is the fundamentals of laminar boundary-layer theory, the details will not be provided about advanced researches. However, readers who are interested in details may check [9–12] for subsonic boundary-layer transition, Ref. [13–21] for supersonic/hypersonic boundary-layer transition, and Ref. [22–24] for flow separation. The other researches where boundary-layer flow is involved are [25–30].

Understanding the aforementioned fundamental flows are crucial for a senior undergraduate student or a graduate student in order to simulate more complex flows. However, modeling these equations in a computer environment requires more than knowledge about canonical flows instead it requires knowledge about programming languages as well. There are several learning modules and papers [31–35] for computational fluid dynamics simulation coding; however, there is not enough publication for Julia language [36]. It is a relatively new coding language among the other coding languages such as Fortran, C/C++, Python, and MATLAB but it is getting popular fast because it is trying to fill the gap between the language of the state-of-art CFD codes, Fortran and C/C++, and straightforward/user-friendly languages, Python and MATLAB. Most of the state-of-art CFD codes are written in Fortran and C/C++ because it is so fast and random access memory usage (RAM) can be reduced drastically. On the other hand, MATLAB and Python have user-friendly and easy syntax which makes them favorites of students, with a price, which is the speed of computation. Julia is a language that tries to fill this gap between two different coding styles. It has a user-friendly environment, while also being fast. It provides a working space that helps users to write clear, high-level, generic code that resembles mathematical formulas. Julia's ability to combine high-performance with productivity makes it a great choice for researchers working in different areas.

In this paper, fundamental flow problems such as Hiemenz flow, Homann flow, Blasius flow, and Falkner–Skan flow will be derived from scratch and modeled in the Julia environment. The finite-difference discretization in space with Thomas algorithm as linear system solver is used. Outputs of the code are provided and they are compared with the literature. Additionally, the advantages of the Julia language over other languages will be discussed. The computer codes and the implementation instructions will help students to understand the fundamental flows which will provide insight for student's course work and researches. Using these examples, they can solve more complex flow types and develop their own codes. We make all these codes available on GitHub and they are accessible to everyone. We provide installation instruction for Julia and the required packages in Appendix A. The GitHub link of the codes also can be found in Appendix A.

2. Laminar Boundary-Layer Theory

Understanding canonical flows is crucial to understanding more complex flows such as flow over a wing or an airfoil. Hosseini et al. [37] studied flow over a wing section with a direct numerical simulation (DNS) study. They created a great video about how flow is formed over the aircraft. The video shows the development of the boundary-layer, how the trailing edge looks like a mixing-layer flow, and how Karman vortex street type of wake structures occurs. If the pre-study work is examined, it can be seen that the mesh structure is built on the flow prediction. Using a finer mesh on the critical regions is a key point of the CFD and it requires predictions about the possible flow behavior. The boundary-layer is the origin of many engineering problems in aerodynamics, including wing stall, the skin friction drag on an object, and the heat transfer that occurs in high-speed flight. In this present paper, boundary-layer theory will be examined under two subsections, which are laminar boundary-layer problems and stagnation point problems. In this present paper, fundamental fluid dynamics problems related to boundary-layer theory will be derived and implemented in a relatively new programming language, Julia. The contribution is employing the Julia language. The governing equations and solution methodologies are already published in the literature. The interested reader should refer to the additional references [6,38] for detailed derivations. The manuscript may enable students to adopt the programming language easily.

2.1. Laminar Boundary-Layer Flow Problems

Velocity distribution over a flat plate can be represented with a similarity solution. In this subsection, Blasius and Falkner–Skan similarity solutions will be derived from scratch and they will be solved numerically in the Julia environment. The Julia codes will be available to shorten the learning curve.

2.1.1. Blasius Flow Problem

Schematic description of flow over a flat plate, in other words, Blasius boundary-layer flow can be illustrated as in Figure 1. The governing equations for Blasius flow are boundary-layer equations. These equations can be obtained by non-dimensionalizing the two-dimensional Navier–Stokes equations that are:

$$\frac{\partial u}{\partial x} + \frac{\partial v}{\partial y} = 0 \tag{1}$$

$$u\frac{\partial u}{\partial x} + v\frac{\partial u}{\partial y} = -\frac{1}{\rho}\frac{\partial p}{\partial x} + \nu\left(\frac{\partial^2 u}{\partial x^2} + \frac{\partial^2 u}{\partial y^2}\right) \tag{2}$$

$$u\frac{\partial v}{\partial x} + v\frac{\partial v}{\partial y} = -\frac{1}{\rho}\frac{\partial p}{\partial y} + \nu\left(\frac{\partial^2 v}{\partial x^2} + \frac{\partial^2 v}{\partial y^2}\right), \tag{3}$$

where u and v are the velocities in the x-direction and y-direction, respectively. p is the pressure, v is the kinematic viscosity, ρ is the density. If Equations (1)–(3) are non-dimensionalized with:

$$u^* = \frac{u}{U_\infty}, \quad v^* = \frac{v}{U_\infty}, \quad p^* = \frac{p}{\rho U_\infty^2}, \tag{4}$$

$$x^* = \frac{x}{L}, \quad y^* = \frac{y}{L}, \tag{5}$$

where L is the plate length and U_∞ is the free-stream velocity. The two-dimensional, incompressible non-dimensional Navier–Stokes equations can be shown as:

$$\frac{\partial u^*}{\partial x^*} + \frac{\partial v^*}{\partial y^*} = 0 \tag{6}$$

$$u^* \frac{\partial u^*}{\partial x^*} + v^* \frac{\partial u^*}{\partial y^*} = -\frac{\partial p^*}{\partial x^*} + \frac{1}{Re_\infty} \left(\frac{\partial^2 u^*}{\partial x^{*2}} + \frac{\partial^2 u^*}{\partial y^{*2}} \right) \tag{7}$$

$$u^* \frac{\partial v^*}{\partial x^*} + v^* \frac{\partial v^*}{\partial y^*} = -\frac{\partial p^*}{\partial y^*} + \frac{1}{Re_\infty} \left(\frac{\partial^2 v^*}{\partial x^{*2}} + \frac{\partial^2 v^*}{\partial y^{*2}} \right). \tag{8}$$

Star superscript ($[\ ]^*$) corresponds to non-dimensional variable. In a boundary-layer flow, some variables are smaller than others. For example, $u^* = O(1)$ and $x^* = O(1)$ so $\frac{\partial u^*}{\partial x^*} = O(1)$. In the continuity equation, two terms must be in the same order to obtain 0 so $\frac{\partial v^*}{\partial y^*} = O(1)$. It can be seen that from Figure 1, $y = O(\delta^*)$ where δ^* is the non-dimensionalized boundary-layer thickness. From here, it can be concluded that $v = O(\delta^*)$. It has to be noted that $\delta^* << 1$. If the same approach is applied to momentum equations, the final system of equations will be:

$$\frac{\partial u^*}{\partial x^*} + \frac{\partial v^*}{\partial y^*} = 0 \tag{9}$$

$$u^* \frac{\partial u^*}{\partial x^*} + v^* \frac{\partial u^*}{\partial y^*} = -\frac{\partial p^*}{\partial x^*} + \frac{1}{Re_\infty} \frac{\partial^2 u^*}{\partial y^{*2}} \tag{10}$$

$$\frac{\partial p^*}{\partial y^*} = 0. \tag{11}$$

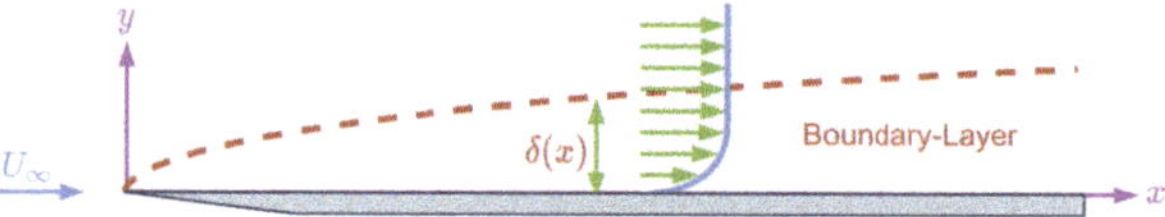

Figure 1. Schematic description of the flow over a flat plate.

For an inviscid flow, where the Reynolds number is so high, the viscous term can be neglected. The pressure is constant as obtained in the Equation (11) and $v = 0$. Equation (10) will be:

$$-\frac{dP_e^*}{dx^*} = U_e^* \frac{dU_e^*}{dx^*}, \tag{12}$$

where U_e and P_e are the inviscid velocity and pressure. Finally, the two-dimensional boundary-layer equations in the dimensionless form can be written as:

$$\frac{\partial u}{\partial x} + \frac{\partial v}{\partial y} = 0 \tag{13}$$

$$u \frac{\partial u}{\partial x} + v \frac{\partial u}{\partial y} = U_e \frac{dU_e}{dx} + v \frac{\partial^2 u}{\partial y^2}. \tag{14}$$

It has to be noted that star superscript is suppressed to avoid confusion. In order to solve this problem, similarity transformation can be used. The similarity parameter (η) can be chosen as:

$$\eta = \frac{y}{\sqrt{\frac{\nu x}{U_\infty}}}.$$ (15)

After that, velocity components can be shown as:

$$u(x,y) = U_\infty F(\eta)$$ (16)

$$v(x,y) = \frac{G(\eta)}{\sqrt{\frac{x}{U_\infty \nu}}}.$$ (17)

If Equations (13) and (14) are rearranged with the given velocities, the boundary-layer equations will be:

$$G' = \frac{\eta}{2}F'$$ (18)

$$F'' = -\frac{1}{2}\eta FF' + GF'.$$ (19)

In order to make the next procedure easier, it can be assumed that $F = f'$. After that, Equation (18) can be integrated and put into Equation (19). The final ordinary differential equation can be obtained as:

$$2f''' + ff'' = 0.$$ (20)

The boundary conditions of the system will be:

$$u(x,0) = U_\infty F(\eta = 0) = 0 \longrightarrow F(0) = f'(0) = 0$$ (21)

$$v(x,0) = \frac{G(\eta = 0)}{\sqrt{\frac{x}{U_\infty \nu}}} = 0 \longrightarrow G(0) = f(0) = 0$$ (22)

$$u(x,y \to \infty) = U_\infty F(\eta \to \infty) = 1 \longrightarrow F(\infty) = f'(\infty) = f(\infty) = 1.$$ (23)

The given assumptions reduced the system of second-order partial differential equations into a third-order ordinary differential equation. The computational approach will be examined in the Julia framework.

2.1.2. Numerical Solution of Blasius Flow Problem

In this computational section, the solution of the third-order ordinary differential equation will be transformed into two equations and they will be solved with Thomas algorithm [39] which is one of the best methods for the tridiagonal matrices. It has to be noted that the same equations can be solved with any other methods such as Runge-Kutta, Runge-Kutta-Fehlberg, compact finite difference, high-order finite-difference; however, in this present paper, authors used the finite difference method for spatial discretization with the Thomas algorithm for the linear system solution. If it is assumed that $f' = h$ and $f = p$ in Equation (20), the system of equations will be:

$$2h'' + ph' = 0$$ (24)

$$p' - h = 0,$$ (25)

with the following boundary conditions:

$$p(0) = 0$$ (26)

$$h(0) = 0$$ (27)

$$h(\infty) = 1.$$ (28)

If the second-order central finite difference method is applied to h variable:

$$h_n'' = \frac{h_{n+1} - 2h_n + h_{n-1}}{(\Delta\eta)^2} + O((\Delta\eta)^2) \tag{29}$$

$$h_n' = \frac{h_{n+1} - h_{n-1}}{2\Delta\eta} + O((\Delta\eta)^2). \tag{30}$$

In this paper, details of the finite difference and derivation of it from Taylor's series will not be covered. If the reader is curious about the derivation of them, the book of Moin [40] can be checked. If the finite difference approach for h is substituted into the Equation (24), the final equation will be:

$$A_n h_{n+1} + B_n h_n + C_n h_{n-1} = 0, \tag{31}$$

where the A_n, B_n, and C_n are:

$$A_n = \frac{2}{\Delta\eta^2} + \frac{p_n}{2\Delta\eta} \tag{32}$$

$$B_n = \frac{-4}{\Delta\eta^2} \tag{33}$$

$$C_n = \frac{2}{\Delta\eta^2} - \frac{p_n}{2\Delta\eta}, \tag{34}$$

with the boundary conditions for the system the tridiagonal system ($Ax = b$) can be shown as:

$$\underbrace{\begin{bmatrix} 1 & 0 & 0 & 0 & \cdots & 0 \\ A_2 & B_2 & C_2 & 0 & \cdots & 0 \\ \vdots & \ddots & \ddots & \ddots & & \vdots \\ 0 & 0 & A_{N-2} & B_{N-2} & C_{N-2} & 0 \\ 0 & 0 & 0 & A_{N-1} & B_{N-1} & C_{N-1} \\ 0 & 0 & 0 & 0 & 0 & 1 \end{bmatrix}}_{A} \underbrace{\begin{bmatrix} h_1 \\ h_2 \\ \vdots \\ h_{N-2} \\ h_{N-1} \\ h_N \end{bmatrix}}_{x} = \underbrace{\begin{bmatrix} 0 \\ 0 \\ \vdots \\ 0 \\ 0 \\ 1 \end{bmatrix}}_{b} \tag{35}$$

It has to be noted that the first and last rows of the matrix are known and they are coming from the boundary conditions. In order to solve this problem, the Thomas algorithm can be applied. If relation between h_n and h_{n-1} is:

$$h_n = G_n + H_n h_{n+1} \tag{36}$$

$$h_{n-1} = G_{n-1} + H_{n-1} h_n. \tag{37}$$

If Equation (37) is substituted into Equation (31), the final equation will be:

$$h_n = \frac{-C_n G_{n-1}}{B_n + C_n H_{n-1}} + \frac{-A_n}{B_n + C_n H_{n-1}} h_{n+1}. \tag{38}$$

If Equations (36) and (38) are compared, G_n and H_n coefficients can be found as:

$$G_n = \frac{-C_n G_{n-1}}{B_n + C_n H_{n-1}} \tag{39}$$

$$H_n = \frac{-A_n}{B_n + C_n H_{n-1}}. \tag{40}$$

The problem here is that one has to know G_1 and H_1 in order to start the calculation. H_1 can be assumed as 0 and the G_1 can be calculated as 0 from the boundary conditions. Once G_1 and H_1 are calculated, G_n and H_n can be calculated from Equations (39) and (40). The given formulations can be implemented as it is shown in Listing 1.

The problem in Listing 1 is the usage of the p and h values which are not defined yet. In order to start to iteration, the initial assumption for h must be given. Most of the time, the linear assumption is the best assumption. Figure 2 shows the schematic of the profile transformation after some iterations. Once the initialization of h is done, p can be calculated from the integration of $p' = h$. The discrete integration of p can be written as:

$$p_n = p_{n-1} + \int_{\eta_{n-1}}^{\eta_n} h d\eta. \tag{41}$$

Listing 1. Implementation of Thomas algorithm for Blasius profile.

```
A = [ 2/Δη² + p[i]/(2*Δη)  for i=1:N]
B = [−4/Δη²                for i=1:N]
C = [ 2/Δη² − p[i]/(2*Δη)  for i=1:N]
D = [ 0 for i=1:N]

for i=2:N−1
G[i] = − ( C[i]*G[i−1] + D[i] )/(B[i] + C[i] * H[i−1])
H[i] = −                  A[i]   /(B[i] + C[i] * H[i−1])
end

for i=N-1:-1:2
h[i] = G[i] + H[i] * h[i+1]
end
```

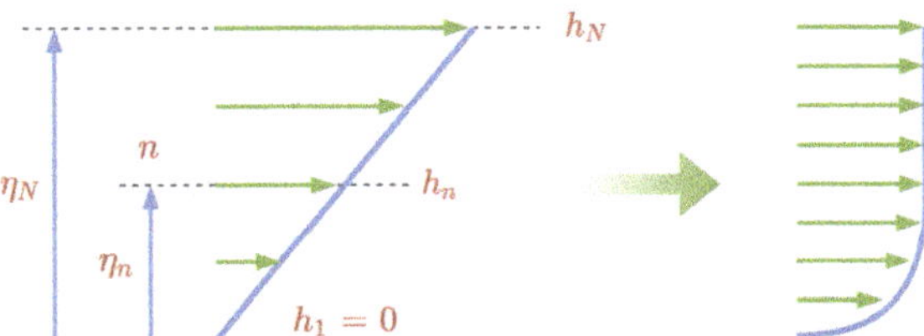

Figure 2. Schematic description of the change in the velocity profile from initial guess to final solution.

If the integration is done numerically with the trapezoidal rule [41], the initial guess for h and p can be calculated in Julia as shown in Listing 2

After the initialization is done, Listing 1 can be run with p calculation of Listing 2, until the change in h is smaller than an arbitrary parameter ϵ which can be taken as 1×10^{-8}. The final profile will be as shown in Figure 3. The profile is also compared with Schlichting [6] in order to validate the results. It has to be noted that, Schlichting used $\eta = \dfrac{y}{\sqrt{\frac{2\nu x}{U_\infty}}}$ as

similarity coordinate so the figure is plotted according to this nondimensionalization.

Listing 2. Implementation of initial velocity guess and initial p for Blasius profile.

```
h = [(i−1)/(N−1)   for i=1:N]

for i = 2:N
p[i] = p[i−1] + (h[i] + h[i−1])*Δη/2
end
```

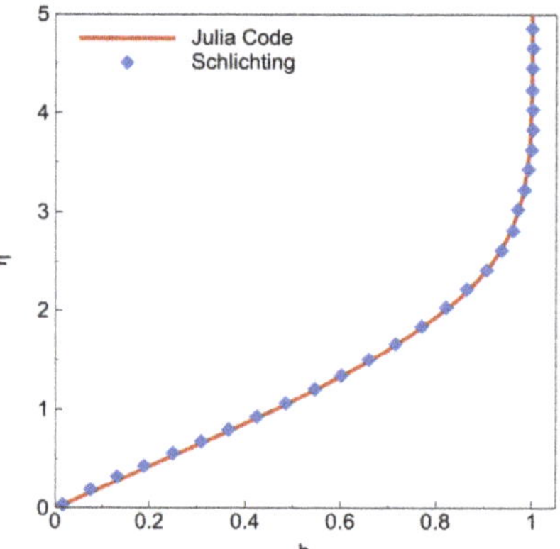

Figure 3. The velocity distribution of the Blasius similarity solution obtained by given Julia code and data digitized from Schlichting [6].

2.1.3. Falkner–Skan Flow Problem

The Falkner–Skan similarity solution can be considered as a family of the similarity solutions since it is a general solution that includes Blasius flow, Hiemenz flow (see Section 2.2.1), and more. Falkner–Skan equation cannot represent Homann flow (see Section 2.2.3) because Homann flow is an axisymmetric flow. On the other hand, Falkner–Skan flow is a two-dimensional flow. The name 'similarity solution' arises from the solutions at two arbitrary stations which are related to one another by means of a scale factor. A flow that can be represented with boundary-layer Equations (13) and (14), and satisfies the following equation:

$$\frac{u(x_1), \frac{y}{g(x_1)}}{U_e(x_1)} = \frac{u(x_2), \frac{y}{g(x_2)}}{U_e(x_2)}, \tag{42}$$

can be considered as self-similar. If boundary-layer Equations (13) and (14) are considered and similarity a transformation is assumed as:

$$u(x,y) = U_e(x)f'(\xi), \quad \xi = \frac{y}{g(x)}. \tag{43}$$

If the continuity equation of boundary-layer Equation (13) is modified with the given transformation, the final equation will be:

$$\frac{\partial v}{\partial \xi} = -U_e'f'g + U_ef''g'\xi. \tag{44}$$

If Equation (44) is integrated over ξ to find the velocity v, the v velocity will be:

$$v(x,\xi) = -(U_e g)'f + U_e\xi g'f' + H(x). \tag{45}$$

$H(x)$ can be calculated from the boundary conditions on the wall where ξ is zero and correspondingly, velocities and $f'(0)$ are zero. $H(x)$ will be 0 if $f(0)$ is chosen as 0. If calculated derivatives and velocities are substituted into Equation (14), after some simplification, the final equation will be:

$$f''' + \frac{g}{\nu}(U_e g)'ff'' + \frac{g^2 U_e'}{\nu}(1 - f') = 0. \tag{46}$$

Since f must be the function of ζ only and must not be a function of x, the coefficients of second and third terms must be constant. The final Falkner–Skan equation for the family of self-similar solutions:

$$f''' + \alpha f f'' + \beta(1 - f'^2) = 0, \tag{47}$$

where $\alpha = \frac{g}{\nu}(U_e g)'$ and $\beta = \frac{g^2 U_e'}{\nu}$. The α and β can be further solved for the velocity scale and length scale. If the following consideration is applied:

$$\alpha^* = g(U_e g)' \tag{48}$$
$$\beta^* = g^2 U_e'. \tag{49}$$

If $(U_e g^2)'$ is written in terms of α^* and β^*, the obtained equation integrated with respect to x, the resultant equation will be:

$$U_e g^2 = (2\alpha^* - \beta^*)x + c. \tag{50}$$

The constant of the integration represents a shift in the origin on x. Hence it doesn't affect the result and it also can be calculated from the stagnation point where $x = 0$ and $U_e = 0$ as a result, $c = 0$. If relation of $\beta^* = g^2 U_e'$ is divided by Equation (50) and integrated with respect to x, U_e can be calculated as:

$$U_e = c_1 x^m, \tag{51}$$

where $m = \frac{\beta^*}{2\alpha^* - \beta^*}$ and c_1 is a positive or negative constant which depends on the sign of U_e. It can be concluded from these calculations that similar solutions exist when the inviscid velocity is proportional to x raised to some power. Next, Equation (50) can be used by taking $c = 0$ to calculate the g which is:

$$g = \sqrt{\frac{2\alpha^*}{c_1(1+m)}} x^{1-m}. \tag{52}$$

Self-similar boundary-layers occur when the external velocity is the simple power law ($U_e = U_0(x/L)^m$), where the arbitrary constants U_0 and L have the same sign as U and x. The similarity variable for these kinds of flows can be written as:

$$\eta = \frac{y}{\delta} = \frac{y}{\sqrt{\pm\frac{\nu x}{U_e}}} = \frac{y}{\sqrt{\pm\frac{\nu L}{U_0}(\frac{x}{L})^{1-m}}}. \tag{53}$$

When U_e and x have the same signs, the Falkner–Skan equation can be written as:

$$f''' + \frac{1}{2}(m+1)f f'' + m(1 - f'^2) = 0. \tag{54}$$

The two arbitrary constants α and β have been reduced to one constant m by fixing the scale for the function $\delta(x)$. The boundary conditions of the equation are:

$$f(0) = 0 \tag{55}$$
$$f'(0) = 0 \tag{56}$$
$$f'(\eta \to \infty) = 1. \tag{57}$$

If the Falkner–Skan Equation (54) is carefully examined, it can be seen that when the constant m is 0, the equation will be Blasius flow. Moreover, if the constant m is 1, the equation will be Hiemenz flow (see Section 2.2.1). This important point is stated before and it is emphasized one more time after the derivation. Another great usage of the Falkner–Skan equation is to simulate the boundary-layer over a wedge with half-angle $\theta = \frac{m\pi}{m+1}$ when the m is between 0 and 1. If the m is in between 1 and 2, the Falkner–Skan equation will solve a corner flow with $\theta > \frac{\pi}{2}$. The visual schematic of four different physical flows that can be calculated from the Falkner–Skan equation can be seen in Figure 4. One interesting point of the equation is that when $m = -0.0904$, the obtained profile will have zero-shear at the wall which corresponds to the verge of the separation point for all x stations.

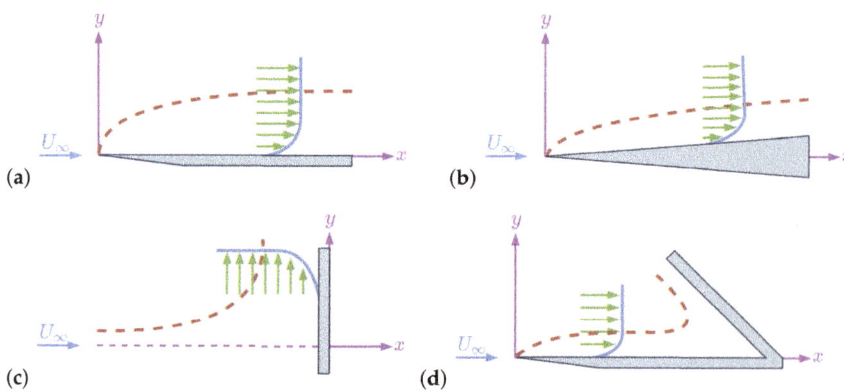

Figure 4. The representation of flow types that can be calculated with the Falkner–Skan equation. (**a**) Blasius flow ($m = 0$), (**b**) wedge flow ($0 < m < 1$), (**c**) Hiemenz flow ($m = 1$), and (**d**) corner flow ($1 < m < 2$).

2.1.4. Numerical Solution of Falkner–Skan Flow Problem

In this section, the Falkner–Skan equation will be solved with the Thomas algorithm and central finite difference scheme. In order to do that, the third-order ordinary differential equation should be reduced to second-order and first-order differential equations as it is done in Blasius flow. If it is assumed that $f' = h$ and $f = p$ in Equation (54), the system of equations will be:

$$h'' + \frac{1}{2}(m+1)ph' + m(1 - h^2) = 0 \tag{58}$$

$$p' - h = 0, \tag{59}$$

where the boundary conditions of the system are:

$$p(0) = 0 \tag{60}$$
$$h(0) = 0 \tag{61}$$
$$h(\infty) = 1 \tag{62}$$

If finite difference scheme is applied to h variable, the final system of equations will be:

$$A_n h_{n+1} + B_n h_n + C_n h_{n-1} + D_n = 0, \tag{63}$$

where the A_n, B_n, C_n, and D_n are:

$$A_n = \frac{1}{\Delta\eta^2} + (m+1)\frac{p_n}{4\Delta\eta} \tag{64}$$

$$B_n = \frac{-2}{\Delta\eta^2} - mh_n \tag{65}$$

$$C_n = \frac{1}{\Delta\eta^2} - (m+1)\frac{p_n}{4\Delta\eta} \tag{66}$$

$$D_n = m. \tag{67}$$

The relation between h_n and h_{n-1} is taken as it is done in Blasius flow. When Equation (37) is substituted into Equation (63), the final G_n and H_n will be:

$$G_n = \frac{-C_n G_{n-1} + D_n}{B_n + C_n H_{n-1}} \tag{68}$$

$$H_n = \frac{-A_n}{B_n + C_n H_{n-1}}. \tag{69}$$

The implementation of these variables in the Julia environment can be seen in Listing 3.

Listing 3. Implementation of the Thomas algorithm for the Falkner–Skan profile.

```julia
m = 0.5
h = [(i−1)/(N−1)   for i=1:N]

for i = 2:N
p[i] = p[i−1] + (h[i] + h[i−1])*Δη/2
end

while 1e-8<=errorProfile

A = [ 1/Δη² + (m+1)*p[i]/(4*Δη) for i=1:N]
B = [−2/Δη² − m*h[i]              for i=1:N]
C = [ 1/Δη² − (m+1)*p[i]/(4*Δη) for i=1:N]
D = [ m for i=1:N]

for i=2:N-1
G[i] = − ( C[i]*G[i−1] + D[i] )/(B[i] + C[i] * H[i−1])
H[i] = −                 A[i]  /(B[i] + C[i] * H[i−1])
end

hp = copy(h)

for i=N−1:−1:2
h[i] = G[i] + H[i] * h[i+1]
end

errorProfile = maximum(abs.(hp-h))

for i = 2:N
p[i] = p[i−1] + (h[i] + h[i−1])*Δη/2
end
end
```

The final profiles for the varying m values can be seen in the Figure 5. The profiles are also compared with Schlichting [6] in order to validate the results. It has to be noted that, Schlichting used $\eta = \sqrt{\frac{m+1}{2}} \dfrac{y}{\sqrt{\frac{\nu x}{U_\infty}}}$ as similarity coordinate so the figure is plotted according to this nondimensionalization.

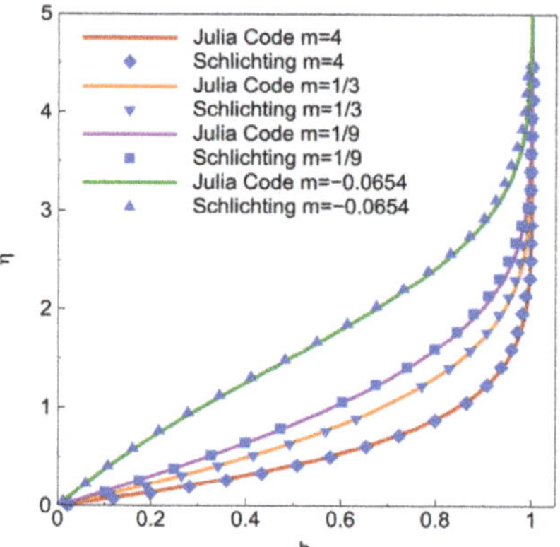

Figure 5. The velocity distribution of the Falkner–Skan similarity solution obtained by Julia code and data digitized from Schlichting [6] for varying m values.

2.2. Stagnation Point Flow Problems

Boundary-layer velocity distribution over a wall that is perpendicular to the flow velocity vector can be represented with a similarity solution. In this subsection, Hiemenz and Homann similarity solutions will be derived from scratch and they will be solved numerically in the Julia environment. The Julia codes will be available to shorten the learning curve.

2.2.1. Hiemenz Flow Problem

This problem is that of a fluid flow that is parallel to the y-axis in the far-field impinging on a wall that coincides with the x-axis. The flow which is perpendicular to a cylinder can be assumed as the Hiemenz flow around the stagnation point. The schematic description of the Hiemenz flow can be seen in Figure 6. In the Hiemenz flow, viscous forces away from the wall become so small in comparison with the inertia forces, particularly when the Reynolds number is large. In this case, inviscid irrotational flow assumption ($\xi = \nabla \times U = 0$) can be done. The velocity can be represented with a scalar function, ϕ, which is the velocity potential. The velocities in the x-direction and y-direction can be written as:

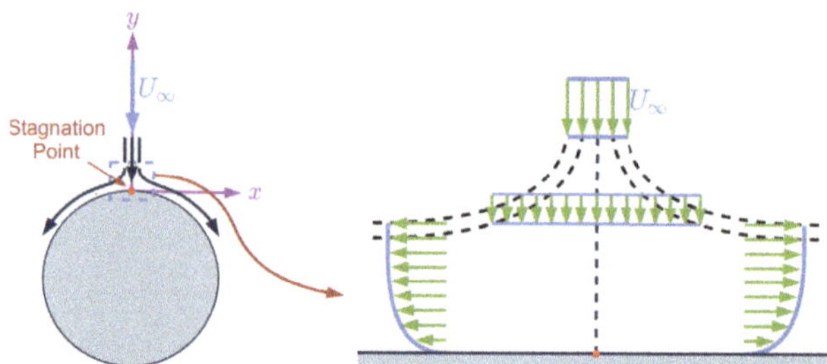

Figure 6. Schematic description of the Hiemenz and Homann flow. **Left** side corresponds to general flow and the **Right** side is the extended vision of the dashed rectangle. The flow represents Hiemenz flow if the circle is the projection of a cylinder and Homann flow if the circle is the projection of a sphere.

$$u = \frac{\partial \phi}{\partial x}, \quad v = \frac{\partial \phi}{\partial y}. \tag{70}$$

If the dimensional Navier–Stokes equations (Equations (1)–(3)) are considered for this flow as well, and the continuity equation becomes:

$$\frac{\partial^2 \phi}{\partial x^2} + \frac{\partial^2 \phi}{\partial y^2} = 0. \tag{71}$$

On the wall, in the absence of viscosity, the flow can slip in the x-direction but in y-direction the velocity must be zero ($\frac{\partial \phi}{\partial y}|_{y=0} = 0$) because of the no-penetration boundary condition. The inviscid flow solution for the potential function was found to be:

$$\phi_{(x,y)} = \frac{a}{2}(x^2 - y^2), \tag{72}$$

where a is a constant that depends on the freestream flow and the body shape. This solution satisfies the governing equations for inviscid, irrotational flow, and the boundary conditions which can be shown as:

$$u = \frac{\partial \phi}{\partial x} = ax, \quad v = \frac{\partial \phi}{\partial y} = -ay. \tag{73}$$

Since potential the function is obtained, the stream-function can be calculated from the potential function. the velocity components, in terms of stream-function, $\psi_{(x,y)}$:

$$u = ax = \frac{\partial \psi}{\partial y}, \quad v = -ay = -\frac{\partial \psi}{\partial x}. \tag{74}$$

If Equation (74) is integrated for x and y, the stream function can be found as:

$$\psi = axy + c. \tag{75}$$

In order to find the pressure in the inviscid flow, Bernoulli equation [42] can be used. The pressure from the Bernoulli equation is:

$$p_0 - p = \frac{a^2 \rho}{2}(x^2 + y^2), \tag{76}$$

where p_0 is the stagnation pressure and ρ is the density. So far, flow is assumed as inviscid flow; however viscous forces will modify the inviscid solution, in particular in the y-direction. Hence, the viscous velocity components can be assumed as:

$$u(x,y) = xg(y) \tag{77}$$
$$v(x,y) = -f(y) \tag{78}$$

If Equations (77) and (78) are substituted in the Navier–Stokes equations (Equations (1)–(3)), the final system of equations will be:

$$g = f' \tag{79}$$
$$xg^2 - xfg' = -\frac{1}{\rho}\frac{\partial p}{\partial x} + vxg'' \tag{80}$$
$$ff' = -\frac{1}{\rho}\frac{\partial p}{\partial y} + vf''. \tag{81}$$

Once Equation (79) is substituted into Equations (80) and (81), two non-linear ordinary differential equations can be obtained as:

$$xf'^2 - xff'' = -\frac{1}{\rho}\frac{\partial p}{\partial x} + vxf''' \tag{82}$$

$$ff' = -\frac{1}{\rho}\frac{\partial p}{\partial y} + vf'' \tag{83}$$

It has to be noted that these two equations are coupled, which means they have to be solved together. If the x-momentum and y-momentum are solved for pressure, it can be seen that the pressure will be in the form of:

$$p(x,y) = \frac{1}{2}x^2(constant) + H(y), \tag{84}$$

where $H(y)$ is a function depends on only y. On the other hand, if the inviscid pressure compared with the viscous pressure as $y \to \infty$, the pressure gradient in x-direction and y-direction can be found as:

$$\frac{\partial p}{\partial x} = -a^2\rho x \tag{85}$$

$$\frac{\partial p}{\partial y} = -\frac{a^2\rho}{2}F'(y). \tag{86}$$

If the pressure gradients are substituted into Equations (82) and (83), the resultant system of ordinary differential equations can be found as:

$$f'^2 - ff'' = a^2 + vf''' \tag{87}$$

$$ff' = \frac{a^2}{2}F' - vf''. \tag{88}$$

Momentum equations are decoupled in this system of equations so once, $f(y)$ is calculated from Equation (87), $F(y)$ in the Equation (88) can be solved. The boundary conditions of the system are:

$$u(x,0) = 0 \to f'_{(y=0)} = 0 \tag{89}$$

$$v(x,0) = 0 \to f_{(y=0)} = 0 \tag{90}$$

$$u(x,y \to \infty) = ax \to f'_\infty = a \tag{91}$$

$$v(x,y \to \infty) = -ay \to f''_\infty = 0 \tag{92}$$

Additionally, $F(y \to \infty) = y^2$ can be obtained from y-momentum equation as $y \to \infty$. The final Equations (87) and (88) can be solved with the given boundary conditions; however, it is possible to get rid of the dependence on a. In order to do that, affine transformation [43] can be used. If it is assumed that $f(y) = A\varphi(\eta)$ and $y = B\eta$ where A and B are constant to be determined, the f derivative functions can be calculated as:

$$f' = \frac{A}{B}\varphi' \tag{93}$$

$$f'' = \frac{A}{B^2}\varphi'' \tag{94}$$

$$f''' = \frac{A}{B^3}\varphi''' \tag{95}$$

If Equations (93)–(95) substituted into the Equations (87) and (88) and assume that $\frac{v}{AB} = 1$ and $\frac{a^2 B^2}{A^2} = 1$, the A and B coefficients can be calculated as:

$$A = \sqrt{av} \tag{96}$$

$$B = \sqrt{\frac{v}{a}} \tag{97}$$

The given transformation allows to reduce two ordinary differential equations into one third-order ordinary differential equation as:

$$\varphi''' + \varphi\varphi'' - \varphi'^2 + 1 = 0 \tag{98}$$

where the boundary conditions are:

$$f'(0) = 0 \rightarrow \varphi'(0) = 0 \tag{99}$$
$$f(0) = 0 \rightarrow \varphi(0) = 0 \tag{100}$$
$$f'(\infty) = a \rightarrow \varphi'(\infty) = 1 \tag{101}$$

and the velocities:

$$u = ax\varphi'(\eta) \tag{102}$$
$$v = -\sqrt{av}\,\varphi(\eta) \tag{103}$$

the resultant equation along with the boundary conditions can be solved with different numerical approaches such as Runge-Kutta, Runge-Kutta-Fehlberg, compact finite difference, high-order finite difference; however, in this present paper, the Thomas algorithm with finite difference discretization will be used to solve the Hiemenz profile in the Julia framework.

2.2.2. Numerical Solution of Hiemenz Flow Problem

The computational approach for the Hiemenz flow is similar to the Blasius solution since the final ordinary equation of the Hiemenz flow (Equation (98)) is similar to the Blasius similarity solution (Equation (20)). If it is assumed that $\varphi' = h$ and $\varphi = p$ in Equation (98), the system of equations will be:

$$h'' + ph' - h^2 + 1 = 0 \tag{104}$$
$$p' - h = 0, \tag{105}$$

where the boundary conditions of the system are:

$$p(0) = 0 \tag{106}$$
$$h(0) = 0 \tag{107}$$
$$h(\infty) = 1. \tag{108}$$

If finite difference scheme is applied to the h variable as in Blasius flow. The final system of equations will be:

$$A_n h_{n+1} + B_n h_n + C_n h_{n-1} + D_n = 0, \tag{109}$$

where the A_n, B_n, C_n, and D_n are:

$$A_n = \frac{1}{\Delta\eta^2} + \frac{p_n}{2\Delta\eta} \tag{110}$$

$$B_n = \frac{-2}{\Delta\eta^2} - h_n \tag{111}$$

$$C_n = \frac{1}{\Delta\eta^2} - \frac{p_n}{2\Delta\eta} \tag{112}$$

$$D_n = 1. \tag{113}$$

The relation between h_n and h_{n-1} is taken as it is done in Blasius flow. When Equation (37) is substituted into the Equation (109), the final G_n and H_n coefficients can be found as:

$$G_n = \frac{-C_n G_{n-1} + D_n}{B_n + C_n H_{n-1}} \tag{114}$$

$$H_n = \frac{-A_n}{B_n + C_n H_{n-1}}. \tag{115}$$

In order to start the calculation, H_1 can be assumed as 0 and the G_1 can be calculated as 0 from the boundary conditions. Once G_1 and H_1 are calculated, G_n and H_n can be calculated from Equations (114) and (115). p can be calculated as it is done for Blasius flow:

$$p_n = p_{n-1} + \int_{\eta_{n-1}}^{\eta_n} h\,d\eta. \tag{116}$$

The system of equations can be implemented in the Julia environment as it is shown in Listing 4. The linear profile assumption is taken as the initial condition of h (see Figure 2) as it is done for the Blasius solution. The reason hp and $errorProfile$ variables are used in the Listing 4 is that the linear profile converges to the Hiemenz profile in each iteration so in order to check the difference between previous and present profiles, the solution vector from the previous iteration is copied and it is compared with the new solution vector. If the difference between these two solution vectors is less than ϵ, which is an arbitrary limit and can be taken as 1×10^{-8}, then it can be said that the solution has converged. It has to be noted that ϵ can be taken as any number; however, if it is small, the solution will be more accurate. The final result of the Hiemenz flow can be seen in Figure 7. The results are also validated with White [44].

2.2.3. Homann Flow Problem

Homann flow is similar to Hiemenz flow. The only difference is that Homann flow is an axisymmetric version of the Hiemenz flow. The same schematic (Figure 6) can represent this flow as well; however, in this flow, the circle is the projection of a sphere. On the other hand, it was the projection of a cylinder in Hiemenz flow. Derivation of the Homann similarity solution has the almost same procedure as well but it uses cylindrical coordinates instead of Cartesian coordinates. The velocity components of the flow can be shown as:

$$v_r = v_r(r, z) \tag{117}$$

$$v_\theta = 0 \tag{118}$$

$$v_z = v_z(r, z) \tag{119}$$

$$p = p(r, z). \tag{120}$$

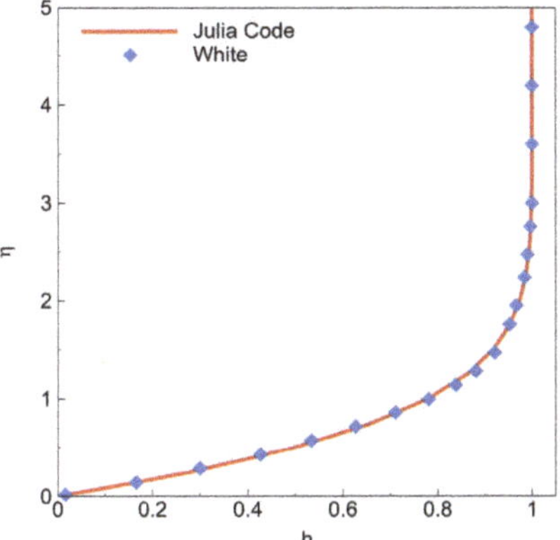

Figure 7. The velocity distribution of Hiemenz similarity solution obtained by Julia code and data digitized from White [44].

As axisymmetric assumptions, derivative with respect to θ ($\frac{\partial}{\partial \theta}$) and v_θ can be assumed as zero. The Navier–Stokes equations in cylindrical coordinates can be written as:

$$\frac{\partial v_r}{\partial r} + \frac{v_r}{r} + \frac{\partial v_z}{\partial z} = 0 \tag{121}$$

$$v_r\frac{\partial v_r}{\partial r} + v_z\frac{\partial v_z}{\partial z} = -\frac{1}{\rho}\frac{\partial p}{\partial r} + \nu\left(\frac{\partial^2 v_r}{\partial r^2} + \frac{1}{r}\frac{\partial v_r}{\partial r} - \frac{v_r}{r^2} + \frac{\partial^2 v_r}{\partial z^2}\right) \tag{122}$$

$$v_r\frac{\partial v_z}{\partial r} + v_z\frac{\partial v_z}{\partial z} = -\frac{1}{\rho}\frac{\partial p}{\partial z} + \nu\left(\frac{\partial^2 v_z}{\partial r^2} + \frac{1}{r}\frac{\partial v_z}{\partial r} + \frac{\partial^2 v_z}{\partial z^2}\right). \tag{123}$$

If the same procedures applied to Navier–Stokes in Cartesian coordinates for Hiemenz flow, are applied in cylindrical coordinates, the final potential function and stream function will be:

$$\phi(r,z) = k\left(\frac{r^2}{2} - z^2\right) \tag{124}$$

$$\psi(r,z) = kzr^2, \tag{125}$$

where k is a constant which depends on the freestream flow and the body shape. The corresponding velocities are:

$$v_r = \frac{\partial \phi}{\partial r} = \frac{1}{r}\frac{\partial \psi}{\partial z} = kr \tag{126}$$

$$v_z = \frac{\partial \phi}{\partial z} = \frac{-1}{r}\frac{\partial \psi}{\partial r} = -2kz. \tag{127}$$

Once velocities are calculated, the pressure from the Bernoulli can be calculated as:

$$p_0 - p = \frac{1}{2}\rho k^2(r^2 + 4z^2). \tag{128}$$

The viscous velocity components and pressure can be calculated as:

$$v_r = rg(z) \tag{129}$$

$$v_z = -2f(z) \tag{130}$$

$$p_0 - p = \frac{1}{2}\rho k^2(r^2 + F(z)). \tag{131}$$

Listing 4. Implementation of Thomas algorithm for Hiemenz profile.

```
h = [(i−1)/(N−1)  for i=1:N]

for i = 2:N
p[i] = p[i−1] + (h[i] + h[i−1])*Δη/2
end

while 1e−8<=errorProfile

A = [ 1/Δη² + p[i]/(2*Δη) for i=1:N]
B = [−2/Δη² − h[i]            for i=1:N]
C = [ 1/Δη² − p[i]/(2*Δη) for i=1:N]
D = [ 1 for i=1:N]

for i=2:N−1
G[i] = − ( C[i]*G[i−1] + D[i] )/(B[i] + C[i] * H[i−1])
H[i] = −                 A[i]   /(B[i] + C[i] * H[i−1])
end

hp = copy(h)

for i=N−1:−1:2
h[i] = G[i] + H[i] * h[i+1]
end

errorProfile = maximum(abs.(hp−h))

for i = 2:N
p[i] = p[i−1] + (h[i] + h[i−1])*Δη/2
end
end
```

After the affine transformation, the final ordinary differential equation will be:

$$\varphi''' + 2\varphi\varphi'' - \varphi'^2 + 1 = 0, \tag{132}$$

where the boundary conditions of the equation is:

$$\varphi'(0) = 0 \tag{133}$$
$$\varphi(0) = 0 \tag{134}$$
$$\varphi'(\infty) = 1 \tag{135}$$

If one compares Hiemenz and Homann similarity solutions (see Equations (98) and (132)) the only difference is the 2 in the second term. In the same manner, the computational process will be the same except for two lines of code.

2.2.4. Numerical Solution of Homann Flow Problem

Since equations of Hiemenz and Homann similarity solutions are the same except one coefficient, procedures for the solution are also the same, except for two lines of code. If it is assumed that $\varphi' = h$ and $\varphi = p$ in Equation (132), the system of equations will be:

$$h'' + 2ph' - h^2 + 1 = 0 \tag{136}$$
$$p' - h = 0, \tag{137}$$

where the boundary conditions of the system are:

$$p(0) = 0 \tag{138}$$
$$h(0) = 0 \tag{139}$$
$$h(\infty) = 1 \tag{140}$$

If the finite difference scheme is applied to the h variable, the final system of equations will be:

$$A_n h_{n+1} + B_n h_n + C_n h_{n-1} + D_n = 0, \tag{141}$$

where the A_n, B_n, C_n, and D_n are:

$$A_n = \frac{1}{\Delta\eta^2} + \frac{p_n}{\Delta\eta} \tag{142}$$

$$B_n = \frac{-2}{\Delta\eta^2} - h_n \tag{143}$$

$$C_n = \frac{1}{\Delta\eta^2} - \frac{p_n}{\Delta\eta} \tag{144}$$

$$D_n = 1. \tag{145}$$

The final G_n and H_n coefficients are the same as Hiemenz flow as well and they are:

$$G_n = \frac{-C_n G_{n-1} + D_n}{B_n + C_n H_{n-1}} \tag{146}$$

$$H_n = \frac{-A_n}{B_n + C_n H_{n-1}}. \tag{147}$$

The final code can be seen in Listing 5. If the Hiemenz code (4) and Homann code (5) are compared, the only difference is in the A and C and it is because of the finite difference approach for the h'.

Listing 5. Implementation of Thomas algorithm for Homann profile.

```
h = [(i-1)/(N-1)   for i=1:N]

for i = 2:N
p[i] = p[i-1] + (h[i] + h[i-1])*Δη/2
end

while 1e-8<=errorProfile

A = [ 1/Δη² + p[i]/Δη for i=1:N]
B = [-2/Δη² - h[i]    for i=1:N]
C = [ 1/Δη² - p[i]/Δη for i=1:N]
D = [ 1 for i=1:N]

for i=2:N-1
G[i] = - ( C[i]*G[i-1] + D[i] )/(B[i] + C[i] * H[i-1])
H[i] = -                  A[i]  /(B[i] + C[i] * H[i-1])
end

hp = copy(h)

for i=N-1:-1:2
h[i] = G[i] + H[i] * h[i+1]
end

errorProfile = maximum(abs.(hp-h))

for i = 2:N
p[i] = p[i-1] + (h[i] + h[i-1])*Δη/2
end
end
```

The final solution profile can be seen in Figure 8. The results are also validated with White [44]. One can use these results to validate their own codes. Reference data will be shared on GitHub as well.

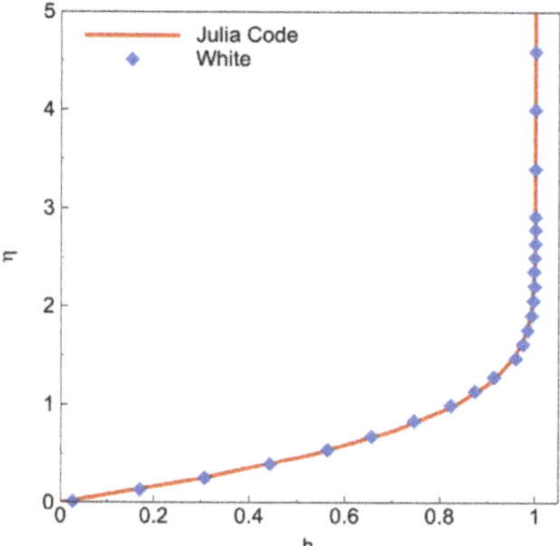

Figure 8. The velocity distribution of Homann similarity solution obtained by Julia code and digitized from White [44].

3. Discussion

Fast computational fluid dynamics solvers are crucial for engineers because the design process requires a lot of simulations to reach the final and optimized design. One of the most important factors that make a solver fast is the language itself. Writing almost the same script in different languages may give different solution central processing unit (CPU) times. For instance, a script that uses long and complex for loops in both Julia and Python environments will result in different solution times because Python is slower with for loops in general. The reasons that cause this slow behavior will not be covered in this paper since it is out of the scope of this paper. However, it is important to state these differences in order to decide which coding language is proper for the simulation that is planned.

Sometimes, some other criteria, such as a user-friendly environment, might be the critical condition. The easy matrix–matrix multiplication and backslash linear system solver can be some of the user-friendly examples. In CFD, the usage of vectors and matrices is so common. Most of the time, it is required to multiply or add vectors or matrices with one another. Fortran, which is one of the fastest languages and also one of the fundamental languages in the CFD industry, does not have a built-in element-wise vector or matrix multiplication feature. This requires the use of for loops for each element-wise matrix and vector operation. As a result of this, every time, one needs to use a for loop to do these operations. This ends up with hard-to-read codes and also excessive usage of indices is another source of possible mistakes that will cause trouble during the debugging. However, in the Julia environment, it can be done in one line without any trouble. Listing 6 is showing the two for loop usage. Both of them use Julia syntax in order to prevent confusion; however, it is required to state that Fortran has a different syntax than this but, logically, the for loops are the same with Fortran logic.

Another user-friendly feature of Julia over Fortran is to plot a vector in the code with built-in functions. However, in Fortran, it is not possible to do it with built-in functions so one needs to extract the solution vector or matrix to an external file to visualize it. As it is stated before, coding language selection can be an important topic. In this section, the strong sides of the Julia environment will be stated. One of the user-friendly features of Julia is compact for loop syntax. Listing 7 shows the traditional way and compact way to use a for loop in the Julia environment.

Listing 6. Comparison of the element-wise and compact matrix-matrix multiplication.

```julia
for l=1:m
for k=1:nz
for j=1:ny
for i=1:nx
A[i,j,k,l] = B[i,j,k,l]*C[i,j,k,l]
end
end
end
end

A .= B.*C
```

Listing 7. Comparison of the traditional and compact way of for loop usage.

```julia
for i=1:N
h[i] = (i−1)/(N−1)
end

h = [(i−1)/(N−1)  for i=1:N]
```

The readability of the code is also extremely important in order to explain the code to others or provide it as open-source code. Julia is so strong in this topic because it allows the usage of LATEX language within the code. It is one of the unique abilities of this language. If you are writing a formula that is full of Greek letters and regular letters, it is inevitable to use complex variable names. Comparison of the usage of Greek letters and regular formula writing can be seen in Listing 8.

Listing 8. Comparison of the Greek letter usage and conventional formula writing in Julia.

```julia
alpha = a*beta−c/(gamma−1)*(d_i−1/(eta^2))

α = a*β−c/(γ−1)*(d_i−1/(η^2))
```

The last advantage of Julia stated in this present paper is the ability to do both dynamic and static RAM allocations. It may be important to manually allocate the variable sizes and types for optimized code. However, sometimes it is easier to use just variables without any initializing. Initializing differences in the Julia environment can be seen in Listing 9.

Listing 9. Different initializing methods in Julia environment.

```julia
nx = 10
ny = 15.2

q = Array{Float64}(undef, nx)
for i=1:nx
q[i] = i
end

a = Array{Float64}(undef, nx)
c = zeros(nx)
a = q
b = q
c = q
```

As it is seen from the example scripts, Julia is a user-friendly, fast, open-source, and free language which can increase productivity drastically [36]. It is a great choice both for those who are new to coding and coding experts. Julia also can call the C, Fortran, and Python libraries so it is great for experienced engineers who think their previous code in other coding languages will be useless.

4. Conclusions

In the computational fluid dynamics industry, it is crucial to have some predictions about the flow that will be simulated. It helps to spot the location at which finer mesh is required. Fundamental knowledge about canonical flows is crucial in this point because most of the complex flow consists of a combination of a couple of canonical flows. For example, an airfoil CFD simulation can consist of boundary-layer flow over the smooth part of it, mixing layer flow where the tail ends, and blunt body flow in the wake region. In other words, one airfoil simulation consists of three different canonical flows. A full understanding on canonical flows is extremely important in order to simulate similar flows accurately and cheaply. In this paper, boundary-layer theory is introduced and boundary-layer flows are derived from scratch. The Blasius flow, Hiemenz flow, Homann flow, Falkner–Skan flow are the focus of this paper. Once the derivations of them are completed, derived forms are implemented in the Julia environment. In order to model the equations, a finite difference scheme for space discretization is used and the Thomas algorithm is used for the linear system solution. It has to be noted that other methods such as Runge-Kutta, Runge-Kutta-Fehlberg, compact finite difference, high-order finite difference can also be used to solve the given ODEs; however, in this present paper, authors preferred the finite difference method and Thomas algorithm. The authors further discussed the advantages of the Julia language over some other coding languages available in the literature. It is shown that Julia syntax is so straightforward, easy, and user-friendly. Strong sides of the Julia environment are stated with the given comparisons as well. The popularity of Julia may drastically increase in the near future because of its potential.

Author Contributions: code generation, F.O. and K.K.; validation, F.O. and K.K.; writing—original draft preparation, F.O.; writing—review and editing, F.O. and K.K.; visualization, F.O. All authors have read and agreed to the published version of the manuscript.

Funding: This research received no external funding.

Institutional Review Board Statement: Not applicable.

Informed Consent Statement: Not applicable.

Data Availability Statement: All the data used and generated in this study is available in the GitHub link provided in Appendix A.

Conflicts of Interest: The authors declare no conflict of interest.

Appendix A

Julia setup files can be downloaded from their website (https://julialang.org/downloads/ (accessed on 1 June 2021)). The website also includes instructions on how to install Julia on Windows, Linux, and mac operating systems. Some of the useful resources for learning Julia are listed below:

- https://docs.julialang.org/en/v1/ (accessed on 1 June 2021)
- https://www.coursera.org/learn/julia-programming (accessed on 1 June 2021)
- https://www.youtube.com/user/JuliaLanguage/featured (accessed on 1 June 2021)
- https://www.youtube.com/user/Parallel Computing and Scientific Machine Learning (accessed on 1 June 2021)
- https://discourse.julialang.org/ (accessed on 1 June 2021)

It is common to use external packages for Julia. In order to do that, Pkg, which is Julia's built-in package manager, can be used. Once Julia is opened, Pkg can be activated with the "]" button in Windows. In Linux, calling "julia" in the terminal will open it. After that "Pkg.add("Pluto")" will trigger the setup process for that package. In here, we used Pluto as an example because, in GitHub, our codes are developed in the Pluto environment. After Pluto is installed. Pluto can be run with "Pluto.run()". This command will open a new tab in the browser which you can run your Julia codes. After that, the "using Pluto" line must be placed to the top of the file. For "Plots" package, the commands will be "Pkg.add("Plots")" and "using Plots". Since the Plots package does not have a GUI, there is not a command called "Plots.run()".

Other than Pluto, JuliaPro which includes Julia and the Juno IDE (https://juliacomputing.com/products/juliapro/ (accessed on 1 June 2021)) can be used as an editor and compiler.

This software contains a set of packages for plotting, optimization, machine learning, database, and much more. Pluto is appropriate for small scripts while JuliaPro is better for more complex codes. The GitHub link of the codes used in this paper is:

- https://github.com/frkanz/A-CFD-Tutorial-in-Julia (accessed on 1 June 2021)

References

1. Prandtl, L. *Über Flüssigkeitsbewegung bei sehr Kleiner Reibung, Verh 3 int*; English Translation; Math-Kongr: Heidelberg, Germanty, 1904. Available online: http://homepage.ntu.edu.tw/~wttsai/Adv_Fluid/NACA_TM-452.pdf (accessed on 1 June 2021).
2. Blasius, H. Grenzschichten in Flüssigkeiten mit kleiner Reibung. *Z. Math. Phys.* **1908**, *60*, 397–398.
3. Hager, W.H. Blasius: A life in research and education. *Exp. Fluids* **2003**, *34*, 566–571. [CrossRef]
4. Cousteix, T.; Cebeci, J. *Modeling and Computation of Boundary-Layer Flows*; Springer: Berlin/Heidelberg, Germany, 2005.
5. Hiemenz, K. Die Grenzschicht an einem in den gleichformigen Flussigkeitsstrom eingetauchten geraden Kreiszylinder. *Dinglers Polytech. J.* **1911**, *326*, 321–324.
6. Schlichting, H.; Gersten, K. *Boundary-Layer Theory*; Springer: Berlin/Heidelberg, Germany, 2016.
7. Howarth, L. *On the Calculation of Steady Flow in the Boundary Layer Near the Surface of a Cylinder in a Stream*; Technical Report; Aeronautical Research Council London: London, UK, 1934.
8. Homann, F. Der Einfluss grosser Zähigkeit bei der Strömung um den Zylinder und um die Kugel. *ZAMM-J. Appl. Math. Mech. Für Angew. Math. Und Mech.* **1936**, *16*, 153–164. [CrossRef]
9. Brennan, G.; Gajjar, J.; Hewitt, R. Tollmien–Schlichting wave cancellation via localised heating elements in boundary layers. *J. Fluid Mech.* **2021**, *909*, A16-1. [CrossRef]
10. Brennan, G.S.; Gajjar, J.S.; Hewitt, R.E. Cancellation of Tollmien–Schlichting waves with surface heating. *J. Eng. Math.* **2021**, *128*, 1–23. [CrossRef]
11. Corelli Grappadelli, M.; Sattler, S.; Scholz, P.; Radespiel, R.; Badrya, C. Experimental investigations of boundary layer transition on a flat plate with suction. In Proceedings of the AIAA Scitech 2021 Forum, Virtual Event, 11–15 and 19–21 January 2021; p. 1452.
12. Rigas, G.; Sipp, D.; Colonius, T. Nonlinear input/output analysis: Application to boundary layer transition. *J. Fluid Mech.* **2021**, *911*, A15. [CrossRef]
13. Haley, C.; Zhong, X. Supersonic mode in a low-enthalpy hypersonic flow over a cone and wave packet interference. *Phys. Fluids* **2021**, *33*, 054104. [CrossRef]
14. Malik, M.R. Numerical methods for hypersonic boundary layer stability. *J. Comput. Phys.* **1990**, *86*, 376–413. [CrossRef]
15. Fedorov, A. Transition and stability of high-speed boundary layers. *Annu. Rev. Fluid Mech.* **2011**, *43*, 79–95. [CrossRef]
16. Long, T.; Dong, Y.; Zhao, R.; Wen, C. Mechanism of stabilization of porous coatings on unstable supersonic mode in hypersonic boundary layers. *Phys. Fluids* **2021**, *33*, 054105. [CrossRef]
17. Fong, K.D.; Wang, X.; Zhong, X. Numerical simulation of roughness effect on the stability of a hypersonic boundary layer. *Comput. Fluids* **2014**, *96*, 350–367. [CrossRef]
18. Kara, K.; Balakumar, P.; Kandil, O. Receptivity of hypersonic boundary layers due to acoustic disturbances over blunt cone. In Proceedings of the 45th AIAA Aerospace Sciences Meeting and Exhibit, Reno, Nevada, 8–11 January 2007; p. 945.
19. Kara, K.; Balakumar, P.; Kandil, O. Effects of wall cooling on hypersonic boundary layer receptivity over a cone. In Proceedings of the 38th Fluid Dynamics Conference and Exhibit, Seattle, WA, USA, 23–26 June 2008; p. 3734.
20. Kara, K.; Balakumar, P.; Kandil, O.A. Effects of nose bluntness on hypersonic boundary-layer receptivity and stability over cones. *AIAA J.* **2011**, *49*, 2593–2606. [CrossRef]
21. Oz, F.; Kara, K. Effects of Local Cooling on Hypersonic Boundary-Layer Stability. In Proceedings of the AIAA Scitech 2021 Forum, Virtual Event, 11–15 and 19–21 January 2021; p. 0940.
22. Drozdz, A.; Niegodajew, P.; Romanczyk, M.; Sokolenko, V.; Elsner, W. Effective use of the streamwise waviness in the control of turbulent separation. *Exp. Therm. Fluid Sci.* **2021**, *121*, 110291. [CrossRef]
23. Iyer, P.S.; Malik, M.R. Wall-modeled LES of flow over a Gaussian bump. In Proceedings of the AIAA Scitech 2021 Forum, Virtual Event, 11–15 and 19–21 January 2021; p. 1438.
24. Mohammed-Taifour, A.; Weiss, J. Periodic forcing of a large turbulent separation bubble. *J. Fluid Mech.* **2021**, *915*, A24. [CrossRef]
25. Hady, F.; Ibrahim, F.; Abdel-Gaied, S.; Eid, M. Effect of heat generation/absorption on natural convective boundary-layer flow from a vertical cone embedded in a porous medium filled with a non-Newtonian nanofluid. *Int. Commun. Heat Mass Transf.* **2011**, *38*, 1414–1420. [CrossRef]
26. Hady, F.M.; Ibrahim, F.S.; Abdel-Gaied, S.M.; Eid, M.R. Radiation effect on viscous flow of a nanofluid and heat transfer over a nonlinearly stretching sheet. *Nanoscale Res. Lett.* **2012**, *7*, 1–13. [CrossRef]
27. Hady, F.; Ibrahim, F.; Abdel-Gaied, S.; Eid, M.R. Boundary-layer non-Newtonian flow over vertical plate in porous medium saturated with nanofluid. *Appl. Math. Mech.* **2011**, *32*, 1577–1586. [CrossRef]
28. Hady, F.; Ibrahim, F.; Abdel-Gaied, S.; Eid, M. Boundary-layer flow in a porous medium of a nanofluid past a vertical cone. In *An Overview of Heat Transfer Phenomena*; Kazi, S.N., Ed.; IntechOpen: London, UK, 2012; pp. 91–104.
29. Sohail, M.; Naz, R.; Abdelsalam, S.I. Application of non-Fourier double diffusions theories to the boundary-layer flow of a yield stress exhibiting fluid model. *Phys. A Stat. Mech. Appl.* **2020**, *537*, 122753. [CrossRef]

30. Bhatti, M.; Alamri, S.Z.; Ellahi, R.; Abdelsalam, S.I. Intra-uterine particle–fluid motion through a compliant asymmetric tapered channel with heat transfer. *J. Therm. Anal. Calorim.* **2020**, *144*, 2259–2267. [CrossRef]

31. Barba, L.; Forsyth, G. CFD Python: The 12 steps to Navier-Stokes equations. *J. Open Source Educ.* **2018**, *2*, 21. [CrossRef]

32. Oliphant, T.E. *A Guide to NumPy*; Trelgol Publishing: Austin, TX, USA, 2006; Volume 1.

33. Ketcheson, D.I. Teaching numerical methods with IPython notebooks and inquiry-based learning. In Proceedings of the 13th Python in Science Conference, Austin, TX, USA, 6–12 July 2014; pp. 19–24.

34. Ketcheson, D.I.; Mandli, K.; Ahmadia, A.J.; Alghamdi, A.; de Luna, M.Q.; Parsani, M.; Knepley, M.G.; Emmett, M. PyClaw: Accessible, extensible, scalable tools for wave propagation problems. *SIAM J. Sci. Comput.* **2012**, *34*, 210–231. [CrossRef]

35. Pawar, S.; San, O. CFD Julia: A learning module structuring an introductory course on computational fluid dynamics. *Fluids* **2019**, *4*, 159. [CrossRef]

36. Bezanson, J.; Edelman, A.; Karpinski, S.; Shah, V.B. Julia: A fresh approach to numerical computing. *SIAM Rev.* **2017**, *59*, 65–98. [CrossRef]

37. Hosseini, M.; Vinuesa, R.; Hanifi, A.; Henningson, D.; Schlatter, P. Turbulent flow around a wing profile, a direct numerical simulation. In Proceedings of the 68th Annual Meeting of the APS Division of Fluid Dynamics, Boston, MA, USA, 22–24 November 2015.

38. Tannehill, J.C.; Pletcher, R.H.; Anderson, D.A. *Computational Fluid Mechanics and Heat Transfer*; Taylor & Francis: Bristol, PA, USA, 1997.

39. Press, W.H.; Teukolsky, S.A.; Vetterling, W.T.; Flannery, B.P. *Numerical Recipes in Fortran 90*; Cambridge University Press: New York, NY, USA, 1996.

40. Moin, P. *Fundamentals of Engineering Numerical Analysis*; Cambridge University Press: New York, NY, USA, 2010.

41. Atkinson, K.E. *An Introduction to Numerical Analysis*; John Wiley & Sons: New York, USA, 1989.

42. Munson, B.R.; Young, D.F.; Okiishi, T.H. Fundamentals of fluid mechanics. *Oceanogr. Lit. Rev.* **1995**, *10*, 831.

43. Rogers, D.F.; Adams, J.A. *Mathematical Elements for Computer Graphics*; McGraw-Hill: New York, NY, USA 1989.

44. White, F.M.; Corfield, I. *Viscous Fluid Flow*; McGraw-Hill: New York, NY, USA, 2006; Volume 3.

Article

PyDA: A Hands-On Introduction to Dynamical Data Assimilation with Python

Shady E. Ahmed, Suraj Pawar and Omer San *

School of Mechanical and Aerospace Engineering, Oklahoma State University, Stillwater, OK 74078, USA; shady.ahmed@okstate.edu (S.E.A.); supawar@okstate.edu (S.P.)
* Correspondence: osan@okstate.edu; Tel.: +1-405-744-2457; Fax: +1-405-744-7873

Received: 1 November 2020; Accepted: 26 November 2020; Published: 29 November 2020

Abstract: Dynamic data assimilation offers a suite of algorithms that merge measurement data with numerical simulations to predict accurate state trajectories. Meteorological centers rely heavily on data assimilation to achieve trustworthy weather forecast. With the advance in measurement systems, as well as the reduction in sensor prices, data assimilation (DA) techniques are applicable to various fields, other than meteorology. However, beginners usually face hardships digesting the core ideas from the available sophisticated resources requiring a steep learning curve. In this tutorial, we lay out the mathematical principles behind DA with easy-to-follow Python module implementations so that this group of newcomers can quickly feel the essence of DA algorithms. We explore a series of common variational, and sequential techniques, and highlight major differences and potential extensions. We demonstrate the presented approaches using an array of fluid flow applications with varying levels of complexity.

Keywords: data assimilation; variational and sequential methods; Kalman filtering; forward sensitivity; measurements fusion

---

1. Introduction

Data assimilation (DA) refers to a class of techniques that lie at the interface between computational sciences and real measurements, and aim at fusing information from both sides to provide better estimates of the system's state. One of the very mature applications that significantly utilize DA is weather forecast, that we rely on in our daily life. In order to predict the weather (or the state of any system) in the future, a model has to be solved, most often by numerical simulations. However, a few problems rise at this point and we refer to only few of them here. First, for accurate predictions, these simulations need to be initiated from the true initial condition, which is never known exactly. For large scale systems, it is almost impossible to experimentally measure the full state of the system at a given time. For example, imagine simulating the atmospheric or oceanic flow, then you need to measure the velocity, temperature, density, etc. at every location corresponding to your numerical grid! Even in the hypothetical case when this is possible, measurements are always contaminated by noise, reducing the fidelity of your estimation. Second, the mathematical model that completely describes all the underlying processes and dynamics of the system is either unknown or hard to deal with. Then, approximate and simplified models are adopted instead. Third, the computational resources always constrain the level of accuracy in the employed schemes and enforce numerical approximations. Luckily, DA appears at the intersection of all these efforts and introduces a variety of approaches to mitigate these problem, or at least reduce their effects. In particular, DA techniques combines possibly incomplete dynamical models, prior information about initial system's state and parameterization, and sparse and corrupted measurement data to yield optimized trajectory in order to describe the system's dynamics and evolution.

As indicated above, dynamical data assimilation techniques have a long history in computational meteorology and geophysical fluid dynamics sciences [1–3]. Then comes the question: why do we write such an introductory tutorial about a historical topic? Before answering this question, we highlight a few points. DA borrows ideas from numerical modeling and analysis, linear algebra, optimization, and control. Although these topics are taught separately in almost every engineering discipline, their combination is rarely presented. We believe that incorporating DA course in engineering curricula is important nowadays as it provides a variety of global tools and ideas that can potentially be applied in many areas, not just meteorology. This was proven while administering a graduate class on "Data Assimilation in Science and Engineering" at Oklahoma State University, as students from different disciplines and backgrounds were astonished by the feasibility and utility of DA techniques to solve numerous inverse problems they are working on, not related to weather forecast. Nonetheless, the availability of beginner-friendly resources has been the major shortage that students suffered from. The majority of textbooks either derives DA algorithms from their very deep roots or surveys their historical developments, without focus on actual implementations. On the other hand, available packages are presented in a sophisticated way that optimizes data storage and handling, computational cost, and convergence. However, this level of sophistication takes a steep learning curve to understand the computational pipeline as well as the algorithmic steps, and a lot of learners fall hopeless during this journey.

Therefore, the main objective of this tutorial paper is to familiarize beginning researchers and practitioners with basic DA ideas along with easy-to-follow pieces of codes to feel the essence of DA and trigger the priceless "aha" moments. With this in mind, we choose Python as the coding language, being a popular, interpreted language, and easy to understand even with minimum programming background, although not the most computationally favored language in high performance computing (HPC) environments. Moreover, whenever possible, we utilize the built-in functions and libraries to minimize the user coding efforts. In other words, the provided codes are presented for demonstrative purposes only using an array of academic test problems, and significant modifications should be incorporated before dealing with complex applications. Meanwhile, this tutorial will give the reader a jump-start that hopefully shortens the learning curve of more advanced packages. A Python-based DA testing suite has been also designed to compare different methodologies [4].

Dynamical data assimilation techniques can be generally classified into variational DA and sequential DA. Variational data assimilation which works by setting an optimization problem defined by a cost functional along with constraints that collectively incorporate our knowledge about the system. The minimizer of the cost functional represents the DA estimate of the unknown system's variables and/or parameters. On the other hand, in sequential methods (also known as statistical methods), the system state is evolved in time using background information until observations become available. At this instant, an update (correction) to the system's variables and/or parameters is estimated and the solver is re-initialized with this new updated information until new measurements are collected, and so on. We give an overview of both approaches as well as basic implementation. In particular, we briefly discuss the three dimensional variational data assimilation (3DVAR) [5,6], the four dimensional variational data assimilation (4DVAR) [7–12], and forward sensitivity method (FSM) [13,14] as examples of variational approaches. Kalman filtering and its variants [15–22] are the most popular applications of sequential methods. We introduce the main ideas behind standard Kalman filter and its extensions for nonlinear and high-dimensional problems. The famous Lorenz 63 is utilized to illustrate the merit of all presented algorithms, being a simple low-order dynamical system that exhibit interesting dynamics. This is to help readers to digest the different pieces of codes and follow the computational pipeline. Then, the paper is concluded with a section that provides the deployment of selected DA approaches for dynamical systems with increasing levels of dimensionality and complexity. We highlight here that the primary purpose of this paper is to provide an introductory tutorial on the data assimilation for educational purposes. All Python implementations of the presented algorithms as well as the test cases are made publicly accessible at our GitHub repository https://github.com/Shady-Ahmed/PyDA.

2. Preliminaries

2.1. Notation

Before we dive into the technical details of dynamical data assimilation approaches, we briefly present and describe our notations and assumptions. In general, we assume that all vector-valued functions or variables are written as a column vector. Unless stated otherwise, boldfaced lowercase letters are used to denote vectors and boldface uppercase letters are reserved to matrices. We suppose that the system state at any time t is denoted as $\mathbf{u}(t) = [u_1(t), u_2(t), \ldots, u_n(t)]^T \in \mathbb{R}^n$, where n is the state-space dimension. The dynamics of the system are governed by the following differential equation

$$\frac{d\mathbf{u}}{dt} = \mathbf{f}(\mathbf{u}; \boldsymbol{\theta}), \tag{1}$$

where $\mathbf{f} : \mathbb{R}^n \times \mathbb{R}^p \to \mathbb{R}^n$ encapsulates the model's dynamics, with $\boldsymbol{\theta} \in \mathbb{R}^p$ being the vector of model's parameters and p being the number of these parameters. With a time-integration scheme applied, the discrete-time model can be written as follows,

$$\mathbf{u}(t_{k+1}) = M(\mathbf{u}(t_k); \boldsymbol{\theta}), \tag{2}$$

where M is the one-time step transition map that evolves the state at time t_k to time $t_{k+1} = t_k + \Delta t$, with Δt being the time step length.

We denote the true value of the state variable as $\mathbf{u}_t$, which is assumed to be unknown and a good approximation of it is sought. Our prior information about the state $\mathbf{u}$ is called the background, with a subscript of b as $\mathbf{u}_b$. This represents our beginning knowledge, which might come from historical data, numerical simulations, or just an intelligent guess. The discrepancy between this background information and true state is denoted as $\boldsymbol{\xi}_b = \mathbf{u}_t - \mathbf{u}_b$, resulting from imperfect model, inaccurate model's initialization, incorrect parameterization, numerical approximations, etc. From probabilistic point of view, we suppose that the background error has a zero mean and a covariance matrix of $\mathbf{B}$. This can be represented as $E[\boldsymbol{\xi}_b] = 0$ and $E[\boldsymbol{\xi}_b \boldsymbol{\xi}_b^T] = \mathbf{B}$, where $\mathbf{B} \in \mathbb{R}^{n \times n}$ is a symmetric and positive-definite matrix and the superscript T refers to the transpose operation. Moreover, we assume that unknown true state has a multivariate Gaussian distribution with a mean $\mathbf{u}_b$ and a covariance matrix $\mathbf{B}$ (i.e., $\mathbf{u}_t = \mathcal{N}(\mathbf{u}_b, \mathbf{B})$).

We define the set of the collected measurements at a specific time t_k as $\mathbf{w}(t_k) \in \mathbb{R}^m$, where m is the dimension of observation-space. We highlight that the observed quantity need not be the same as the state variable. For instance, if the state variable that we are trying to resolve is the temperature of sea surface, we may have access only to radiance measurements by satellites. However, those case be related to each other through Planch–Stefan's law, for instance. Formally, we can relate the observables and the state variables as

$$\mathbf{w}(t_k) = h(\mathbf{u}_t(t_k)) + \boldsymbol{\xi}_m, \tag{3}$$

where $h : \mathbb{R}^n \to \mathbb{R}^m$ defines the mapping from state-space to measurement-space and $\boldsymbol{\xi}_m \in \mathbb{R}^m$ denotes the measurement noise. The mapping h can refer to the sampling (and probably interpolation) of state variables at the measurements locations, relating different quantities of interest (e.g., relating see surface temperature to emitted radiance), or both! The model's map M and observation operator h can be linear, nonlinear, or a combination of them. Similar to the background error, the observation noise $\boldsymbol{\xi}_m$ is assumed to possess a multivariate normal distribution, with a zero mean and a covariance matrix $\mathbf{R} \in \mathbb{R}^{m \times m}$, i.e., $\boldsymbol{\xi}_m = \mathcal{N}(0, \mathbf{R})$. An extra grounding assumption is that the measurement noise and the state variables (either true or background) are uncorrelated. Furthermore, all noises are assumed to be temporally uncorrelated (i.e., white noise). Even though we consider only Gaussian distribution for the background error, and observation noise, we emphasize that there has been a lot of studies dealing with non-Gaussian data assimilation [23–26].

The objective of data assimilation is to provide an algorithm that fuses our prior information $\mathbf{u}_b$ and measurement data $\mathbf{w}$ to yield a better approximation of the unknown true state. This better approximation is called the analysis, and denoted as $\mathbf{u}_a$. The difference between this better approximation and the true state is denoted as $\xi_a = \mathbf{u}_t - \mathbf{u}_a$.

2.2. Twin Experiment Framework

In a realistic situation, the true state values are unknown and noisy measurements are collected by sensing devices. However, for testing ideas, the ground truth need to be known beforehand such that the convergence and accuracy of the developed algorithm can be evaluated. In this sense, the concept of twin experiment has been popular in data assimilation (and inverse problems, in general) studies. First, a prototypical test case (all called toy problems!) is selected based on the similarities between its dynamics and real situations. Similar to your first "Hello World!" program, the Lorenz 63 and Lorenz 96 are often used in numerical weather forecast investigations, the one-dimensional Burgers equation is explored in computational fluid dynamics developments, the two-dimensional Kraichnan turbulence and three-dimensional Taylor–Green vortex are analyzed in turbulence studies, and so on. A reference true trajectory is computed by fixing all parameters and running the forward solver until some final time is reached. Synthetic measurements are then collected by sampling the true trajectory at some points in space and time. A mapping can be applied on the true state variables and arbitrary random noise is artificially added (e.g., a white Gaussian noise). Finally, the data assimilation technique of interest is implemented starting from false values of the state variables or the model's parameters along with the synthetic measurement data. The output trajectory of the algorithm is thus compared against the reference solution, and the performance can be evaluated. It is always recommended that researchers get familiar with twin experiment frameworks as they provide well-structured and controlled environments for testing ideas. For instance, the influence of different measurement sparsity and/or level of noise can be cheaply assessed, without the need to locate or modify sensors.

3. Three Dimensional Variational Data Assimilation

The three dimensional variational data assimilation (3DVAR) framework can be derived from either an optimal control or Bayesian analysis points of view. The interested readers can be referred to other resources for mathematical foundations (e.g., [27]). In order to compute a good approximation of the system state, the following cost functional can be defined,

$$J(\mathbf{u}) = \frac{1}{2}(\mathbf{w} - h(\mathbf{u}))^T \mathbf{R}^{-1}(\mathbf{w} - h(\mathbf{u})) + \frac{1}{2}(\mathbf{u} - \mathbf{u}_b)^T \mathbf{B}^{-1}(\mathbf{u} - \mathbf{u}_b), \tag{4}$$

where the first term penalizes the discrepancy between the actual measurement $\mathbf{w}$ and the state variable mapped into the observation space $h(\mathbf{u})$ (also called the model predicted measurement). The second term aims at incorporating the prior information, weighted by the inverse of the covariance matrix to reflect our confidence in this background. We highlight that all terms in Equation (4) are evaluated at the same time, and thus the 3DVAR can be referred to as a stationary case.

The minimizer of $J(\mathbf{u})$ (i.e., the analysis) can be obtained by setting the gradient of the cost functional to zero as follows,

$$\nabla J(\mathbf{u}) = -\mathbf{D}_h^T(\mathbf{u}_a)\mathbf{R}^{-1}(\mathbf{w} - h(\mathbf{u}_a)) + \mathbf{B}^{-1}(\mathbf{u}_a - \mathbf{u}_b) = 0, \tag{5}$$

where $\mathbf{D}_h(\mathbf{u}) \in \mathbb{R}^{m \times n}$ is the Jacobian matrix of the operator $h(\mathbf{u})$. The difficulty of solving Equation (5) depends on the form of $h(\mathbf{u_a})$ as it can either be linear, or highly nonlinear.

3.1. Linear Case

For linear observation operator (i.e., $h(\mathbf{u}) = \mathbf{H}\mathbf{u}$, and $\mathbf{D}_h(\mathbf{u}) = \mathbf{H}$, where $\mathbf{H}$ is an $m \times n$ matrix), the evaluation of the analysis $\mathbf{u}_a$ in Equation (5) reduces to solving the following linear system of equations

$$(\mathbf{B}^{-1} + \mathbf{H}^T \mathbf{R}^{-1} \mathbf{H})\mathbf{u}_a = (\mathbf{B}^{-1}\mathbf{u}_b + \mathbf{H}^T \mathbf{R}^{-1}\mathbf{w}). \tag{6}$$

We note that $(\mathbf{B}^{-1} + \mathbf{H}^T \mathbf{R}^{-1} \mathbf{H})$ on the left-hand side is an $n \times n$ matrix, and hence this is called the model-space approach to 3DVAR. Furthermore, a popular incremental form can be derived from Equation (6) by adding and subtracting $\mathbf{H}^T \mathbf{R}^{-1} \mathbf{H}\mathbf{u}_a$ to/from the right-hand side and rearranging to get the following form,

$$\mathbf{u}_a = \mathbf{u}_b + (\mathbf{B}^{-1} + \mathbf{H}^T \mathbf{R}^{-1} \mathbf{H})^{-1}\mathbf{H}^T \mathbf{R}^{-1}(\mathbf{w} - \mathbf{H}\mathbf{u}_b). \tag{7}$$

Moreover, the Sherman–Morrison–Woodbury inversion formula can be used to derive an observation-space solution to the 3DVAR problem (for details, see [27], page 327) as follows,

$$\mathbf{u}_a = \mathbf{u}_b + \mathbf{B}\mathbf{H}^T(\mathbf{R} + \mathbf{H}\mathbf{B}\mathbf{H}^T)^{-1}(\mathbf{w} - \mathbf{H}\mathbf{u}_b). \tag{8}$$

Note that $(\mathbf{R} + \mathbf{H}\mathbf{B}\mathbf{H}^T)$ is an $m \times m$ matrix, compared to $(\mathbf{B}^{-1} + \mathbf{H}^T \mathbf{R}^{-1} \mathbf{H})$ being an $n \times n$ matrix. Thus, Equation (7) or Equation (8) might be computationally favored based on the values of n and m. We also highlight that in either cases, matrix inversion is rarely (almost never) computed directly, and efficient linear system solvers should be utilized, instead. An example of a Python function for the implementation of the 3DVAR algorithm with a linear operator is shown in Listing 1.

Listing 1. Implementation of 3DVAR for with a linear observation operator.

```python
import numpy as np
def Lin3dvar(ub,w,H,R,B,opt):

    # The solution of the 3DVAR problem in the linear case requires
    # the solution of a linear system of equations.
    # Here, we utilize the built-in numpy function to do this.
    # Other schemes can be used, instead.
    if opt == 1: #model-space approach
    Bi = np.linalg.inv(B)
    Ri = np.linalg.inv(R)
    A = Bi + (H.T)@Ri@H
    b = Bi@ub + (H.T)@Ri@w
    ua = np.linalg.solve(A,b) #solve a linear system

    elif opt == 2: #model-space incremental approach
    Bi = np.linalg.inv(B)
    Ri = np.linalg.inv(R)
    A = Bi + (H.T)@Ri@H
    b = (H.T)@Ri@(w-H@ub)
    ua = ub + np.linalg.solve(A,b) #solve a linear system

    elif opt == 3: #observation-space incremental approach
    A = R + H@B@(H.T)
    b = (w-H@ub)
    ua = ub + B@(H.T)@np.linalg.solve(A,b) #solve a linear system

    return ua
```

3.2. Nonlinear Case

On the other hand, if $h(\mathbf{u})$ is a nonlinear function, Equation (5) implies the solution of a system of nonlinear equations. Unlike linear systems, few algorithms are available to directly solve nonlinear systems and their convergence and stability are usually questionable. Alternatively, we can use Taylor series to expand $h(\mathbf{u})$ around an initial estimate of $\mathbf{u}_a$, denoted as $\mathbf{u}_c$, where $\mathbf{u}_a = \mathbf{u}_c + \Delta\mathbf{u}$. The first-order approximation of $h(\mathbf{u}_a)$ can be written as

$$h(\mathbf{u}_a) \approx h(\mathbf{u}_c) + \mathbf{D}_h(\mathbf{u}_c)\Delta\mathbf{u}, \tag{9}$$

and Equation (5) can be approximated as

$$\mathbf{D}_h^T(\mathbf{u}_c)\mathbf{R}^{-1}(\mathbf{w} - h(\mathbf{u}_c) - \mathbf{D}_h(\mathbf{u}_c)\Delta\mathbf{u}) = \mathbf{B}^{-1}(\mathbf{u}_c + \Delta\mathbf{u} - \mathbf{u}_b). \tag{10}$$

Thus, the correction to the initial guess of $\mathbf{u}_a$ can be computed by solving the following system of linear equations

$$\left(\mathbf{B}^{-1} + \mathbf{D}_h^T(\mathbf{u}_c)\mathbf{R}^{-1}\mathbf{D}_h(\mathbf{u}_c)\right)\Delta\mathbf{u} = \left(\mathbf{B}^{-1}(\mathbf{u}_b - \mathbf{u}_c) + \mathbf{D}_h(\mathbf{u}_c)^T\mathbf{R}^{-1}(\mathbf{w} - h(\mathbf{u}_c))\right), \tag{11}$$

and a new guess of $\mathbf{u}_a$ is estimated as $\mathbf{u}_c + \Delta\mathbf{u}$, which is then plugged back into Equation (11) and the computations are repeated until convergence is reached. Python implementation of the 3DVAR in nonlinear observation operator is presented in Listing 2 Although we only present the first order approximation of $h(\mathbf{u})$, higher order expansions can be utilized for increased accuracy [27].

Listing 2. Implementation of the 3DVAR for with a nonlinear observation operator, using first-order approximation.

```python
import numpy as np
def NonLin3dvar(ub,w,ObsOp,JObsOp,R,B):

    # The solution of the 3DVAR problem in the nonlinear case requires
    # the solution of a linear system of equations.
    # Here, we utilize the built-in numpy function to do this.
    # Other schemes can be used, instead.
    Bi = np.linalg.inv(B)
    Ri = np.linalg.inv(R)
    ua = np.copy(ub)
    for iter in range(100):
        Dh = JObsOp(ua)
        A = Bi + (Dh.T)@Ri@Dh
        b = Bi@(ub-ua) + (Dh.T)@Ri@(w-ObsOp(ua))
        du = np.linalg.solve(A,b) #solve a linear system
        ua = ua + du
        if np.linalg.norm(du) <= 1e-4:
            break
    return ua
```

3.3. Example: Lorenz 63 System

The Lorenz 63 equations have been utilized as a toy problem in data assimilation studies, capturing some of the interesting mechanisms of weather systems. The three-equation model can be written as

$$
\begin{aligned}
\frac{dx}{dt} &= \sigma(y - x), \\
\frac{dy}{dt} &= x(\rho - z) - y, \\
\frac{dz}{dt} &= xy - \beta z,
\end{aligned}
\tag{12}
$$

where the values of $\sigma = 10, \beta = 8/4, \rho = 28$ are usually used to exhibit a chaotic behavior. If we like to put Equation (12) with the notations introduced in Section 2, we can write $\mathbf{u} = [x, y, z]^T$ with $n = 3$, and $\boldsymbol{\theta} = [\sigma, \beta, \rho]^T$ with $p = 3$. A Python function describing the dynamics of the Lorenz 63 system is given in Listing 3.

Listing 3. A Python function for the Lorenz 63 dynamics.

```python
import numpy as np
def Lorenz63(state,*args): #Lorenz 96 model
sigma = args[0]
beta = args[1]
rho = args[2]
x, y, z = state #Unpack the state vector
f = np.zeros(3) #Derivatives
f[0] = sigma * (y - x)
f[1] = x * (rho - z) - y
f[2] = x * y - beta * z
return f
```

Equation (12) describe the continuous-time evolution of the Lorenz system. In order to obtain the discrete-time mapping $M(\cdot; \cdot)$, a temporal integration scheme has to be applied. In Listing 4, one-step time integration functions are provided in Python using the first-order Euler and the fourth-order Runge–KuKutta schemes. Note that these functions requires a right-hand side function as input, this is mainly the continuous-time model $f(\mathbf{u})$ (e.g., Listing 3).

Listing 4. Python functions for the time integration using the 1st Euler and the 4th Runge–Kutta schemes.

```python
import numpy as np
def euler(rhs,state,dt,*args):
k1 = rhs(state,*args)
new_state = state + dt*k1
return new_state

def RK4(rhs,state,dt,*args):
k1 = rhs(state,*args)
k2 = rhs(state+k1*dt/2,*args)
k3 = rhs(state+k2*dt/2,*args)
k4 = rhs(state+k3*dt,*args)
new_state = state + (dt/6)*(k1+2*k2+2*k3+k4)
return new_state
```

For twin experiment testing, we suppose a true initial condition of $\mathbf{u}_t(0) = [1, 1, 1]^T$ and measurements are collected each 0.2 time units for a total time of 2. We suppose that we measure the full system state (i.e., $h(\mathbf{u}) = \mathbf{u}$, $m = 3$, and $\mathbf{H} = \mathbf{I}_3$, where $\mathbf{I}_3$ is the 3×3 identity matrix). Measurements are considered to be contaminated by a white Gaussian noise with a zero mean and a covariance matrix $\mathbf{R} = \mathrm{Diag}(\sigma_1^2, \sigma_2^2, \sigma_3^2)$. For simplicity, we let $\sigma_1 = \sigma_2 = \sigma_3 = 0.15$. For data assimilation testing, we assume that we begin with a perturbed initial condition of $\mathbf{u}(0) = [2, 3, 4]^T$. Then, background state values are computed at $t = 0.2$ by time integration of Equation (12) starting from this false initial condition. Observations at $t = 0.2$ are assimilated to provide the analysis at $t = 0.2$. After that, background state values are computed at $t = 0.4$ by time integration of Equation (12) starting from the analysis at $t = 0.2$, and so on. A sample implementation of the 3DVAR framework is presented in Listing 5, where a fixed background covariance matrix $\mathbf{B} = \mathrm{Diag}(0.01, 0.01, 0.01)$ is assumed. Solution trajectories are presented in Figure 1 for a total time of 10, where observations are only available up to $t = 2$.

Listing 5. Implementation of the 3DVAR for the Lorenz 63 system.

```python
import numpy as np
import matplotlib.pyplot as plt

#%% Application: Lorenz 63
# parameters
sigma = 10.0
beta = 8.0/3.0
rho = 28.0
dt = 0.01
tm = 10
nt = int(tm/dt)
t = np.linspace(0,tm,nt+1)

u0True = np.array([1,1,1]) # True initial conditions

######################### Twin experiment ###############################
np.random.seed(seed=1)
sig_m= 0.15 # standard deviation for measurement noise
R = sig_m**2*np.eye(3) #covariance matrix for measurement noise
H = np.eye(3) #linear observation operator

dt_m = 0.2 #time period between observations
tm_m = 2 #maximum time for observations
nt_m = int(tm_m/dt_m) #number of observation instants

#t_m = np.linspace(dt_m,tm_m,nt_m) #np.where( (t<=2) & (t%0.1==0) )[0]
ind_m = (np.linspace(int(dt_m/dt),int(tm_m/dt),nt_m)).astype(int)
t_m = t[ind_m]

#time integration
uTrue = np.zeros([3,nt+1])
uTrue[:,0] = u0True
km = 0
w = np.zeros([3,nt_m])
for k in range(nt):
uTrue[:,k+1] = RK4(Lorenz63,uTrue[:,k],dt,sigma,beta,rho)
if (km<nt_m) and (k+1==ind_m[km]):
w[:,km] = H@uTrue[:,k+1] + np.random.normal(0,sig_m,[3,])
km = km+1

plt.plot(t,uTrue[0,:])
```

```python
plt.plot(t_m,w[0,:],'o')


######################### Data Assimilation ##############################
u0b = np.array([2.0,3.0,4.0])
sig_b= 0.1
B = sig_b**2*np.eye(3)

#time integration
ub = np.zeros([3,nt+1])
ub[:,0] = u0b
ua = np.zeros([3,nt+1])
ua[:,0] = u0b
km = 0
for k in range(nt):
ub[:,k+1] = RK4(Lorenz63,ub[:,k],dt,sigma,beta,rho)
ua[:,k+1] = RK4(Lorenz63,ua[:,k],dt,sigma,beta,rho)

if (km<nt_m) and (k+1==ind_m[km]):
ua[:,k+1] = Lin3dvar(ua[:,k+1],w[:,km],H,R,B,3)
km = km+1

############################# Plotting ###################################
import matplotlib as mpl
mpl.rc('text', usetex=True)
mpl.rcParams['text.latex.preamble']=[r"\usepackage{amsmath}"]
mpl.rcParams['text.latex.preamble'] = [r'\boldmath']
font = {'family' : 'normal',
'weight' : 'bold',
'size'   : 20}
mpl.rc('font', **font)

fig, ax = plt.subplots(nrows=3,ncols=1, figsize=(10,8))
ax = ax.flat

for k in range(3):
ax[k].plot(t,uTrue[k,:], label=r'\bf{True}', linewidth = 3)
ax[k].plot(t,ub[k,:], ':', label=r'\bf{Background}', linewidth = 3)
ax[k].plot(t[ind_m],w[k,:], 'o', fillstyle='none', \
label=r'\bf{Observation}', markersize = 8, markeredgewidth = 2)
ax[k].plot(t,ua[k,:], '--', label=r'\bf{Analysis}', linewidth = 3)
ax[k].set_xlabel(r'$t$',fontsize=22)
ax[k].axvspan(0, tm_m, color='y', alpha=0.4, lw=0)

ax[0].legend(loc="center", bbox_to_anchor=(0.5,1.25),ncol =4,fontsize=15)

ax[0].set_ylabel(r'$x(t)$')
ax[1].set_ylabel(r'$y(t)$')
fig.subplots_adjust(hspace=0.5)
```

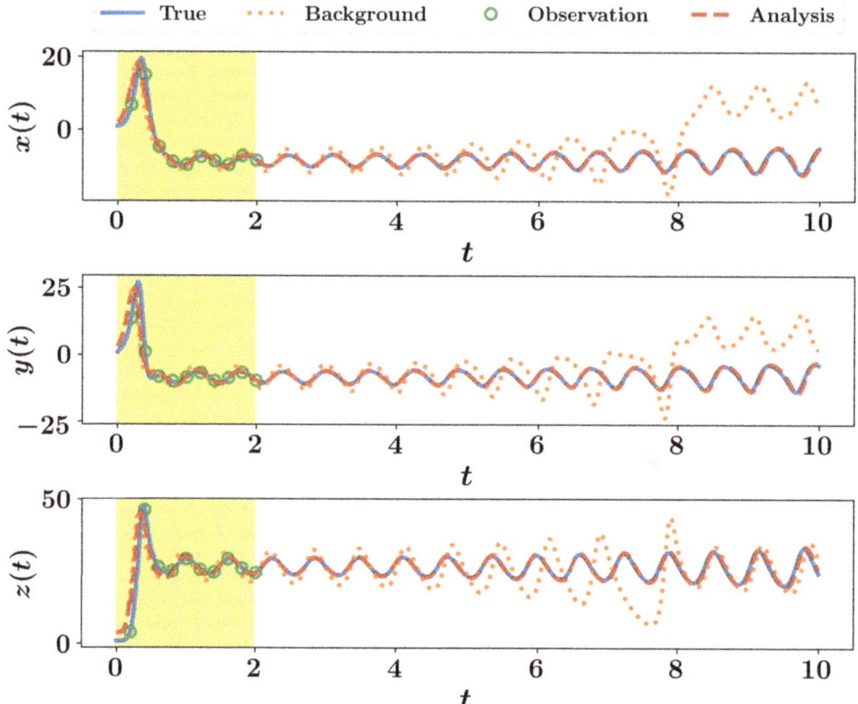

Figure 1. Results of 3DVAR implementation for the Lorenz 63 system.

4. Four Dimensional Variational Data Assimilation

We highlighted in Section 3 that the 3DVAR can be referred to as a stationary case since the observations, background, and analysis all correspond to a fixed time instant. In other words, the optimization problem that minimizes Equation (4) takes place in the spatial state-space only. As an extension, the four dimensional variational data assimilation (4DVAR) aims to solve the optimization in both space and time, proving a non-stationary framework. In particular, the model's dynamics are incorporated into the optimization problem to relate different points in time to each other. The cost functional for the 4DVAR Can be written as follows,

$$J(\mathbf{u}(t_0)) = \sum_{t_k \in \mathcal{T}} \frac{1}{2}(\mathbf{w}(t_k) - h(\mathbf{u}(t_k)))^T \mathbf{R}^{-1}(t_k)(\mathbf{w}(t_k) - h(\mathbf{u}(t_k))), \tag{13}$$

where $\mathbf{w}(t_k)$ is the measurement at time t_k and $\mathcal{T}$ defines the set of time instants where observations are available. Note that the argument of this cost functional is the initial condition $\mathbf{u}(t_0)$. In other words, the purpose of the 4DVAR algorithm is to evaluate an initial state estimate, which if evolved in time, would produce a trajectory that is as close to the collected measurements as possible (weighted by the inverse of the covariance matrix of interfering noise). This is the place where the model's dynamics comes into play to relate initial condition to future predictions when measurements are accessible. In other words, the values of $\mathbf{u}(t_k)$) are constrained by the underlying model. Instead of presenting the linear and nonlinear mappings separately, we will focus on the general case of both nonlinear model mapping and nonlinear observation operator, where simplification to linear cases should be straightforward.

In Equation (2), we introduced the one-step transition map and here, we can extend it to the k-step transition case by applying Equation (2) recursively as

$$\mathbf{u}(t_k) = M^{(k)}(\mathbf{u}(t_0); \boldsymbol{\theta}) = M(M^{(k-1)}(\mathbf{u}(t_0); \boldsymbol{\theta}); \boldsymbol{\theta}), \tag{14}$$

where $M^{(1)}(\mathbf{u}(t_0); \boldsymbol{\theta}) = M(\mathbf{u}(t_0); \boldsymbol{\theta})$. Now, we consider a base trajectory given by $\overline{\mathbf{u}}(t_k)$ for $k = 1, 2, \ldots$ generated from an initial condition of $\overline{\mathbf{u}}(t_0)$. A perturbed trajectory ($\mathbf{u}(t_k)$ for $k = 1, 2, \ldots,$) can be obtained by correcting the initial condition as $\mathbf{u}(t_0) = \overline{\mathbf{u}}(t_0) + \Delta \mathbf{u}_0$ and the difference between the perturbed and based trajectories can be written as

$$\mathbf{u}(t_k) - \overline{\mathbf{u}}(t_k) = M^{(k)}(\overline{\mathbf{u}}(t_0) + \Delta \mathbf{u}_0; \boldsymbol{\theta}) - M^{(k)}(\overline{\mathbf{u}}(t_0); \boldsymbol{\theta}). \tag{15}$$

A first-order Taylor expansion of $M(\overline{\mathbf{u}}(t_0) + \Delta \mathbf{u}_0; \boldsymbol{\theta})$ around $\overline{\mathbf{u}}(t_0)$ can be given as follows

$$M(\overline{\mathbf{u}}(t_0) + \Delta \mathbf{u}_0; \boldsymbol{\theta}) \approx M(\overline{\mathbf{u}}(t_0); \boldsymbol{\theta}) + \mathbf{D}_M(\overline{\mathbf{u}}(t_0))\Delta \mathbf{u}_0, \tag{16}$$

where $\mathbf{D}_M(\mathbf{u}(t_k))$ is the Jacobian of the model $M(\mathbf{u}; \boldsymbol{\theta})$, evaluated at $\mathbf{u}(t_k)$, also known as the tangent linear operator. Note that $M(\overline{\mathbf{u}}(t_0); \boldsymbol{\theta}) = \overline{\mathbf{u}}(t_1)$ and $M(\overline{\mathbf{u}}(t_0) + \Delta \mathbf{u}_0; \boldsymbol{\theta}) = \mathbf{u}(t_1)$, thus $\Delta \mathbf{u}_1 = \mathbf{u}(t_1) - \overline{\mathbf{u}}(t_1) \approx \mathbf{D}_M(\overline{\mathbf{u}}(t_0))\Delta \mathbf{u}_0$. Similarly, we can expand $M(\overline{\mathbf{u}}(t_1) + \Delta \mathbf{u}_1; \boldsymbol{\theta})$ around $\overline{\mathbf{u}}(t_1)$ as follows,

$$M(\overline{\mathbf{u}}(t_1) + \Delta \mathbf{u}_1; \boldsymbol{\theta}) \approx M(\overline{\mathbf{u}}(t_1); \boldsymbol{\theta}) + \mathbf{D}_M(\overline{\mathbf{u}}(t_1))\Delta \mathbf{u}_1, \tag{17}$$

where $\mathbf{u}(t_2) = M(\mathbf{u}(t_1); \boldsymbol{\theta}) \approx M(\overline{\mathbf{u}}(t_1) + \Delta \mathbf{u}_1; \boldsymbol{\theta})$ and $\mathbf{u}(t_2) = M(\overline{\mathbf{u}}(t_1); \boldsymbol{\theta})$. Consequently, $\Delta \mathbf{u}_2 = \mathbf{u}(t_2) - \overline{\mathbf{u}}(t_2) \approx \mathbf{D}_M(\overline{\mathbf{u}}(t_1))\Delta \mathbf{u}_1$, which can be generalized as,

$$\Delta \mathbf{u}_{k+1} \approx \mathbf{D}_M(\overline{\mathbf{u}}(t_k))\Delta \mathbf{u}_k, \tag{18}$$

with $\mathbf{u}(t_k) \approx \Delta \mathbf{u}_k + \overline{\mathbf{u}}(t_k)$. It is customary to call Equation (18) as the perturbation equation, or the tangent linear system (TLS). Equation (18) can be related to $\Delta \mathbf{u}_0$ by recursion as follows,

$$\begin{aligned}
\Delta \mathbf{u}_{k+1} &\approx \mathbf{D}_M(\overline{\mathbf{u}}(t_k))\Delta \mathbf{u}_k \\
&\approx \mathbf{D}_M(\overline{\mathbf{u}}(t_k))\mathbf{D}_M(\overline{\mathbf{u}}(t_{k-1}))\Delta \mathbf{u}_{k-1} \\
&\approx \mathbf{D}_M(\overline{\mathbf{u}}(t_k))\mathbf{D}_M(\overline{\mathbf{u}}(t_{k-1}))\mathbf{D}_M(\overline{\mathbf{u}}(t_{k-2}))\Delta \mathbf{u}_{k-2} \\
&\approx \mathbf{D}_M(\overline{\mathbf{u}}(t_k))\mathbf{D}_M(\overline{\mathbf{u}}(t_{k-1}))\mathbf{D}_M(\overline{\mathbf{u}}(t_{k-2}))\mathbf{D}_M(\overline{\mathbf{u}}(t_{k-3}))\ldots\mathbf{D}_M(\overline{\mathbf{u}}(t_0))\Delta \mathbf{u}_0,
\end{aligned}$$

which can be short-handed as $\Delta \mathbf{u}_{k+1} \approx \mathbf{D}_M(\overline{\mathbf{u}}(t_{k:0}))\Delta \mathbf{u}_0$ (please, notice the order of matrix multiplication and the subscript "$k:0$").

Now, we investigate the first order variation ΔJ of the cost functional $J(\overline{\mathbf{u}}(t_0))$ induced by the perturbation $\Delta \mathbf{u}_0$ in the initial condition. This can be approximated as below,

$$\begin{aligned}
\Delta J &= \Delta \mathbf{u}_0^T \nabla J(\overline{\mathbf{u}}(t_0)) \tag{19} \\
&= -\sum_{t_k \in \mathcal{T}} \Delta \mathbf{u}_k^T \mathbf{D}_h^T(\overline{\mathbf{u}}(t_k))\mathbf{R}^{-1}(t_k)(\mathbf{w}(t_k) - h(\overline{\mathbf{u}}(t_k))). \tag{20}
\end{aligned}$$

Given that $\Delta \mathbf{u}_k \approx \mathbf{D}_M(\overline{\mathbf{u}}(t_{k-1:0}))\Delta \mathbf{u}_0$, then $\Delta \mathbf{u}_k^T \approx \Delta \mathbf{u}_0^T \mathbf{D}_M^T(\overline{\mathbf{u}}(t_0))\mathbf{D}_M^T(\overline{\mathbf{u}}(t_1))\ldots \mathbf{D}_M^T(\overline{\mathbf{u}}(t_{k-1})) = \Delta \mathbf{u}_0^T \mathbf{D}_M^T(\overline{\mathbf{u}}(t_{0:k-1}))$ and Equation (20) can be rewritten as

$$\Delta J = -\sum_{t_k \in \mathcal{T}} \Delta \mathbf{u}_0^T \mathbf{D}_M^T(\overline{\mathbf{u}}(t_{0:k-1}))\mathbf{D}_h^T(\overline{\mathbf{u}}(t_k))\mathbf{R}^{-1}(t_k)(\mathbf{w}(t_k) - h(\overline{\mathbf{u}}(t_k))). \tag{21}$$

By comparing Equations (19) and (21), the gradient of the cost functional can be approximated as

$$\nabla J(\overline{\mathbf{u}}(t_0)) = - \sum_{t_k \in \mathcal{T}} \mathbf{D}_M^T(\overline{\mathbf{u}}(t_{0:k-1})) \mathbf{D}_h^T(\overline{\mathbf{u}}(t_k)) \mathbf{R}^{-1}(t_k)(\mathbf{w}(t_k) - h(\overline{\mathbf{u}}(t_k))) \tag{22}$$

$$= - \sum_{t_k \in \mathcal{T}} \mathbf{D}_M^T(\overline{\mathbf{u}}(t_{0:k-1})) \mathbf{f}(t_k), \tag{23}$$

where $\mathbf{f}(t_k) = \mathbf{D}_h^T(\overline{\mathbf{u}}(t_k)) \mathbf{R}^{-1}(t_k)(\mathbf{w}(t_k) - h(\overline{\mathbf{u}}(t_k)))$. If we denote the time instants at which measurements are available as $\mathcal{T} = \{t_{O1}, t_{O2}, \ldots, t_{ON}\}$, Equation (23) can be expanded as

$$\nabla J(\overline{\mathbf{u}}(t_0)) = -\left\{ \mathbf{D}_M^T(\overline{\mathbf{u}}(t_{0:O1-1})) \mathbf{f}(t_{O1}) + \mathbf{D}_M^T(\overline{\mathbf{u}}(t_{0:O2-1})) \mathbf{f}(t_{O2}) + \cdots + \mathbf{D}_M^T(\overline{\mathbf{u}}(t_{0:ON-1})) \mathbf{f}(t_{ON}) \right\}. \tag{24}$$

Now, defining a sequence of $\lambda_k \in \mathbb{R}^n$ as below,

$$\lambda_k = \begin{cases} \mathbf{f}_k, & \text{if } t_k = t_{ON} \\ \mathbf{D}_M^T(\overline{\mathbf{u}}(t_k))\lambda_{k+1} + \mathbf{f}_k, & \text{if } t_k \in \{t_{O1}, t_{O2}, \ldots, t_{ON-1}\} \\ \mathbf{D}_M^T(\overline{\mathbf{u}}(t_k))\lambda_{k+1}, & \text{otherwise.} \end{cases} \tag{25}$$

It can be verified that $\nabla J(\overline{\mathbf{u}}(t_0)) = -\lambda_0$ (assume some numbers and you can see this relation holds!). Therefore, in order to obtain the gradient of the cost functional, λ_0 has to be computed, which depends on the evaluation of λ_1. In turn, the computation of λ_1 requires λ_2 and so on. Equation (25) is known as the first-order adjoint equation, as it implies the evaluation of λ_k sequence from t_{k+1} to t_k (i.e., reverse order).

Therefore, the first-order approximation of the 4DVAR works as follows. Starting from a prior guess of the initial condition, the base trajectory is computed by solving the model forward in time until the final time corresponding the last observation point (i.e., $k = 0, 1, 2, \ldots, ON$). Then, the value of $\lambda_{ON} = \mathbf{f}_{ON}$ is evaluated at final time. After that, λ_k is evolved backward in time using Equation (25) until $\nabla J(\overline{\mathbf{u}}(t_0)) = -\lambda_0$ is obtained. This value of the gradient is thus utilized to update the initial condition and a new base trajectory is generated. The solution of the 4DVAR problem requires the solution of the model dynamics forward in time and adjoint problem backward in time until to compute the gradient of the cost functional and update the initial condition. The process is thus repeated until convergence takes place. Listing 6 shows a sample function to compute the gradient of the cost functional $\nabla J(\overline{\mathbf{u}}(t_0))$ corresponding to a base trajectory generated from a guess of the initial condition $\overline{\mathbf{u}}(t_0)$. We highlight that, in practice, the storage of the base trajectory as well as the λ sequence at every time instant might be overwhelming. However, we are not addressing such issues in these introductory tutorials to data assimilation techniques.

Listing 6. Computation of the gradient of the cost functional with the 4DVAR using the first-order adjoint algorithm.

```python
def Adj4dvar(rhs,Jrhs,ObsOp,JObsOp,t,ind_m,u0b,w,R,opt,*args):

    # The solution of the 4DVAR problem requires the evaluation of
    # the forward model to generate base trajectory and
    # the Jacobian of the model to solve the adjoint problem.
    # Inputs:
    #rhs: defines the right-hand side of the continuous time forward model f
    #Jrhs: defines the Jacobian matrix of rhs D_f(u)
    #ObsOp: defines the observation operator h(u)
    #JObsOp: defines the Jacobian of the observation operator D_h(u)
    #t: vector of time
    #ind_m: indices of measurement instants
```

```python
#u0b: initial condition for base trajectory
#w: matrix of measurements
#R: covariance matrix of measurement noise
#opt: [0=euler] or [1=RK4] defines the time integration scheme to
#comptue the discrete-time forward map and its Jacobian
# Output: The Jacobian of the cost functional

n = len(u0b)
#determine the assimilation window
t = t[:ind_m[-1]+1] #cut the time till the last observatino point
nt = len(t)-1
dt = t[1] - t[0]
ub = np.zeros([n,nt+1]) #base trajectory
lam = np.zeros([n,nt+1]) #lambda sequence
fk = np.zeros([n,len(ind_m)])

Ri = np.linalg.inv(R)

ub[:,0] = u0b
if opt == 0: #Euler
#forward model
for k in range(nt):
ub[:,k+1] = euler(rhs,ub[:,k],dt,*args)

#backward adjoint
k = ind_m[-1]
fk[:,-1] = (JObsOp(ub[:,k])).T @ Ri @ (w[:,-1]-ObsOp(ub[:,k]))
lam[:,k] = fk[:,-1] #lambda_N = f_N

km = len(ind_m)-2
for k in range(ind_m[-1],0,-1):
DM = Jeuler(rhs,Jrhs,ub[:,k-1],dt,*args)
lam[:,k-1] =   (DM).T @ lam[:,k]
if k-1 == ind_m[km]:
fk[:,km] =(JObsOp(ub[:,k-1])).T @ Ri @ (w[:,km]-ObsOp(ub[:,k-1]))
lam[:,k-1] = lam[:,k-1] + fk[:,km]
km = km - 1

elif opt == 1: #RK4
# forward model
for k in range(nt):
ub[:,k+1] = RK4(rhs,ub[:,k],dt,*args)

#backward adjoint
k = ind_m[-1]
fk[:,-1] = (JObsOp(ub[:,k])).T @ Ri @ (w[:,-1]-ObsOp(ub[:,k]))
lam[:,k] = fk[:,-1] #lambda_N = f_N

km = len(ind_m)-2
for k in range(ind_m[-1],0,-1):
DM = JRK4(rhs,Jrhs,ub[:,k-1],dt,*args)
lam[:,k-1] =   (DM).T @ lam[:,k]
if k-1 == ind_m[km]:
fk[:,km] = (JObsOp(ub[:,k-1])).T @ Ri @(w[:,km]-ObsOp(ub[:,k-1]))
lam[:,k-1] = lam[:,k-1] + fk[:,km]
```

```
km = km - 1

dJ0 = -lam[:,0]
return dJ0
```

The gradient $\nabla J(\overline{\mathbf{u}}(t_0))$ should be used in a minimization algorithm to update the initial condition for the next iteration. One simple algorithm is the simple gradient descent where an updated value of the initial state is computed as $\overline{\mathbf{u}}(t_0))^{new} = \overline{\mathbf{u}}(t_0))^{old} - \beta_n \nabla J(\overline{\mathbf{u}}(t_0)^{old})$, where β_n is some step parameter. This can be normalized as $\overline{\mathbf{u}}(t_0))^{new} = \overline{\mathbf{u}}(t_0))^{old} - \beta \dfrac{\nabla J(\overline{\mathbf{u}}(t_0)^{old})}{\|\nabla J(\overline{\mathbf{u}}(t_0)^{old})\|}$. The value of β might be predefined, or more efficiently updated at each iteration using an additional optimization algorithm (e.g., line-search). For the sake of completeness, we present a line-search routine in Listing 7 using the Golden search algorithm. This is based on the definition of the cost functional in Listing 8.

Listing 7. A line-search Python function using the Golden search method.

```python
def GoldenAlpha(p,rhs,ObsOp,t,ind_m,u0,w,R,opt,*args):

    # p is the optimization direction
    a0=0
    b0=1
    r=(3-np.sqrt(5))/2

    uncert = 1e-5 # Specified uncertainty

    a1= a0 + r*(b0-a0);
    b1= b0 - r*(b0-a0);
    while (b0-a0) > uncert:

    if loss(rhs,ObsOp,t,ind_m,u0+a1*p,w,R,opt,*args)  < loss(rhs,ObsOp,t,\
    ind_m,u0+b1*p,w,R,opt,*args):
    b0=b1;
    b1=a1;
    a1= a0 + r*(b0-a0);
    else:
    a0=a1;
    a1=b1;
    b1= b0 - r*(b0-a0);
    alpha = (b0+a0)/2

    return alpha
```

Listing 8. Computation of the cost functional defined in Equation (13).

```python
# cost functional (w-h(u))^T * R^{-1} * (w-h(u))
def loss(rhs,ObsOp,t,ind_m,u0,w,R,opt,*args):

    n = len(u0)
    #determine the assimilation window
    t = t[:ind_m[-1]+1] #cut the time till the last observation point
    nt = len(t)-1
    dt = t[1] - t[0]
    u = np.zeros([n,nt+1]) #trajectory

    u[:,0] = u0
```

```python
Ri = np.linalg.inv(R)
floss = 0
km = 0
nt_m = len(ind_m)
if opt == 0: #Euler
#forward model
for k in range(nt):
u[:,k+1] = euler(rhs,u[:,k],dt,*args)

if (km<nt_m) and (k+1==ind_m[km]):
tmp = w[:,km] - ObsOp(u[:,k+1])
tmp = tmp.reshape(-1,1)
floss = floss + np.linalg.multi_dot(( tmp.T, Ri , tmp ))
km = km + 1

elif opt == 1: #RK4
# forward model
for k in range(nt):
u[:,k+1] = RK4(rhs,u[:,k],dt,*args)
if (km<nt_m) and (k+1==ind_m[km]):
tmp = w[:,km] - ObsOp(u[:,k+1])
tmp = tmp.reshape(-1,1)
floss = floss + np.linalg.multi_dot(( tmp.T, Ri , tmp ))
km = km + 1

floss = floss[0,0]/2
return floss
```

Example: Lorenz 63 System

Similar to the 3DVAR demonstration, we apply the described 4DVAR using the first-order adjoint method on the Lorenz 63 system. We also begin with the same erroneous initial condition of $\mathbf{u}(0) = [2,3,4]^T$ and observations are collected each 0.2 time units, contaminated with a Gaussian noise with diagonal covariance matrix defined as $\mathbf{R} = \sigma_m^2 \mathbf{I}_3$, where $\sigma_m = 0.15$ is the standard deviation for measurement noise. Moreover, we simply define a linear observation operator defined as $h(\mathbf{u}) = \mathbf{u}$, with a Jacobian of identity matrix. We utilize the simple gradient descent for minimizing the cost functional, equipped by a Golden search method for learning rate optimization. A maximum number of iterations is set to 1000, but we highlight that this is highly dependent on the adopted minimization algorithm as well as the line-search technique. In practice, the evaluation of each iteration might be too computationally expensive, so the number of iterations need to be as low as possible. We define two criteria for convergence, and iterations stop whenever any one of them is achieved. The first one is based on the change in the value of the cost or loss functional and the second one is based on the magnitude of its gradient. Extra criteria might be supplied as well.

Results for running Listing 9 is shown in Figure 2, where we can notice the significant improvement of predictions, compared to the background trajectories. Moreover, we highlight the correction to the initial conditions in Figure 2 which resulted in the analysis trajectory. This is opposed to the 3DVAR implementation, where correction is applied locally at measurements instants only as seen in Figure 1.

Listing 9. Implementation of 4DVAR using the first-order adjoint method for the Lorenz 63 system.

```python
import numpy as np
import matplotlib.pyplot as plt

#%% Application: Lorenz 63
# parameters
sigma = 10.0
beta = 8.0/3.0
rho = 28.0
dt = 0.01
tm = 10
nt = int(tm/dt)
t = np.linspace(0,tm,nt+1)


########################### Twin experiment ###############################
def h(u): # Observation operator
w = u
return w

def Dh(u): #Jacobian of observation operator
n = len(u)
D = np.eye(n)
return D

u0True = np.array([1,1,1]) # True initial conditions
np.random.seed(seed=1)
sig_m= 0.15   # standard deviation for measurement noise
R = sig_m**2*np.eye(3) #covariance matrix for measurement noise

dt_m = 0.2 #time period between observations
tm_m = 2 #maximum time for observations
nt_m = int(tm_m/dt_m) #number of observation instants

ind_m = (np.linspace(int(dt_m/dt),int(tm_m/dt),nt_m)).astype(int)
t_m = t[ind_m]

#time integration
uTrue = np.zeros([3,nt+1])
uTrue[:,0] = u0True
km = 0
w = np.zeros([3,nt_m])
for k in range(nt):
uTrue[:,k+1] = RK4(Lorenz63,uTrue[:,k],dt,sigma,beta,rho)
if (km<nt_m) and (k+1==ind_m[km]):
w[:,km] = h(uTrue[:,k+1]) + np.random.normal(0,sig_m,[3,])
km = km+1

########################### Data Assimilation ###############################
u0b = np.array([2.0,3.0,4.0])
u0a = u0b
J0 = loss(Lorenz63,h,t,ind_m,u0a,w,R,1,sigma,beta,rho)
for iter in range(1000):

#computing the gradient of cost functional with base trajectory
```

```python
dJ = Adj4dvar(Lorenz63,JLorenz63,h,Dh,t,ind_m,u0a,w,R,1,sigma,beta,rho)
#minimization direction
p = -dJ/np.linalg.norm(dJ)
#Golden method for linesearch
alpha = GoldenAlpha(p,Lorenz63,h,t,ind_m,u0a,w,R,1,sigma,beta,rho)
#update initial condition with gradient descent
u0a = u0a + alpha*p

J = loss(Lorenz63,h,t,ind_m,u0a,w,R,1,sigma,beta,rho)

if np.abs(J0-J) < 1e-2:
print('Convergence: loss function')
break
else:
J0=J
if np.linalg.norm(dJ) < 1e-4:
print('Convergence: gradient of loss function')
break

#################### Time Integration [Comparison] ########################
ub = np.zeros([3,nt+1])
ub[:,0] = u0b
ua = np.zeros([3,nt+1])
ua[:,0] = u0a
km = 0
for k in range(nt):
ub[:,k+1] = RK4(Lorenz63,ub[:,k],dt,sigma,beta,rho)
ua[:,k+1] = RK4(Lorenz63,ua[:,k],dt,sigma,beta,rho)

#%%
############################## Plotting ###################################
import matplotlib as mpl
mpl.rc('text', usetex=True)
mpl.rcParams['text.latex.preamble']=[r"\usepackage{amsmath}"]
mpl.rcParams['text.latex.preamble'] = [r'\boldmath']
font = {'family' : 'normal',
'weight' : 'bold',
'size'   : 20}
mpl.rc('font', **font)

fig, ax = plt.subplots(nrows=3,ncols=1, figsize=(10,8))
ax = ax.flat

for k in range(3):
ax[k].plot(t,uTrue[k,:], label=r'\bf{True}', linewidth = 3)
ax[k].plot(t,ub[k,:], ':', label=r'\bf{Background}', linewidth = 3)
ax[k].plot(t[ind_m],w[k,:], 'o', fillstyle='none', \
label=r'\bf{Observation}', markersize = 8, markeredgewidth = 2)
ax[k].plot(t,ua[k,:], '--', label=r'\bf{Analysis}', linewidth = 3)
ax[k].set_xlabel(r'$t$',fontsize=22)
ax[k].axvspan(0, tm_m, color='y', alpha=0.4, lw=0)

ax[0].legend(loc="center", bbox_to_anchor=(0.5,1.25),ncol =4,fontsize=15)
ax[0].set_ylabel(r'$x(t)$')
ax[1].set_ylabel(r'$y(t)$')
```

```
ax[2].set_ylabel(r'$z(t)$')
fig.subplots_adjust(hspace=0.5)
```

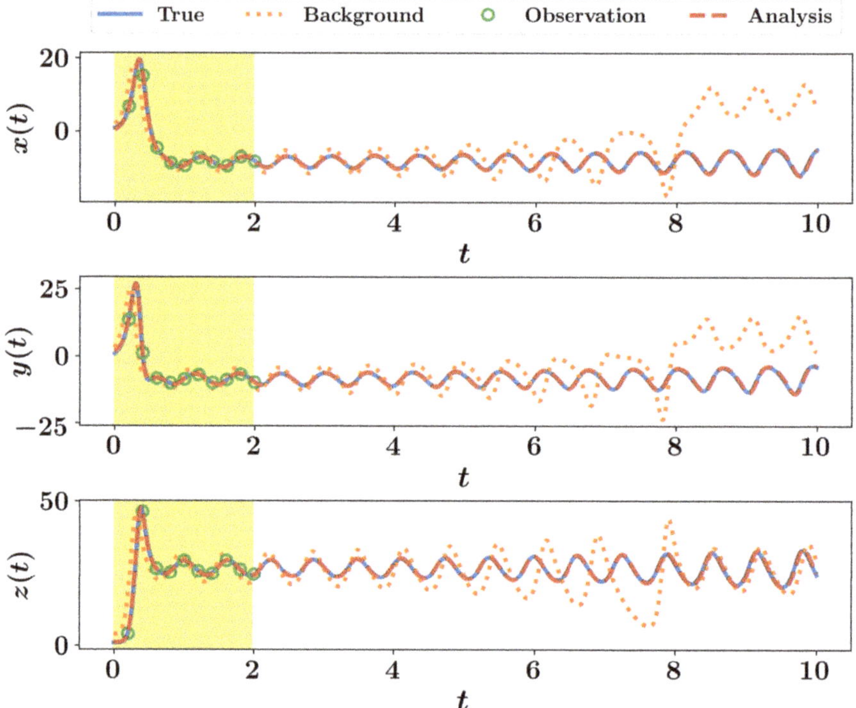

Figure 2. Results of 4DVAR implementation for the Lorenz 63 system.

Before we move to other data assimilation techniques, we highlight a few remarks regarding our presentation of the 4DVAR

- In Listing 9, we utilize the gradient descent approach to minimize the cost function. Readers are encouraged to apply other optimization techniques (e.g., conjugate gradient) that achieve higher convergence rate.
- The determination of the learning rate can be further optimized using more efficient line-search methods, rather than the simple Golden search.
- The Lagrangian multiplier method can be applied to solve the 4DVAR problem instead of the adjoint method, similar results should be obtained.
- The presented algorithm relies on the definition of the cost functional given in Equation (13), based on the discrepancy between measurements and model's predictions. When extra information is available, it can be incorporated into the cost functional. For instance, similar to Equation (4), a term that penalizes the correction magnitude can be added, weighted by the background covariance matrix. Furthermore, symmetries or other physical knowledge can be enforced as hard or weak constraints.
- The first-order adjoint algorithm requires the computation of the Jacobian $\mathbf{D}_M(\mathbf{u})$ of the discrete-time model map $M(\mathbf{u}; \theta)$. This can be computed by plugging the model $f(\mathbf{u}; \theta)$ in a time integration scheme and rearranging everything to rewrite $M(\mathbf{u}; \theta)$ as explicit function of $\mathbf{u}$ and differentiating with respect to components of $\mathbf{u}$. For Lorenz 63 and 1st Euler scheme,

this can be an easy task. However, for a higher dimensional system and more accurate time integrators, this would be cumbersome. Instead, the chain rule can be utilized to compute $\mathbf{D}_M(\mathbf{u})$ as presented in Listing 10, which takes as input the right-hand side of the continuous-time dynamics $f(\cdot;\cdot)$ (described in Listing 3 for Lorenz 63 system) as well as its Jacobian (given in Listing 11 for Lorenz 63).

Listing 10. Python functions for computing the Jacobian $\mathbf{D}_M(\mathbf{u})$ of the discrete-time model map $M(\mathbf{u};\boldsymbol{\theta})$ using the 1st Euler and the 4th Runge–Kutta schemes with chain rule.

```python
import numpy as np
def Jeuler(rhs,Jrhs,state,dt,*args):
n = len(state)
k1 = rhs(state,*args)
dk1 = Jrhs(state,*args)
DM = np.eye(n) + dt*dk1
return DM

def JRK4(rhs,Jrhs,state,dt,*args):
n = len(state)
k1 = rhs(state,*args)
k2 = rhs(state+k1*dt/2,*args)
k3 = rhs(state+k2*dt/2,*args)
dk1 = Jrhs(state,*args)
dk2 = Jrhs(state+k1*dt/2,*args) @ (np.eye(n)+dk1*dt/2)
dk3 = Jrhs(state+k2*dt/2,*args) @ (np.eye(n)+dk2*dt/2)
dk4 = Jrhs(state+k3*dt,*args) @ (np.eye(n)+dk3*dt)
DM = np.eye(n) + (dt/6) * (dk1+2*dk2+2*dk3+dk4)
return DM
```

Listing 11. A Python function for the Jacobian of the continuous-time Lorenz 63 dynamics.

```python
import numpy as np
def JLorenz63(state,*args): #Jacobian of Lorenz 96 model
sigma = args[0]
beta = args[1]
rho = args[2]
x, y, z = state #Unpack the state vector
df = np.zeros([3,3]) #Derivatives

df[0,0] = sigma * (-1)
df[0,1] = sigma * (1)
df[0,2] = sigma * (0)

df[1,0] = 1 * (rho - z)
df[1,1] = -1
df[1,2] = x * (-1)

df[2,0] = 1 * y
df[2,1] = x * 1
df[2,2] = - beta
return df
```

5. Forward Sensitivity Method

We have seen in Section 4 that the minimization of the cost functional via the 4DVAR algorithm requires the solution of the adjoint problem at each iteration, which incurs a significant computational burden for high dimensional systems. Alternatively, Lakshmivarahan and Lewis [13] proposed the forward sensitivity method (FSM) to derive an expression for the correction vector in terms of the forward sensitivity matrices [14]. In their development, simultaneous correction to the initial condition $\mathbf{u}(t_0)$ and the model parameters $\boldsymbol{\theta}$ is treated. For conciseness and consistency with the methods introduced here, we only consider erroneous initial conditions and assume model parameters are perfectly known. Given the discrete-time model map $M(\mathbf{u}; \boldsymbol{\theta})$ in Equation (2), the forecast sensitivity at time t_{k+a} to the initial conditions $\mathbf{u}(t_0)$ can be defined as follows,

$$\frac{\partial u_i(t_{k+1})}{\partial u_j(t_0)} = \sum_{q=1}^{n} \left(\frac{\partial M_i(\mathbf{u}(t_k); \boldsymbol{\theta})}{\partial u_q(t_k)} \right) \left(\frac{\partial u_q(t_k)}{\partial u_j(t_0)} \right), \qquad 1 \leqslant i, j \leqslant n, \tag{26}$$

where $M(\mathbf{u}(t_k); \boldsymbol{\theta}) = [M_1(\mathbf{u}(t_k); \boldsymbol{\theta}), M_2(\mathbf{u}(t_k); \boldsymbol{\theta}), \ldots, M_n(\mathbf{u}(t_k); \boldsymbol{\theta})]^T$. Recall that the Jacobian of the model $M(\mathbf{u}(t_k); \boldsymbol{\theta})$ is defined by the matrix $\mathbf{D}_M(\mathbf{u}(t_k)) \in \mathbb{R}^{n \times n}$ whose $(i, j)^{th}$ entry is defined as $\frac{\partial M_i(\mathbf{u}(t_k); \boldsymbol{\theta})}{\partial u_j(t_k)}$. We also define $\mathbf{U}(t_k)$ as the forward sensitivity matrix of $\mathbf{u}(t_k) \in \mathbb{R}^{n \times n}$ with respect to initial state $\mathbf{u}(t_0)$, where $[\mathbf{U}(t_k)]_{i,j} = \frac{\partial u_i(t_k)}{\partial u_j(t_0)}$ for $1 \leqslant i, j \leqslant n$. Thus, Equation (26) can be rewritten in matrix form as,

$$\mathbf{U}(t_{k+1}) = \mathbf{D}_M(\mathbf{u}(t_k))\mathbf{U}(t_k). \tag{27}$$

Equation (27) provides the dynamic evolution of the forward sensitivity matrix in a recursive manner, initialized by $\mathbf{U}(t_0) = \mathbf{I}_n$, that can be used to relate the prediction error at any time step to the initial condition.

Given the measurement $\mathbf{w}(t_k)$ at time $t_k \in \mathcal{T}$, the forecast error $\mathbf{e}(t_k)$ is defined as the difference between the model forecast and measurements as

$$\mathbf{e}(t_k) = \mathbf{w}(t_k) - \mathbf{h}(\mathbf{u}(t_k)). \tag{28}$$

This is commonly called the innovation in DA terminology. The cost functional in Equation (13) can be rewritten as

$$J(\mathbf{u}(t_0)) = \sum_{t_k \in \mathcal{T}} \frac{1}{2} \|\mathbf{e}(t_k)\|^2_{\mathbf{R}^{-1}(t_k)} = \sum_{t_k \in \mathcal{T}} \frac{1}{2} \mathbf{e}(t_k)^T \mathbf{R}^{-1}(t_k)\mathbf{e}(t_k). \tag{29}$$

With the assumption that the dynamical model is perfect (i.e., correctly encapsulates all the relevant processes) and the model parameters are known, the deterministic part of the forecast error can be attributed to the inaccuracy in the initial condition $\mathbf{u}(t_0)$, defined as $\Delta \mathbf{u}_0 = \mathbf{u}_t(t_0) - \mathbf{u}_b(t_0)$, where $\mathbf{u}_t(t_0)$ denotes the true initial conditions.

Considering a base trajectory $\overline{\mathbf{u}}(t_k)$ for $k = 1, 2, \ldots$ generated from the initial condition of $\overline{\mathbf{u}}(t_0)$, related to the corrected trajectory $(\mathbf{u}(t_k)$ for $k = 1, 2, \ldots,)$ obtained by correcting the initial condition as $\mathbf{u}(t_0) = \overline{\mathbf{u}}(t_0) + \Delta \mathbf{u}_0$, we define the difference between both trajectories at any time t_k as $\Delta \mathbf{u}_k = \mathbf{u}(t_k) - \overline{\mathbf{u}}(t_k)$. We highlight that $\mathbf{u}(t_k)$ is a function of both the initial condition (with the model parameters being known), the first-order Taylor expansion of $\mathbf{u}(t_k)$ around the base trajectory can be written as $\mathbf{u}(t_k) \approx \overline{\mathbf{u}}(t_k) + \mathbf{U}(t_k)\Delta \mathbf{u}_0$, leading to the following relation

$$\Delta \mathbf{u}_k \approx \mathbf{U}(t_k)\Delta \mathbf{u}_0. \tag{30}$$

If we let the perturbed (corrected) trajectory to be the sought true trajectory, Equation (3) can be rewritten as

$$\mathbf{w}(t_k) = h(\overline{\mathbf{u}}(t_k) + \Delta\mathbf{u}_k) + \xi_m, \tag{31}$$

and a first order expansion of $\mathbf{w}(t_k)$ (neglecting the measurement noise) will be as follows,

$$\mathbf{w}(t_k) \approx h(\overline{\mathbf{u}}(t_k)) + \mathbf{D}_h(\overline{\mathbf{u}}(t_k))\Delta\mathbf{u}_k, \tag{32}$$

and the forecast error at the base trajectory can be approximated as

$$\mathbf{e}(t_k) = \mathbf{D}_h(\overline{\mathbf{u}}(t_k))\Delta\mathbf{u}_k. \tag{33}$$

Equations (30) and (33) can be combined to yield the following,

$$\mathbf{e}(t_k) = \mathbf{D}_h(\overline{\mathbf{u}}(t_k))\mathbf{U}(t_k)\Delta\mathbf{u}_0, \tag{34}$$

which relates the forecast error at any time t_k and the discrepancy between the true and erroneous initial condition in a linear relationship. In order to account for all the time instants at which observations are available ($\mathcal{T} = \{t_{O1}, t_{O2}, \ldots, t_{ON}\}$), Equation (34) can be concatenated at different times and written as a linear system of equations as follows,

$$\mathbf{Q}\Delta\mathbf{u}_0 = \mathbf{e}_F, \tag{35}$$

where the matrix $\mathbf{Q}^{Nm \times n}$ and the vector $\mathbf{e}_F \in \mathbb{R}^{Nm}$ are computed as,

$$\mathbf{Q} = \begin{bmatrix} \mathbf{D}_h(\overline{\mathbf{u}}(t_{O1}))\mathbf{U}(t_{O1}) \\ \mathbf{D}_h(\overline{\mathbf{u}}(t_{O2}))\mathbf{U}t_{O2}) \\ \vdots \\ \mathbf{D}_h(\overline{\mathbf{u}}(t_{ON}))\mathbf{U}(t_{ON}) \end{bmatrix}, \quad \mathbf{e}_F = \begin{bmatrix} \mathbf{e}(t_{O1}) \\ \mathbf{e}(t_{O2}) \\ \vdots \\ \mathbf{e}(t_{ON}) \end{bmatrix}. \tag{36}$$

Depending on the value of Nm relative to n, Equation (35) can give rise to either an over-determined or an under-determined linear inverse problem. In either case, the inverse problem can be solved in a weighted least squares sense to find a first-order estimation of the optimal correction or perturbation to the initial condition $\Delta\mathbf{u}_0$, with $\mathbf{R}^{-1}$ being the weighting matrix, where $\mathbf{R}$ is a block-diagonal matrix constructed as follows,

$$\mathbf{R} = \begin{bmatrix} \mathbf{R}(t_{O1}) & & & \\ & \mathbf{R}(t_{O2}) & & \\ & & \ddots & \\ & & & \mathbf{R}(t_{ON}) \end{bmatrix}, \tag{37}$$

and the solution of Equation (35) can be written as $\Delta\mathbf{u}_0 = \left(\mathbf{Q}^T\mathbf{R}^{-1}\mathbf{Q}\right)^{-1}\mathbf{Q}^T\mathbf{R}^{-1}\mathbf{e}_F$ for the over-determined case. This first order approximation progressively yields better results by repeating the entire process for multiple iterations until convergence with certain tolerance [13]. In essence, the FSM is an alternative to the 4DVAR algorithm, replacing the solution of the adjoint problem backward in time (i.e., Equation (25)), by the successive matrix evaluation in Equation (27). However, we highlight that in the 4DVAR approach, the actual forecast error is computed as $\mathbf{e}_k = \mathbf{w}(t_k) - h(\overline{\mathbf{u}}(t_k))$

while its first-order approximation is utilized in the FSM development. The duality between the two approaches is further discussed in [13].

A Python implementation of the FSM approach is presented in Listing 12. We note that we solve the Equation (35) using the built-in numpy least-squares function. However, more efficient iterative schemes can be adopted in practice.

Listing 12. Python function for computing the correction vector $\Delta \mathbf{u}_0$ using the forward sensitivity method with first-order approximation.

```python
import numpy as np
from scipy.linalg import block_diag
from scipy.linalg import sqrtm

def fsm1st(rhs,Jrhs,ObsOp,JObsOp,t,ind_m,u0b,w,R,opt,*args):

# Implementation of the first-order forward sensitivity method (FSM) to
# correct the initial conditions based on the forecast sensitivity matrices
# Inputs:
#rhs: defines the right-hand side of the continuous time forward model f
#Jrhs: defines the Jacobian matrix of rhs D_f(u)
#ObsOp: defines the observation operator h(u)
#JObsOp: defines the Jacobian of the observation operator D_h(u)
#t: vector of time
#ind_m: indices of measurement instants
#u0b: initial condition for base trajectory
#w: matrix of measurements
#R: covariance matrix of measurement noise
#opt: [0=euler] or [1=RK4] defines the time integration scheme to
#comptue the discrete-time forward map and its Jacobian
# Output: the correction vector du0

n = len(u0b)
#determine the assimilation window
t = t[:ind_m[-1]+1] #cut the time till the last observation point
nt = len(t)-1
dt = t[1] - t[0]
ub = np.zeros([n,nt+1]) #base trajectory
Ri = np.linalg.inv(R)

ub[:,0] = u0b
U = np.eye(n,n) #Initialization of U
Q = np.zeros((1,n))   #Dh*U
ef = np.zeros((1,1)) #w-h(u)
W = np.zeros((1,1))   #weighting matrix
km = 0
nt_m = len(ind_m)
if opt == 0: #Euler
#forward model
for k in range(nt):
ub[:,k+1] = euler(rhs,ub[:,k],dt,*args)
DM = Jeuler(rhs,Jrhs,ub[:,k],dt,*args)
U = DM @ U
if (km<nt_m) and (k+1==ind_m[km]):
tmp = w[:,km] - ObsOp(ub[:,k+1])
ek = tmp.reshape(-1,1)
```

```python
ef = np.vstack((ef,ek))
Qk = JObsOp(ub[:,k+1]) @ U
Q = np.vstack((Q,Qk))
W = block_diag(W,Ri)
km = km + 1
elif opt == 1: #RK4
# forward model
for k in range(nt):
ub[:,k+1] = RK4(rhs,ub[:,k],dt,*args)
DM = JRK4(rhs,Jrhs,ub[:,k],dt,*args)
U = DM @ U
if (km<nt_m) and (k+1==ind_m[km]):
tmp = w[:,km] - ObsOp(ub[:,k+1])
ek = tmp.reshape(-1,1)
ef = np.vstack((ef,ek))
Qk = JObsOp(ub[:,k+1]) @ U
Q = np.vstack((Q,Qk))
W = block_diag(W,Ri)
km = km + 1
Q = np.delete(Q, (0), axis=0)
ef = np.delete(ef, (0), axis=0)
W = np.delete(W, (0), axis=0)
W = np.delete(W, (0), axis=1)

# solve weighted least-squares
W1 = sqrtm(W)
du0 = np.linalg.lstsq(W1@Q, W1@ef, rcond=None)[0]

return du0.ravel()
```

Example: Lorenz 63 System

We apply the described FSM to estimate the initial conditions for the Lorenz 63 system, using the same parameters and setup as described in Section 4. Sample code is presented in Listing 13. We highlight that instead of adding the correction vector $\Delta\mathbf{u}_0$ directly to the base value $\overline{\mathbf{u}}(t_0)$, we multiply it with a learning rate to mitigate the effects of first-order approximations. We utilize the golden search method to update this learning rate at each iteration.

Listing 13. Implementation of the FSM for the Lorenz 63 system.

```python
#%% Application: Lorenz 63
######################### Data Assimilation #############################
u0b = np.array([2.0,3.0,4.0])
u0a = u0b
J0 = loss(Lorenz63,h,t,ind_m,u0a,w,R,1,sigma,beta,rho)
for iter in range(200):

#computing the correction vector
du0 = fsm1st(Lorenz63,JLorenz63,h,Dh,t,ind_m,u0a,w,R,1,sigma,beta,rho)
#minimization direction
p = du0#/np.linalg.norm(du0)
#Golden method for linesearch
alpha = GoldenAlpha(p,Lorenz63,h,t,ind_m,u0a,w,R,1,sigma,beta,rho)
#update initial condition with gradient descent
u0a = u0a + alpha*p
```

```python
J = loss(Lorenz63,h,t,ind_m,u0a,w,R,1,sigma,beta,rho)
if np.abs(J0-J) < 1e-2:
print('Convergence: loss function')
break
#else:
J0=J
if np.linalg.norm(du0) < 1e-4:
print('Convergence: correction vector')
break


#################### Time Integration [Comparison] #########################

ub = np.zeros([3,nt+1])
ub[:,0] = u0b
ua = np.zeros([3,nt+1])
ua[:,0] = u0a
km = 0
for k in range(nt):
ub[:,k+1] = RK4(Lorenz63,ub[:,k],dt,sigma,beta,rho)
ua[:,k+1] = RK4(Lorenz63,ua[:,k],dt,sigma,beta,rho)
```

Prediction results are provided in Figure 3, where we notice a large discrepancy at the estimated initial conditions. However, the predicted trajectory perfectly match the true one for the rest of the testing time window. This is largely affected by the nature of the Lorenz system itself and the attachment to its attractor. Furthermore, this can be partially attributed to the lack of background information and its contribution to the cost functional. Moreover, this can be highly improved by adding more observations close to the initial time since the correction vector is estimated based on the forecast error computed at observation times. Anyhow, we see that the analysis trajectory is significantly more accurate than the background one, with iterative first-order approximations of the forward sensitivity method, even for long time predictions.

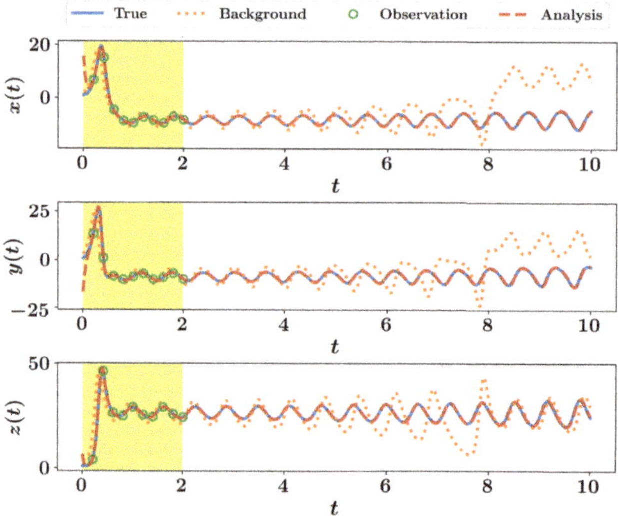

Figure 3. Results of FSM implementation for the Lorenz 63 system.

6. Kalman Filtering

The idea behind Kalman filtering techniques is to propagate the mean as well as the covariance matrix of the system's state sequentially in time. That is, in addition to providing an improved state estimate (i.e., the analysis), it also gives some information about the statistical properties of this state estimate. This is one main difference between Kalman filtering and variational methods, which often assumes a fixed (stationary) background covariance matrices. Kalman filters are also very popular in systems engineering, robotics, navigation, and control. Almost all modern control systems use the Kalman filter. It assisted the guidance of the Apollo 11 lunar module to the moon's surface, and most probably will do the same for next generations of aircraft as well.

Although most application in fluid dynamics involve nonlinear systems, we first describe the standard Kalman filter developed for the linear dynamical system case with linear observation operator described as

$$\mathbf{u}_t(t_{k+1}) = \mathbf{M}_k \mathbf{u}_t(t_k) + \boldsymbol{\xi}_p(t_{k+1}), \tag{38}$$

$$\mathbf{w}(t_k) = \mathbf{H}_k \mathbf{u}_t(t_k) + \boldsymbol{\xi}_m(t_k) \tag{39}$$

where $\mathbf{M} \in \mathbb{R}^{n \times n}$ is a non-singular system matrix defining the underlying governing processes and $\boldsymbol{\xi}_p \in \mathbb{R}^n$ describes the process noise (or model error). $\mathbf{H} \in \mathbb{R}^{m \times n}$ represents the measurement system with a measurement noise of $\boldsymbol{\xi}_m \in \mathbb{R}^m$.

As presented in Section 2, the true state $\mathbf{u}_t(t_k)$ is assumed to be a random variable with known mean $E[\mathbf{u}_t(t_k)] = \mathbf{u}_b(t_k)$ and covariance matrix of $E[(\mathbf{u}_t(t_k) - \mathbf{u}_b(t_k))(\mathbf{u}_t(t_k) - \mathbf{u}_b(t_k))^T] = \mathbf{B}_k$. In Kalman filtering, we note that the covariance matrix evolves in time, and thus appears the subscript. We also assume that the process noise is unbiased with zero mean and a covariance matrix $\mathbf{Q}$. That is $E(\boldsymbol{\xi}_p(t_k)) = 0$ and $E(\boldsymbol{\xi}_p(t_k)\boldsymbol{\xi}_p(t_k)^T) = \mathbf{Q}_k$.

Thus, the goal of the filtering problem is to find a good estimate (analysis) $\mathbf{u}_a(t_k)$ of the true system's state $\mathbf{u}_t(t_k)$ given a dynamical model a set of noisy observation $\{\mathbf{w}(t_i)\}$ collected at some time instants $t_i \in (0, t_k]$. The optimality of the estimate $\mathbf{u}_a(t_k)$ is defined as the one which minimizes $E[(\mathbf{u}_t(t_k) - \mathbf{u}_a(t_k))^T(\mathbf{u}_t(t_k) - \mathbf{u}_a(t_k))]$. This filtering process generally consists of two steps: the forecast step and the data assimilation step.

The forecast step is performed using the predictable part of the given dynamical model starting from the best known information at time t_k (denoted as $\widehat{\mathbf{u}}_b(t_k)$) to produce a forecast or background estimate $\mathbf{u}_b(t_{k+1}) = \mathbf{M}_k \widehat{\mathbf{u}}_b(t_k)$. The difference between the background forecast and true state at t_{k+1} can be written as follows,

$$\begin{aligned}
\boldsymbol{\xi}_b(t_{k+1}) &= \mathbf{u}_t(t_{k+1}) - \mathbf{u}_b(t_{k+1}) \\
&= (\mathbf{M}_k \mathbf{u}_t(t_k) + \boldsymbol{\xi}_p(t_{k+1})) - \mathbf{M}_k \widehat{\mathbf{u}}_b(t_k) \\
&= \mathbf{M}_k(\mathbf{u}_t(t_k) - \widehat{\mathbf{u}}_b(t_k)) + \boldsymbol{\xi}_p(t_{k+1}) \\
&= \mathbf{M}_k \widehat{\boldsymbol{\xi}}_b(t_k) + \boldsymbol{\xi}_p(t_{k+1}),
\end{aligned}$$

where $\widehat{\boldsymbol{\xi}}_b(t_k) = (\mathbf{u}_t(t_k) - \widehat{\mathbf{u}}_b(t_k))$ is the error estimate at t_k, with zero mean and covariance matrix of $\widehat{\mathbf{B}}_k$.

The covariance matrix of the background estimate at t_{k+1} can be evaluated as $\mathbf{B}_{k+1} = E[\boldsymbol{\xi}_b(t_{k+1})\boldsymbol{\xi}_b(t_{k+1})^T] = E\left[\left(\mathbf{M}_k \widehat{\boldsymbol{\xi}}_b(t_k) + \boldsymbol{\xi}_p(t_{k+1})\right)\left(\mathbf{M}_k \widehat{\boldsymbol{\xi}}_b(t_k) + \boldsymbol{\xi}_p(t_{k+1})\right)^T\right]$. Since, $\widehat{\boldsymbol{\xi}}_b(t_k)$ and $\boldsymbol{\xi}_p(t_{k+1})$ are assumed to be uncorrelated (i.e., $E[\widehat{\boldsymbol{\xi}}_b(t_k)\boldsymbol{\xi}_p(t_{k+1})^T] = 0$), the background covariance matrix at t_{k+1} can be computed as follows,

$$\mathbf{B}_{k+1} = \mathbf{M}_k \widehat{\mathbf{B}}_k \mathbf{M}_k^T + \mathbf{Q}_{k+1}. \tag{40}$$

Now, with the forecast step, we have a background estimate at t_{k+1} defined as $\mathbf{u}_b(t_{k+1})$ with a covariance matrix $\mathbf{B}_{k+1}$. Then, measurements $\mathbf{w}(t_{k+1})$ are collected at t_{k+1} with a linear operator $\mathbf{H}_{k+1}$ and measurement noise $\boldsymbol{\xi}_m(t_{k+1})$ with zero mean a covariance matrix of $\mathbf{R}_{k+1}$. Thus, we would like to fuse these pieces of information to create an optimal unbiased estimate (analysis) $\mathbf{u}_a(t_{k+1})$ with a covariance matrix $\mathbf{P}_{k+1}$. This can be defines as linear function of $\mathbf{u}_b(t_{k+1})$ and $\mathbf{w}(t_{k+1})$ as follows,

$$\mathbf{u}_a(t_{k+1}) = \mathbf{u}_b(t_{k+1}) + \mathbf{K}_{k+1}\left(\mathbf{w}(t_{k+1}) - \mathbf{H}_{k+1}\mathbf{u}_b(t_{k+1})\right), \tag{41}$$

where $\left(\mathbf{w}(t_{k+1}) - \mathbf{H}_{k+1}\mathbf{u}_b(t_{k+1})\right)$ is the innovation vector and $\mathbf{K} \in \mathbb{R}^{n \times m}$ is called the Kalman gain matrix. We highlight that Kalman gain matrix is defined in such a way to minimize $E[(\mathbf{u}_t(t_{k+1}) - \mathbf{u}_a(t_{k+1}))^T(\mathbf{u}_t(t_{k+1}) - \mathbf{u}_a(t_{k+1}))] = \text{tr}(\mathbf{P}_{k+1})$. This can be written as [27]

$$\mathbf{K}_{k+1} = \mathbf{B}_{k+1}\mathbf{H}_{k+1}^T\left(\mathbf{H}_{k+1}\mathbf{B}_{k+1}\mathbf{H}_{k+1}^T + \mathbf{R}_{k+1}\right)^{-1}, \tag{42}$$

resulting in an analysis covariance matrix defined as

$$\mathbf{P}_{k+1} = (\mathbf{I}_n - \mathbf{K}_{k+1}\mathbf{H}_{k+1})\mathbf{B}_{k+1}, \tag{43}$$

where $\mathbf{I}_n$ is the $n \times n$ identity matrix. The resulting analysis $\mathbf{u}_a(t_{k+1})$ is known as the best linear unbiased estimate (BLUE). We highlight that information at t_k might correspond to the analysis (i.e., $\widehat{\mathbf{u}}_b(t_k) = \mathbf{u}_a(t_k)$) obtained from the last data assimilation implementation, or just from previous forecast if no other information is available. Thus, the Kalman filtering process can be summarized as follows,

$$
\begin{aligned}
\text{Inputs:} \quad & \widehat{\mathbf{u}}_b(t_k), \widehat{\mathbf{B}}_k, \mathbf{M}_k, \mathbf{Q}_{k+1}, \mathbf{w}(t_{k+1}), \mathbf{R}_{k+1}, \mathbf{H}_{k+1} \\
\text{Forecast:} \quad & \mathbf{u}_b(t_{k+1}) = \mathbf{M}_k\widehat{\mathbf{u}}_b(t_k) \\
& \mathbf{B}_{k+1} = \mathbf{M}_k\widehat{\mathbf{B}}_k\mathbf{M}_k^T + \mathbf{Q}_{k+1} \\
\text{Kalman gain:} \quad & \mathbf{K}_{k+1} = \mathbf{B}_{k+1}\mathbf{H}_{k+1}^T\left(\mathbf{H}_{k+1}\mathbf{B}_{k+1}\mathbf{H}_{k+1}^T + \mathbf{R}_{k+1}\right)^{-1} \\
\text{Analysis:} \quad & \mathbf{u}_a(t_{k+1}) = \mathbf{u}_b(t_{k+1}) + \mathbf{K}_{k+1}\left(\mathbf{w}(t_{k+1}) - \mathbf{H}_{k+1}\mathbf{u}_b(t_{k+1})\right) \\
& \mathbf{P}_{k+1} = (\mathbf{I}_n - \mathbf{K}_{k+1}\mathbf{H}_{k+1})\mathbf{B}_{k+1},
\end{aligned}
$$

where the inputs at t_k are defined as

$$
(\widehat{\mathbf{u}}_b(t_k), \widehat{\mathbf{B}}_k) = \begin{cases} (\mathbf{u}_a(t_k), \mathbf{P}_k) & \text{if } \mathbf{w}(t_k), \text{ is available,} \\ (\mathbf{u}_b(t_k), \mathbf{B}_k) & \text{otherwise.} \end{cases}
$$

Listing 14 describes a basic Python implementation of the data assimilation step using the KF algorithm described before. Although efficient matrix inversion routines that benefit from specific matrix properties can be utilized, we use the standard built-in Numpy matrix inversion function.

Listing 14. Implementation of the KF with linear dynamics and observation operator.

```python
import numpy as np
def KF(ub,w,H,R,B):

# The analysis step for the Kalman filter in the linear case
# i.e., linear model M and linear observation operator H

n = ub.shape[0]
```

```python
# compute Kalman gain
D = H@B@H.T + R
K = B @ H @ np.linalg.inv(D)

# compute analysis
ua = ub + K @ (w-H@ub)
P = (np.eye(n) - K@H) @ B
return ua, P
```

Different forms for evaluating the Kalman gain and the covariance matrices are presented in literature. Some of them are favored for computational cost aspects, while others maintain desirable properties (e.g., symmetry and positive definiteness) for numerically stable implementation [27]. Since we are more interested in nonlinear dynamical models, we shall discuss extensions for standard Kalman filters to account for nonlinearity in the following sections.

7. Extended Kalman Filter

Instead of dealing with linear stochastic dynamics, we look at the nonlinear case with general (nonlinear) observation operator written as

$$\mathbf{u}_t(t_{k+1}) = M(\mathbf{u}_t(t_k); \boldsymbol{\theta}) + \boldsymbol{\xi}_p(t_{k+1}), \tag{44}$$

$$\mathbf{w}(t_k) = h(\mathbf{u}_t(t_k)) + \boldsymbol{\xi}_m(t_k). \tag{45}$$

The first challenge of applying Kalman filter for this system is the propagation of the background covariance matrix in the forecast step. The main clue behind the extended Kalman filter (EKF) to address this issue is to locally linearize $M(\mathbf{u}(t_k))$ by expanding it around the estimate $\widehat{\mathbf{u}}_b(t_k)$ at t_k using the first-order Taylor series as follows,

$$M(\mathbf{u}_t(t_k); \boldsymbol{\theta}) \approx M(\widehat{\mathbf{u}}_b(t_k); \boldsymbol{\theta}) + \mathbf{D}_M(\widehat{\mathbf{u}}_b(t_k))\widehat{\boldsymbol{\xi}}_b(t_k), \tag{46}$$

where $\mathbf{D}_M(\widehat{\mathbf{u}}_b(t_k))$ is the Jacobian (also known as the tangent linear operator) of the forward model $M(\cdot; \cdot)$ evaluated at $\widehat{\mathbf{u}}_b(t_k)$ and $\widehat{\boldsymbol{\xi}}_b(t_k) = (\mathbf{u}_t(t_k) - \widehat{\mathbf{u}}_b(t_k))$ defining the error estimate at t_k, with zero mean and covariance matrix of $\widehat{\mathbf{B}}_k$. Thus, the difference between the background forecast and true state at t_{k+1} can be written as follows,

$$\begin{aligned}
\boldsymbol{\xi}_b(t_{k+1}) &= \mathbf{u}_t(t_{k+1}) - \mathbf{u}_b(t_{k+1}) \\
&= M(\mathbf{u}_t(t_k); \boldsymbol{\theta}) + \boldsymbol{\xi}_p(t_{k+1}) - M(\widehat{\mathbf{u}}_b(t_k); \boldsymbol{\theta}) \\
&\approx M(\widehat{\mathbf{u}}_b(t_k); \boldsymbol{\theta}) + \mathbf{D}_M(\widehat{\mathbf{u}}_b(t_k))\widehat{\boldsymbol{\xi}}_b(t_k) + \boldsymbol{\xi}_p(t_{k+1}) - M(\widehat{\mathbf{u}}_b(t_k); \boldsymbol{\theta}) \\
&\approx \mathbf{D}_M(\widehat{\mathbf{u}}_b(t_k))\widehat{\boldsymbol{\xi}}_b(t_k) + \boldsymbol{\xi}_p(t_{k+1}).
\end{aligned}$$

Similar to the derivation in the linear case, with the assumption of uncorrelation between $\widehat{\boldsymbol{\xi}}_b(t_k)$ and $\boldsymbol{\xi}_p(t_{k+1})$, the background covariance matrix at t_{k+1} can be computed as follows,

$$\mathbf{B}_{k+1} = \mathbf{D}_M(\widehat{\mathbf{u}}_b(t_k))\widehat{\mathbf{B}}_k\mathbf{M}_k^T + \mathbf{Q}_{k+1}. \tag{47}$$

The next challenge regarding the analysis step is the computation of the Kalman gain in case of nonlinear observation operator. Again, $h(\mathbf{u}_t(t_{k+1}))$ is linearized using Talylor series expansion around $\mathbf{u}_b(t_{k+1})$ (i.e., the background forecast) as follows,

$$h(\mathbf{u}_t(t_{k+1})) \approx h(\mathbf{u}_b(t_{k+1})) + \mathbf{D}_h(\mathbf{u}_b(t_{k+1}))\boldsymbol{\xi}_b(t_{k+1}), \tag{48}$$

where $\mathbf{D}_h(\mathbf{u}_b(t_{k+1}))$ is the Jacobian of the observation operator h, computed with the forecast $\mathbf{u}_b(t_{k+1})$. The Kalman gain is thus computed using this first-order approximation of h as follows,

$$\mathbf{K}_{k+1} = \mathbf{B}_{k+1}\mathbf{D}_h(\mathbf{u}_b(t_{k+1}))^T\left(\mathbf{D}_h(\mathbf{u}_b(t_{k+1}))\mathbf{B}_{k+1}\mathbf{D}_h(\mathbf{u}_b(t_{k+1}))^T + \mathbf{R}_{k+1}\right)^{-1}, \tag{49}$$

with an analysis estimate and analysis covariance matrix defined as

$$\mathbf{u}_a(t_{k+1}) = \mathbf{u}_b(t_{k+1}) + \mathbf{K}_{k+1}\left(\mathbf{w}(t_{k+1}) - h(\mathbf{u}_b(t_{k+1}))\right), \tag{50}$$

$$\mathbf{P}_{k+1} = \left(\mathbf{I}_n - \mathbf{K}_{k+1}\mathbf{D}_h(\mathbf{u}_b(t_{k+1}))\right)\mathbf{B}_{k+1}. \tag{51}$$

A summary of the EKF algorithm is described as follows,

Inputs: $\quad \widehat{\mathbf{u}}_b(t_k), \widehat{\mathbf{B}}_k, M(\cdot;\cdot), \mathbf{Q}_{k+1}, \mathbf{w}(t_{k+1}), \mathbf{R}_{k+1}, h(\cdot)$

Forecast: $\quad \mathbf{u}_b(t_{k+1}) = M(\widehat{\mathbf{u}}_b(t_k);\boldsymbol{\theta})$

$\qquad\qquad \mathbf{B}_{k+1} = \mathbf{D}_M(\widehat{\mathbf{u}}_b(t_k))\widehat{\mathbf{B}}_k\mathbf{D}_M(\widehat{\mathbf{u}}_b(t_k))^T + \mathbf{Q}_{k+1}$

Kalman gain: $\quad \mathbf{K}_{k+1} = \mathbf{B}_{k+1}\mathbf{D}_h(\mathbf{u}_b(t_{k+1}))^T\left(\mathbf{D}_h(\mathbf{u}_b(t_{k+1}))_{k+1}\mathbf{B}_{k+1}\mathbf{D}_h(\mathbf{u}_b(t_{k+1}))^T + \mathbf{R}_{k+1}\right)^{-1}$

Analysis: $\quad \mathbf{u}_a(t_{k+1}) = \mathbf{u}_b(t_{k+1}) + \mathbf{K}_{k+1}\left(\mathbf{w}(t_{k+1}) - h(\mathbf{u}_b(t_{k+1}))\right)$

$\qquad\qquad \mathbf{P}_{k+1} = (\mathbf{I}_n - \mathbf{K}_{k+1}\mathbf{D}_h(\mathbf{u}_b(t_{k+1})))\mathbf{B}_{k+1},$

and a Python implementation of the data assimilation step is presented in Listing 15.

Listing 15. Implementation of the (first-order) EKF with nonlinear dynamics and nonlinear observation operator.

```python
import numpy as np
def EKF(ub,w,ObsOp,JObsOp,R,B):
# The analysis step for the extended Kalman filter with nonlinear dynamics
# and nonlinear observation operator
n = ub.shape[0]
# compute Jacobian of observation operator at ub
Dh = JObsOp(ub)
# compute Kalman gain
D = Dh@B@Dh.T + R
K = B @ Dh.T @ np.linalg.inv(D)

# compute analysis
ua = ub + K @ (w-ObsOp(ub))
P = (np.eye(n) - K@Dh) @ B
return ua, P
```

Example: Lorenz 63 System

The first-order approximation of the Kalman filter in nonlinear case, known as extended Kalman filter, is applied for the test case of Lorenz 63 system. The computation of model Jacobian $\mathbf{D}_M(\cdot)$ is presented in Listings 10 and 11 in Section 4. We use the same parameters and initial conditions for the twin experiment framework as before. The sequential implementation of the forecast and analysis steps is shown in Listing 16 and results are illustrated in Figure 4. We adopt the 4th order Runge–Kutta scheme for time integration. For demonstration purposes, we consider zero process noise (i.e., $\mathbf{Q} = 0$). However, we have found that assuming non-zero process noise (e.g., $\mathbf{Q} = 0.01\mathbf{I}_3$) yields better performance.

Listing 16. Implementation of the EKF for the Lorenz 63 system.

```python
#%% Application: Lorenz 63
######################## Data Assimilation ###############################
u0b = np.array([2.0,3.0,4.0])
sig_b= 0.1
B = sig_b**2*np.eye(3)
Q = 0.0*np.eye(3)
#time integration
ub = np.zeros([3,nt+1])
ub[:,0] = u0b
ua = np.zeros([3,nt+1])
ua[:,0] = u0b
km = 0
for k in range(nt):
# Forecast Step
#background trajectory [without correction]
ub[:,k+1] = RK4(Lorenz63,ub[:,k],dt,sigma,beta,rho)
#EKF trajectory [with correction at observation times]
ua[:,k+1] = RK4(Lorenz63,ua[:,k],dt,sigma,beta,rho)
#compute model Jacobian at t_k
DM = JRK4(Lorenz63,JLorenz63,ua[:,k],dt,sigma,beta,rho)
#propagate the background covariance matrix
B = DM @ B @ DM.T + Q
if (km<nt_m) and (k+1==ind_m[km]):
# Analysis Step
ua[:,k+1],B = EKF(ua[:,k+1],w[:,km],h,Dh,R,B)
km = km+1
```

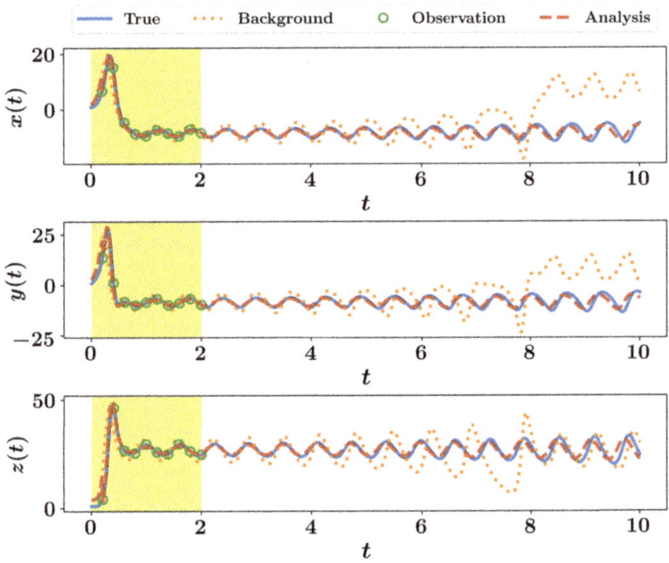

Figure 4. EKF results for the Lorenz 63 system with the assumption of zero process noise.

8. Ensemble Kalman Filter

Despite the sound mathematical and theoretical foundation of Kalman filters (both linear and nonlinear cases), they are not widely utilized in geophysical sciences. The major bottleneck in the computational pipeline of Kalman filtering is the update of background covariance matrix. In typical implementation, the cost of this step is $O(n^3)$, where n is the size of the state vector. For systems governed by ordinary differential equations (ODES), n can be manageable (e.g., 3 in the Lorenz 63 model). However, fluid flows are often governed by partial differential equations. Thus, spatial discretization schemes (e.g., finite difference, finite volume, and finite element) are applied, resulting in a semi-discrete system of ODEs. In geophysical flow dynamics applications (e.g., weather forecast), a dimension of millions or even billions is not uncommon, which hinders the feasible implementation of standard Kalman filtering techniques.

Alternatively, reduced rank algorithms that provides low-order approximation of the covariance matrices are usually adopted. A very popular approach is the ensemble Kalman filter (EnKF), introduced by Evensen [17,28,29] based on the Monte Carlo estimation methods.

The main procedure for these methods is to create an ensemble of size N of the system state denoted as $\{\mathbf{u}(t_k)^{(i)}|1 \leq i \leq N\}$ and apply the filtering algorithm to each member of the established ensemble. The statistical properties of the forecast and analysis are thus extracted from the ensemble using the standard Monte Carlo framework. In the previous discussions, the forecast (background) and analysis covariances are defined as follows,

$$\mathbf{B} = E[\boldsymbol{\xi}_b \boldsymbol{\xi}_b^T] = E[(\mathbf{u}_t - \mathbf{u}_b)(\mathbf{u}_t - \mathbf{u}_b)^T],$$
$$\mathbf{P} = E[\boldsymbol{\xi}_a \boldsymbol{\xi}_a^T] = E[(\mathbf{u}_t - \mathbf{u}_a)(\mathbf{u}_t - \mathbf{u}_a)^T].$$

Alternatively, those can be approximated by the ensemble covariances, given as

$$\mathbf{B} \approx \frac{1}{N-1} \sum_{i=1}^{N} (\mathbf{u}_b^{(i)} - \overline{\mathbf{u_b}})(\mathbf{u}_b^{(i)} - \overline{\mathbf{u_b}})^T,$$

$$\mathbf{P} \approx \frac{1}{N-1} \sum_{i=1}^{N} (\mathbf{u}_a^{(i)} - \overline{\mathbf{u_a}})(\mathbf{u}_a^{(i)} - \overline{\mathbf{u_a}})^T,$$

where the bar denotes the ensemble average defined as $\overline{\mathbf{u}} = \frac{1}{N} \sum_{i=1}^{N} \mathbf{u}^{(i)}$. Thus, an interpretation of EnKF is that the ensemble mean is the best estimate of the state and the spreading of the ensemble around the mean is a definition of the error in this estimate. A larger ensemble size N yields a better approximation of the state estimate and its covariance matrix. In the following, we describe the typical steps for applying the EnKF algorithm.

We begin by creating an initial ensemble $\{\widehat{\mathbf{u}}_b^{(i)}(t_k)|1 \leq i \leq N\}$ at time t_k drawn from the distribution $\mathcal{N}(\widehat{\mathbf{u}}_b(t_k), \widehat{\mathbf{B}}_k)$, where $\widehat{\mathbf{u}}_b(t_k)$ represents our best-known estimate at t_k. It can be verified that the ensemble mean and covariance converge to $\widehat{\mathbf{u}}_b(t_k)$ and $\mathbf{B}_k$ as $N \to \infty$. Then, the forecast step is applied to each member of the enesemble as follows,

$$\mathbf{u}_b^{(i)}(t_{k+1}) = M(\widehat{\mathbf{u}}_b^{(i)}(t_k); \boldsymbol{\theta}) + \boldsymbol{\xi}_p^{(i)}(t_{k+1}), \tag{52}$$

where $\boldsymbol{\xi}_p^{(i)}(t_{k+1})$ is drawn from the multivariate Gaussian distribution with zero mean and covariance matrix of $\mathbf{Q}_{k+1}$ representing the process noise applied to each member. The sample mean of the forecast ensemble can be thus computed, along with the corresponding covaiance matrix as,

$$\mathbf{u}_b(t_{k+1}) \approx \overline{\mathbf{u_b}}(t_{k+1}) = \frac{1}{N} \sum_{i=1}^{N} \mathbf{u}_b^{(i)}(t_{k+1}), \tag{53}$$

$$\boldsymbol{\xi}_b^{(i)}(t_{k+1}) = \mathbf{u}_b^{(i)}(t_{k+1}) - \overline{\mathbf{u_b}}(t_{k+1}), \tag{54}$$

$$\mathbf{B}_{k+1} \approx \frac{1}{N-1} \sum_{i=1}^{N} \boldsymbol{\xi}_b^{(i)}(t_{k+1}) \boldsymbol{\xi}_b^{(i)}(t_{k+1})^T, \tag{55}$$

which provides an approximation for the background covariance at t_{k+1} without actually propagating the covariance matrix, as is the case in standard Kalman filtering.

An enesmble of observations $\{\mathbf{w}^{(i)}(t_{k+1}) | 1 \leq i \leq N\}$, also called virtual observations, is created assuming a Gaussian distribution with a mean equal to the actual observation $\mathbf{w}(t_{k+1})$ and a covariance matrix $\mathbf{R}_{k+1}$. In other words, random Gaussian perturbations with zero mean and covariance matrix $\mathbf{R}_{k+1}$ are added to the actual measurements to create perturbed measurements. The Kalman gain matrix can be computed as before (repeated here for completeness),

$$\mathbf{K}_{k+1} = \mathbf{B}_{k+1} \mathbf{D}_h(\mathbf{u}_b(t_{k+1}))^T \left(\mathbf{D}_h(\mathbf{u}_b(t_{k+1})) \mathbf{B}_{k+1} \mathbf{D}_h(\mathbf{u}_b(t_{k+1}))^T + \mathbf{R}_{k+1} \right)^{-1}. \tag{56}$$

Then, the analysis step is applied for each member in the ensemble cloud as below,

$$\mathbf{u}_a^{(i)} = \mathbf{u}_b^{(i)}(t_{k+1}) + \mathbf{K}_{k+1} \left(\mathbf{w}^{(i)}(t_{k+1}) - h(\mathbf{u}_b^{(i)}(t_{k+1})) \right), \tag{57}$$

and the analyzed estimate at t_{k+1} is computed as the sample mean of the analysis ensemble along with its covariance matrix as follows,

$$\mathbf{u}_a(t_{k+1}) \approx \overline{\mathbf{u_a}}(t_{k+1}) = \frac{1}{N} \sum_{i=1}^{N} \mathbf{u}_a^{(i)}(t_{k+1}), \tag{58}$$

$$\boldsymbol{\xi}_a^{(i)}(t_{k+1}) = \mathbf{u}_a^{(i)}(t_{k+1}) - \overline{\mathbf{u_a}}(t_{k+1}), \tag{59}$$

$$\mathbf{P}_{k+1} \approx \frac{1}{N-1} \sum_{i=1}^{N} \boldsymbol{\xi}_a^{(i)}(t_{k+1}) \boldsymbol{\xi}_a^{(i)}(t_{k+1})^T. \tag{60}$$

We observe that the EnKF algorithm provides approximations of the background and analysis covariance matrices, without the need to evaluate the computationally expensive propagation equations. This comes with the expense of having to evolve an ensemble of system's states. However, the size of the ensemble N is usually smaller than the system's dimension n. Moreover, with parallelization and high performance computing (HPC) frameworks, the forecast step can be distributed efficiently and computational speed-ups can be achieved. A summary of the EnKF algorithm is described as follows,

$$
\begin{aligned}
\text{Inputs:} \quad & \widehat{\mathbf{u}}_b(t_k), \widehat{\mathbf{B}}_k, M(\cdot;\cdot), \mathbf{Q}_{k+1}, \mathbf{w}(t_{k+1}), \mathbf{R}_{k+1}, h(\cdot) \\[4pt]
\text{Ensemble initialization:} \quad & \widehat{\mathbf{u}}_b^{(i)}(t_k) = \widehat{\mathbf{u}}_b(t_k) + \mathbf{e}_b^{(i)}, \quad \mathbf{e}_b^{(i)} \sim \mathcal{N}(\mathbf{0}, \widehat{\mathbf{B}}_k) \\[4pt]
\text{Virtual observations:} \quad & \mathbf{w}^{(i)}(t_{k+1}) = \mathbf{w}^{(i)}(t_{k+1}) + \mathbf{e}_m^{(i)}, \quad \mathbf{e}_m^{(i)} \sim \mathcal{N}(\mathbf{0}, \mathbf{R}_{k+1}) \\[4pt]
\text{Forecast:} \quad & \mathbf{u}_b^{(i)}(t_{k+1}) = M(\widehat{\mathbf{u}}_b^{(i)}(t_k); \boldsymbol{\theta}) + \boldsymbol{\xi}_p^{(i)}(t_{k+1}) \\[4pt]
& \boldsymbol{\xi}_b^{(i)}(t_{k+1}) = \mathbf{u}_b^{(i)}(t_{k+1}) - \overline{\mathbf{u}_b}(t_{k+1}) \\[4pt]
& \mathbf{B}_{k+1} \approx \frac{1}{N-1} \sum_{i=1}^{N} \boldsymbol{\xi}_b^{(i)}(t_{k+1}) \boldsymbol{\xi}_b^{(i)}(t_{k+1})^T \\[8pt]
\text{Kalman gain:} \quad & \mathbf{K}_{k+1} = \mathbf{B}_{k+1} \mathbf{D}_h(\mathbf{u}_b(t_{k+1}))^T \left(\mathbf{D}_h(\mathbf{u}_b(t_{k+1}))_{k+1} \mathbf{B}_{k+1} \mathbf{D}_h(\mathbf{u}_b(t_{k+1}))^T + \mathbf{R}_{k+1} \right)^{-1} \\[8pt]
\text{Analysis:} \quad & \mathbf{u}_a^{(i)}(t_{k+1}) = \mathbf{u}_b^{(i)}(t_{k+1}) + \mathbf{K}_{k+1} \left(\mathbf{w}^{(i)}(t_{k+1}) - h(\mathbf{u}_b^{(i)}(t_{k+1})) \right) \\[4pt]
& \mathbf{u}_a(t_{k+1}) \approx \overline{\mathbf{u_a}}(t_{k+1}) \\[4pt]
& \boldsymbol{\xi}_a^{(i)}(t_{k+1}) = \mathbf{u}_a^{(i)}(t_{k+1}) - \overline{\mathbf{u_a}}(t_{k+1}) \\[4pt]
& \mathbf{P}_{k+1} \approx \frac{1}{N-1} \sum_{i=1}^{N} \boldsymbol{\xi}_a^{(i)}(t_{k+1}) \boldsymbol{\xi}_a^{(i)}(t_{k+1})^T
\end{aligned}
$$

and Listing 17 provides a Python execution of the presented EnKF approach.

Listing 17. Implementation of the EnKF with virtual observations.

```python
import numpy as np
def EnKF(ubi,w,ObsOp,JObsOp,R,B):

    # The analysis step for the (stochastic) ensemble Kalman filter
    # with virtual observations

    n,N = ubi.shape # n is the state dimension and N is the size of ensemble
    m = w.shape[0] # m is the size of measurement vector

    # compute the mean of forecast ensemble
    ub = np.mean(ubi,1)
    # compute Jacobian of observation operator at ub
    Dh = JObsOp(ub)
    # compute Kalman gain
    D = Dh@B@Dh.T + R
    K = B @ Dh @ np.linalg.inv(D)


    wi = np.zeros([m,N])
    uai = np.zeros([n,N])
    for i in range(N):
        # create virtual observations
        wi[:,i] = w + np.random.multivariate_normal(np.zeros(m), R)
        # compute analysis ensemble
        uai[:,i] = ubi[:,i] + K @ (wi[:,i]-ObsOp(ubi[:,i]))

    # compute the mean of analysis ensemble
    ua = np.mean(uai,1)
    # compute analysis error covariance matrix
    P = (1/(N-1)) * (uai - ua.reshape(-1,1)) @ (uai - ua.reshape(-1,1)).T
    return uai, P
```

We highlight a few remarks regarding the EnKF as below,

- Ensemble methods have gained significant popularity because of their simple conceptual formulation and relative ease of implementation. No optimization problem is required to be solved. They are considered non-intrusive in the sense that current solvers can be easily incorporated with minimal modification, as there is no need to derive model Jacobians or adjoint equations.
- The analysis ensemble can be used as initial ensemble for the next assimilation cycle (in which case, we need not compute $\mathbf{P}_{k+1}$). Alternatively, new ensemble can be built, by sampling from multivariate Gaussian distribution with a mean of $\mathbf{u}_a(t_{k+1})$ and covariance matrix of $\mathbf{P}_{k+1}$ (i.e., using $\mathbf{u}_a(t_{k+1})$ and $\mathbf{P}_{k+1}$ in lieu of $\widehat{\mathbf{u}}_b(t_{k+1})$ and $\widehat{\mathbf{B}}_{k+1}$, respectively).
- After virtual observations are made-up, an ensemble measurement error covariance matrix can be arbitrarily computed as an alternative to the actual one [17]. This is especially valuable when the actual measurement noise covariance matrix is poorly known.
- Perturbed observations are needed in EnKF derivation and guarantees that the posterior (analysis) covariance is not underestimated. For instance, in case of small corrections to the forecast, the traditional EnKF without virtual observations yields a error covariance that is about twice smaller than that is needed to match Kalman filter [30]. In other words, the use of virtual observations forces the ensemble posterior covariance to be the same as that of the standard Kalman filter in the limit of very large N. Thus, the same Kalman gain matrix relation is borrowed from standard Kalman filter.
- Instead of assuming virtual observations, alternative formulations of ensemble Kalman filters have been proposed in literature, giving a family of deterministic ensemble Kalman filter (DEnKF), as opposed to the aforementioned (stochastic) ensemble Kalman filer (EnKF). One such variant is briefly discussed in Section 8.1.

8.1. Deterministic Ensemble Kalman Filter

The use of an ensemble of perturbed observations in the EnKF leads to a match between the analysis error covariance and its theoretical value given by Kalman filter. However, this is in a statistical sense only when the ensemble size is large. Unfortunately, this perturbation introduces sampling error, which renders the filter suboptimal, particularly for small ensembles [31]. Alternative formulations that do not require virtual observations can be found in literature, including ensemble square root filters [31,32]. We focus here on a simple formulation proposed by Sakov and Oke [30] that maintains the numerical effectiveness and simplicity EnKF without the need to virtual observations, denoted as deterministic ensemble Kalman filter (DEnKF).

Without measurement perturbation, it can be derived that the resulting analysis error covariance matrix is given as follows,

$$\begin{aligned}
\mathbf{P}_{k+1} &= (\mathbf{I}_n - \mathbf{K}_{k+1}\mathbf{D}_h(\mathbf{u}_b(t_{k+1})))\mathbf{B}_{k+1}(\mathbf{I} - \mathbf{K}\mathbf{D}_h(\mathbf{u}_b(t_{k+1})))^T \\
&= \mathbf{B}_{k+1} - \mathbf{K}_{k+1}\mathbf{D}_h(\mathbf{u}_b(t_{k+1}))\mathbf{B}_{k+1} - \mathbf{B}_{k+1}\mathbf{D}_h(\mathbf{u}_b(t_{k+1}))^T\mathbf{K}_{k+1}^T \\
&\quad + \mathbf{K}_{k+1}\mathbf{D}_h(\mathbf{u}_b(t_{k+1}))\mathbf{B}_{k+1}\mathbf{D}_h(\mathbf{u}_b(t_{k+1}))^T\mathbf{K}_{k+1}^T.
\end{aligned}$$

With the definition of the Kalman gain, it can be seen that $\mathbf{K}_{k+1}\mathbf{D}_h(\mathbf{u}_b(t_{k+1}))\mathbf{B}_{k+1} = \mathbf{B}_{k+1}\mathbf{D}_h(\mathbf{u}_b(t_{k+1}))^T\mathbf{K}_{k+1}^T$. Thus,

$$\mathbf{P}_{k+1} = \mathbf{B}_{k+1} - 2\mathbf{K}_{k+1}\mathbf{D}_h(\mathbf{u}_b(t_{k+1}))\mathbf{B}_{k+1} + \mathbf{K}_{k+1}\mathbf{D}_h(\mathbf{u}_b(t_{k+1}))\mathbf{B}_{k+1}\mathbf{D}_h(\mathbf{u}_b(t_{k+1}))^T\mathbf{K}_{k+1}^T.$$

For small values of $\mathbf{K}_{k+1}\mathbf{D}_h(\mathbf{u}_b(t_{k+1}))$, this form converges to $\mathbf{P}_{k+1} = \mathbf{B}_{k+1} - 2\mathbf{K}_{k+1}\mathbf{D}_h(\mathbf{u}_b(t_{k+1}))\mathbf{B}_{k+1}$ up to the quadratic term. It can be seen that this asymptotically match the theoretical value of analysis covariance matrix in standard Kalman filtering (i.e., $\mathbf{P}_{k+1} = (\mathbf{I}_n - \mathbf{K}_{k+1}\mathbf{D}_h(\mathbf{u}_b(t_{k+1})))\mathbf{B}_{k+1} = \mathbf{B}_{k+1} - \mathbf{K}_{k+1}\mathbf{D}_h(\mathbf{u}_b(t_{k+1})))$ by dividing the Kalman gain by two. Therefore, it can be argued that the DEnKF linearly recovers the theoretical analysis error covariance matrix. This is achieved by applying the analysis equation separately to the forecast mean $\mathbf{u}_b(t_{k+1}) \approx \overline{\mathbf{u_b}}(t_{k+1})$ with the Kalman gain matrix and ensemble of anomalies $\boldsymbol{\zeta}_b^{(i)}(t_{k+1}) = \mathbf{u}_b^{(i)}(t_{k+1}) - \overline{\mathbf{u_b}}(t_{k+1})$ using half of the standard Kalman gain matrix. These steps are summarized as follows,

$$
\begin{aligned}
\text{Inputs:} \quad & \widehat{\mathbf{u}}_b(t_k), \widehat{\mathbf{B}}_k, M(\cdot;\cdot), \mathbf{Q}_{k+1}, \mathbf{w}(t_{k+1}), \mathbf{R}_{k+1}, h(\cdot) \\[4pt]
\text{Ensemble initialization:} \quad & \widehat{\mathbf{u}}_b^{(i)}(t_k) = \widehat{\mathbf{u}}_b(t_k) + \mathbf{e}_b^{(i)}, \quad \mathbf{e}_b^{(i)} \sim \mathcal{N}(\mathbf{0}, \widehat{\mathbf{B}}_k) \\[4pt]
\text{Forecast:} \quad & \mathbf{u}_b^{(i)}(t_{k+1}) = M(\widehat{\mathbf{u}}_b^{(i)}(t_k); \boldsymbol{\theta}) + \boldsymbol{\zeta}_p^{(i)}(t_{k+1}) \\[4pt]
& \mathbf{u}_b(t_{k+1}) \approx \overline{\mathbf{u_b}}(t_{k+1}) \\[4pt]
& \boldsymbol{\zeta}_b^{(i)}(t_{k+1}) = \mathbf{u}_b^{(i)}(t_{k+1}) - \overline{\mathbf{u_b}}(t_{k+1}) \\[4pt]
& \mathbf{B}_{k+1} \approx \frac{1}{N-1} \sum_{i=1}^{N} \boldsymbol{\zeta}_b^{(i)}(t_{k+1}) \boldsymbol{\zeta}_b^{(i)}(t_{k+1})^T \\[6pt]
\text{Kalman gain:} \quad & \mathbf{K}_{k+1} = \mathbf{B}_{k+1} \mathbf{D}_h(\mathbf{u}_b(t_{k+1}))^T \left(\mathbf{D}_h(\mathbf{u}_b(t_{k+1}))_{k+1} \mathbf{B}_{k+1} \mathbf{D}_h(\mathbf{u}_b(t_{k+1}))^T + \mathbf{R}_{k+1} \right)^{-1} \\[6pt]
\text{Analysis:} \quad & \mathbf{u}_a(t_{k+1}) = \mathbf{u}_b(t_{k+1}) + \mathbf{K}_{k+1} \left(\mathbf{w}(t_{k+1}) - h(\mathbf{u}_b(t_{k+1})) \right) \\[4pt]
& \boldsymbol{\zeta}_a^{(i)}(t_{k+1}) = \boldsymbol{\zeta}_b^{(i)}(t_{k+1}) - \frac{1}{2} \mathbf{K}_{k+1} \left(h(\boldsymbol{\zeta}_b^{(i)}(t_{k+1})) \right) \\[4pt]
& \mathbf{u}_a^{(i)}(t_{k+1}) = \mathbf{u}_a(t_{k+1}) + \boldsymbol{\zeta}_a^{(i)}(t_{k+1}) \\[4pt]
& \mathbf{P}_{k+1} \approx \frac{1}{N-1} \sum_{i=1}^{N} \boldsymbol{\zeta}_a^{(i)}(t_{k+1}) \boldsymbol{\zeta}_a^{(i)}(t_{k+1})^T
\end{aligned}
$$

and Listing 18 provides a Python execution of the presented DEnKF approach. Note that the ensemble of observations is not created in this case, compared to the EnKF.

Listing 18. Implementation of DEnKF without virtual observations.

```python
import numpy as np
def DEnKF(ubi,w,ObsOp,JObsOp,R,B):

# The analysis step for the (stochastic) ensemble Kalman filter
# with virtual observations


n,N = ubi.shape # n is the state dimension and N is the size of ensemble
m = w.shape[0] # m is the size of measurement vector

# compute the mean of forecast ensemble
ub = np.mean(ubi,1)
# compute Jacobian of observation operator at ub
Dh = JObsOp(ub)
# compute Kalman gain
D = Dh@B@Dh.T + R
K = B @ Dh @ np.linalg.inv(D)

# compute analysis of mean
ua = ub + K @ (w-ObsOp(ub))

xbi = np.zeros([n,N]) #ensemble of forecast anomalies
xai = np.zeros([n,N]) #ensemble of analysis anomalies

for i in range(N):
# forecast anomalies
```

```python
xbi[:,i] = ubi[:,i] - ub
# analysis of anomalies
xai[:,i] = xbi[:,i] - (1/2) * K @ ObsOp(xbi[:,i])

# compute analysis ensemble
uai = xai + ua.reshape(-1,1)

# compute analysis error covariance matrix
P = (1/(N-1)) * (xai) @ (xai).T
return uai, P
```

8.2. Example: Lorenz 63 System

The same Lorenz 63 system is used to showcase the performance of both the EnKF and DEnKF. In Listing 19, we show the Python application of the EnKF algorithm. In general, the size of ensemble N is much smaller than the state dimension n for the implementation of EnKF to be computationally feasible. However, the state dimension in the Lorenz 63 is 3, and an ensemble of size 3 or less is trivial. The uncertainty in the covariance approximation via the Monte Carlo framework with such small ensemble becomes very high and resulting predictions are unreliable. There exists various approaches that help to increase the fidelity of small ensembles, including localization and inflation. In Section 9.2, we describe simple application of inflation factor and its impact with small ensembles. Here, we stick with the basic implementation with an ensemble size of 10 for both EnKF and DEnKF.

Listing 19. Implementation of EnKF for the Lorenz 63 system.

```python
#%% Application: Lorenz 63
######################### Data Assimilation #############################
u0b = np.array([2.0,3.0,4.0])
sig_b= 0.1
B = sig_b**2*np.eye(3)
Q = 0.0*np.eye(3)
#time integration
ub = np.zeros([3,nt+1])
ub[:,0] = u0b
ua = np.zeros([3,nt+1])
ua[:,0] = u0b
n = 3 #state dimension
m = 3 #measurement dimension
# ensemble size
N = 10
#initialize ensemble
uai = np.zeros([3,N])
for i in range(N):
uai[:,i] = u0b + np.random.multivariate_normal(np.zeros(n), B)

km = 0
for k in range(nt):
# Forecast Step
#background trajectory [without correction]
ub[:,k+1] = RK4(Lorenz63,ub[:,k],dt,sigma,beta,rho)
#EnKF trajectory [with correction at observation times]
for i in range(N): # forecast ensemble
uai[:,i] = RK4(Lorenz63,uai[:,i],dt,sigma,beta,rho) \
+ np.random.multivariate_normal(np.zeros(n), Q)
# compute the mean of forecast ensemble
```

```python
ua[:,k+1] = np.mean(uai,1)
# compute forecast error covariance matrix
B = (1/(N-1))*(uai-ua[:,k+1].reshape(-1,1))@(uai-ua[:,k+1].reshape(-1,1)).T
if (km<nt_m) and (k+1==ind_m[km]):
# Analysis Step
uai,B = EnKF(uai,w[:,km],h,Dh,R,B)
# compute the mean of analysis ensemble
ua[:,k+1] = np.mean(uai,1)
km = km+1
```

Figure 5 shows the EnKF results for the Lorenz 63 system. We see that the analysis trajectory is close to the true one and more accurate than the background. Readers are encouraged to play with the codes to explore the effect of increasing or decreasing the ensemble size with different levels of noise. Furthermore, different observation operators can be defined (for instance, observe only 1 or 2 variables, or assume some nonlinear function $h(\cdot)$).

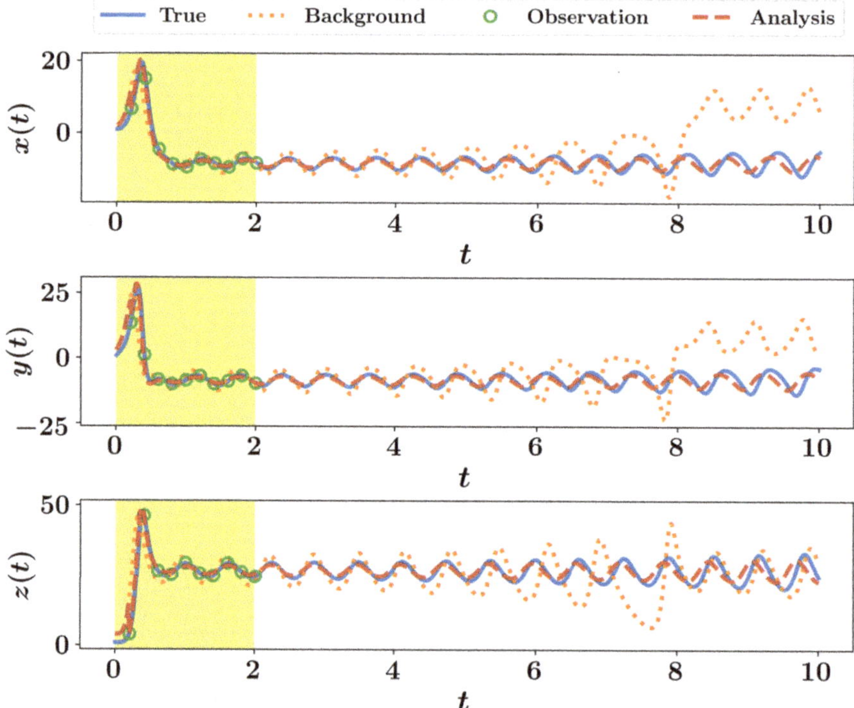

Figure 5. EnKF results for the Lorenz 63 system with virtual observations.

For the sake of completeness, we also sketch the DEnKF predictions in Figure 6. The same implementation in Listing 19 can be adopted, but calling the DEnKF function framed in Listing 18 instead of EnKF. Although different approaches might give slightly dissimilar results, we are not trying to benchmark them in this introductory presentation since we are only showing very simple implementation, with idealized twin experiments.

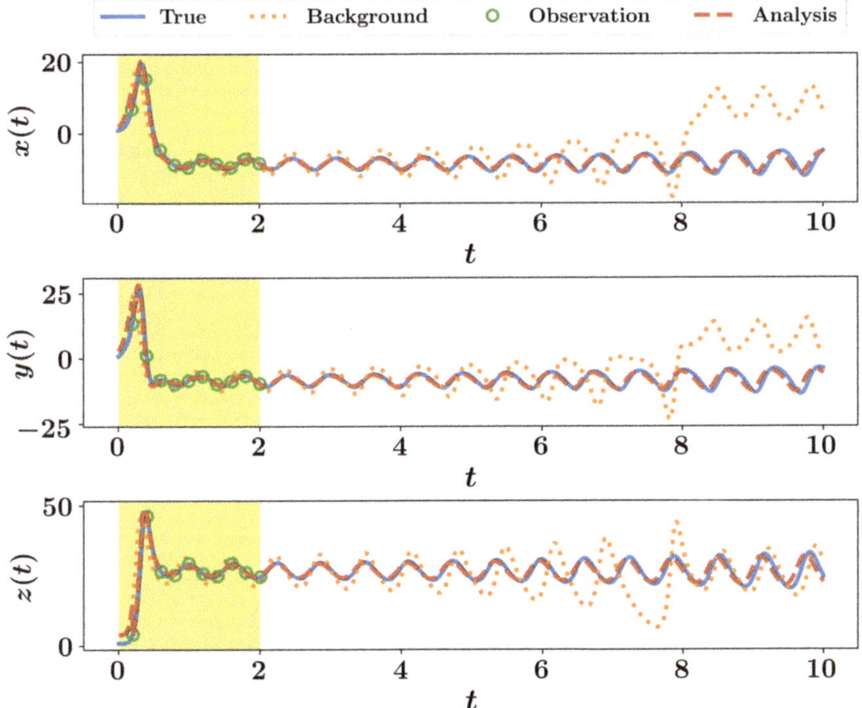

Figure 6. DEnKF results for the Lorenz 63 system without virtual observations.

9. Applications

In this section, we gradually increase the complexity of the test cases using fluid dynamics applications. In Section 9.1, we slightly increase the dimensionality of the system from 3 (as in Lorenz 63 system) to 36 using the Lorenz 96 system and demonstrate the capability of DA algorithms to treat uncertainty in initial conditions. This is further extended in Section 9.2, where we show that DA can recover the hidden underlying processes and provide closure effects using the two-level variant of the Lorenz 96 model. We also introduce the utilization of an inflation factor and its impact to mitigate ensemble collapse and account for a slight under-representation of covariance due to the use of a small ensemble in EnKF and DEnKF. In Section 9.3, we illustrate the application of DA on systems governed by partial differential equations (PDEs) using the Kuramoto–Sivashinsky equation. We highlight that for each application, we only show results for a few selected algorithms and extensions to other approaches covered in this tutorial are left to readers as computer projects. We also emphasize that we are demonstrating the implementation and capabilities of the presented DA algorithms, not assessing their performance nor benchmarking different approaches against each other.

9.1. Lorenz 96 System

The Lorenz 96 model [33] is a system of ordinary differential equations that describes an arbitrary atmospheric quantity as it evolves on a circular array of sites, undergoing forcing, dissipation, and rotation invariant advection [34]. The Lorenz 96 dynamical model can be written as

$$\frac{\mathrm{d}X_i}{\mathrm{d}t} = (X_{i+1} - X_{i-2})X_{i-1} - X_i + F, \quad i = 1, 2, \ldots, n,$$ (61)

where X_i is the state of the system at the ith location and F represents a forcing constant. Periodicity is enforced by assuming that $X_{-1} = X_{n-1}$, $X_0 = X_n$, and $X_{n+1} = X_1$. In the present study, we use $n = 36$, and $F = 8$ defining a forcing term. In order to obtain a valid initial condition, we begin at $t = -5$ using equilibrium conditions defined as ($X_i = F$ for $i = 1, 2, \ldots, n$) and adding a small perturbation to the 20th state variable as $X_{20} = F + 0.01$. Then, ODE integrator is run up to $t = 0$ and solution at $t = 0$ is treated at the true initial conditions for our twin experiment. We assume a background erroneous initial condition by contaminating the true one with Gaussian noise with zero mean and standard deviation of 1.

A total time window of 20 time units is considered, with a time step of $\Delta t = 0.01$ and the RK4 schemes is adopted for time integration. Synthetic measurements are collected at every 0.2 time unit (i.e., each 20 time integration steps) sampled at 9 equidistant locations (i.e., at $i \in \{4, 8, 12, \ldots, 36\}$) from true trajectory assuming that sensors add a white noise with zero mean and a standard deviation of 0.1. We also assume a process noise drawn from a multivariate Gaussian distribution with zero mean and covariance matrix $\mathbf{Q}$ defined as $\mathbf{Q} = 0.1^2 \mathbf{I}_{36}$. We first apply the EKF approach to correct the solution trajectory, which yields very good results as shown in Figure 7. For visualization, we only plot the time evolution of X_9, X_{18}, and X_{36}. We see that the Lorenz 96 is sensitive to the initial conditions and small perturbation is sufficient to produce a very different trajectory (e.g., background solution).

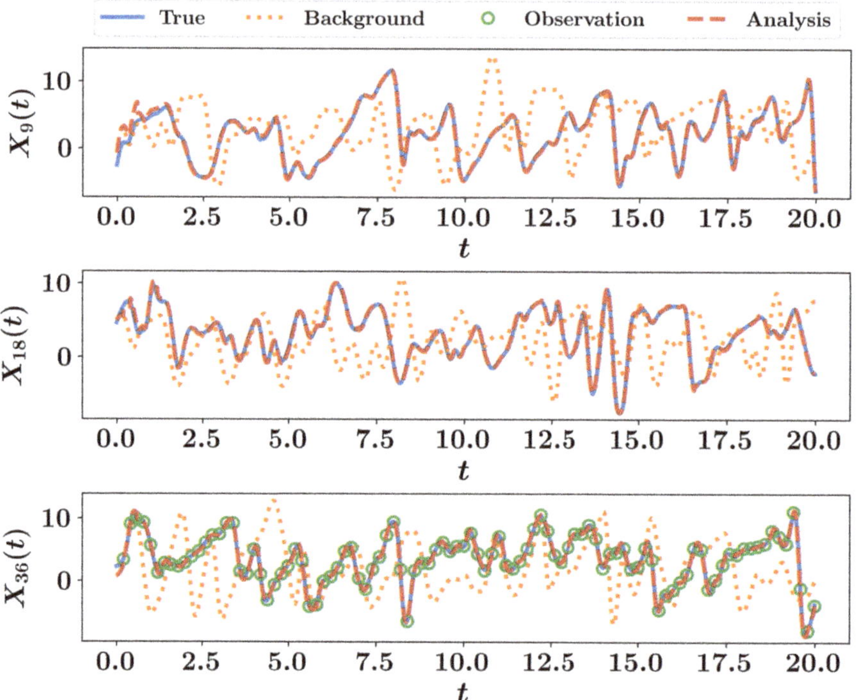

Figure 7. EKF results for the Lorenz 96 system. The trajectories of X_9, X_{18}, and X_{36} are shown.

The second approach to test is the stochastic version of EnKF. We create an ensemble of 50 members to approximate the covariance matrices. Results are depicted in Figure 8 for X_9, X_{18}, and X_{36}. We highlight that observations appear only in the X_{36} plot because observations are collected at $i = 4, 8, 12, \ldots, 36$ (neither X_9 nor X_{18} are measured). We highlight that, generally speaking, increasing the size of ensemble improves the predictions. However, this comes on the expense of the solution of the forward nonlinear model for each added member. Thus, a compromise between the accuracy and computational burden is in place.

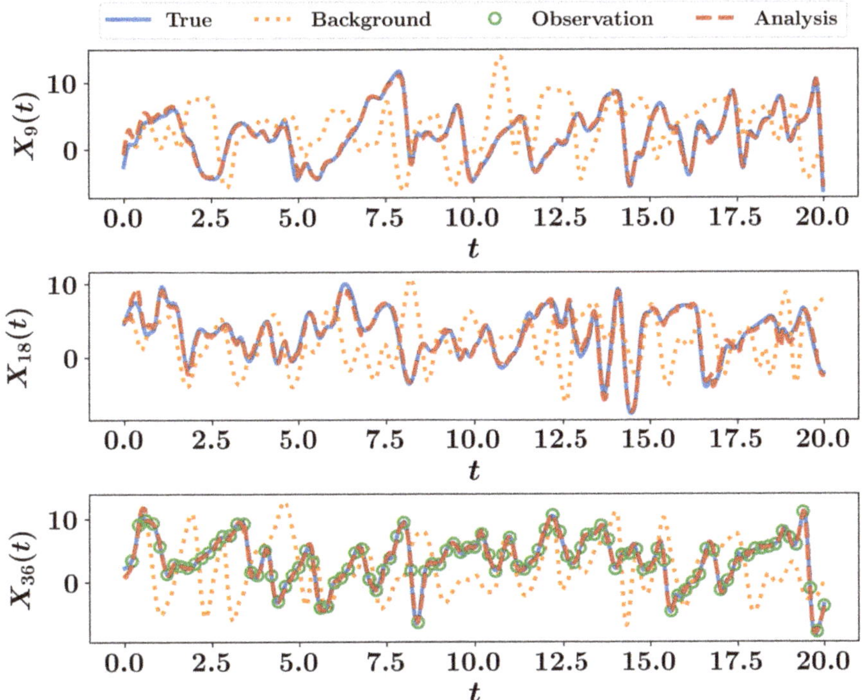

Figure 8. EnKF results for the Lorenz 96 system using an ensemble of 50 members.

9.2. Two-Level Lorenz 96 System

In this section, we describe the two-level variant of the Lorenz 96 model proposed by Lorenz [33]. The two-level Lorenz 96 model can be written as

$$\frac{dX_i}{dt} = -X_{i-1}(X_{i-2} - X_{i+1}) - X_i - \frac{hc}{b}\sum_{j=1}^{J} Y_{j,i} + F, \tag{62}$$

$$\frac{dY_{j,i}}{dt} = -cbY_{j+1,i}(Y_{j+2,i} - Y_{j-1,i}) - cY_{j,i} + \frac{hc}{b}X_i, \tag{63}$$

where Equation (62) represents the evolution of slow, high-amplitude variables X_i ($i = 1,\ldots,n$), and Equation (63) provides the evolution of a coupled fast, low-amplitude variable $Y_{j,i}$ ($j = 1,\ldots,J$). We use $n = 36$ and $J = 10$ in our computational experiments. We utilize $c = 10$ and $b = 10$, which implies that the small scales fluctuate 10 times faster than the larger scales. Furthermore, the coupling coefficient h between two scales is equal to 1 and the forcing is set at $F = 10$ to make both variables exhibit the chaotic behavior.

We utilize the fourth-order Runge–Kutta numerical scheme with a time step $\Delta t = 0.001$ for temporal integration of the Lorenz 96 model. We apply the periodic boundary condition for the slow variables, i.e., $X_{i-n} = X_{i+n} = X_i$. The fast variables are extended by letting $Y_{j,i-n} = Y_{j,i+n} = Y_{j,i}$, $Y_{j-J,i} = Y_{j,i-1}$, and $Y_{j+J,i} = Y_{j,i+1}$. The physical initial condition is computed by starting with an equilibrium condition at time $t = -5$ for slow variables. The equilibrium condition for slow variables is $X_i = F$ for $i \in 1,2,\ldots,n$. We perturb the equilibrium solution for the 18th state variable as

$X_{18} = F + 0.01$. At the time $t = -5$, the fast variables are assigned with random numbers between $-F/10$ to $F/10$. We integrate a two-level Lorenz 96 model by solving both Equations (62) and (63) in a coupled manner up to time $t = 0$. The solution at time $t = 0$ represent the true initial condition. For our twin experiment, we obtain observations by adding noise drawn from the Gaussian distribution with zero mean and $\sigma_o^2 = 1.0$. We assume that observations are sparse in space and are collected at every 10th time step.

The motivation behind this example is to demonstrate how covariance inflation can be utilized to account for the model error. Usually the imperfections in the forecast model is taken into account by adding a Gaussian noise to the forecast model. Another method to account for model error is covarinace inflation. It also helps in alleviating the effect finite number of ensemble members in practical data assimilation and addresses the problem of covariance underestimation in the EnKF algorithm. We use the multiplicative inflation [35] where the ensemble members are pushed away from the ensemble mean by a given inflation factor and mathematically it can be expressed as

$$\mathbf{u}_b^{(i)}(t_{k+1}) \leftarrow \mathbf{u}_a(t_{k+1}) + \lambda \cdot (\mathbf{u}_b^{(i)}(t_{k+1}) - \mathbf{u}_a^{(i)}(t_{k+1})), \tag{64}$$

where λ is the inflation factor. The inflation factor can be a constant scalar over the entire domain at all time step or it can space and time dependent.

In this example, we discard parameterizations of fast variables in the forecast model. The forecast model for two-level Lorenz system with no parameterizations is equivalent to setting the coupling coefficient $h = 0$ in Equation (62) and it reduces to one-level Lorenz 96 model as presented in Section 9.1. We note here that the observations used for data assimilation are obtained by solving a two-level Lorenz 96 model in a coupled manner (i.e., without discarding fast-variables). Therefore, the effect of unresolved scales is embedded in observations. The parameterization of fast variables (i.e., $\frac{hc}{b} \sum_{j=1}^{J} Y_{j,i}$ term in Equation (62)) can be considered as an added noise to the true state of the system for a one-level Lorenz 96 model. Figure 9 displays the RMSE for a two-level Lorenz system when 18 observations are used for DA with different number of ensemble members and inflation factors for EnKF and DEnKF algorithms. Figure 10 shows the full state trajectory of two-level Lorenz system corresponding to minimum RMSE, which is obtained with 50 ensemble members and inflation factor $\lambda = 1.04$ for the EnKF algorithm. The parameters corresponding to minimum RMSE for the DEnKF algorithm are 45 ensemble members and $\lambda = 1.05$.

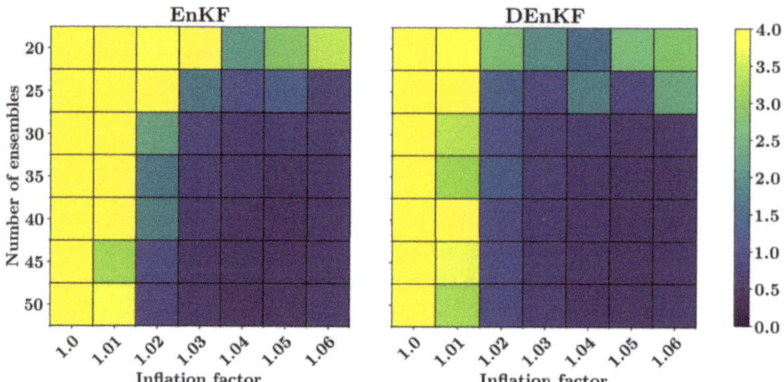

Figure 9. RMSE for a two-level Lorenz model for different combinations of number of ensembles and inflation factor.

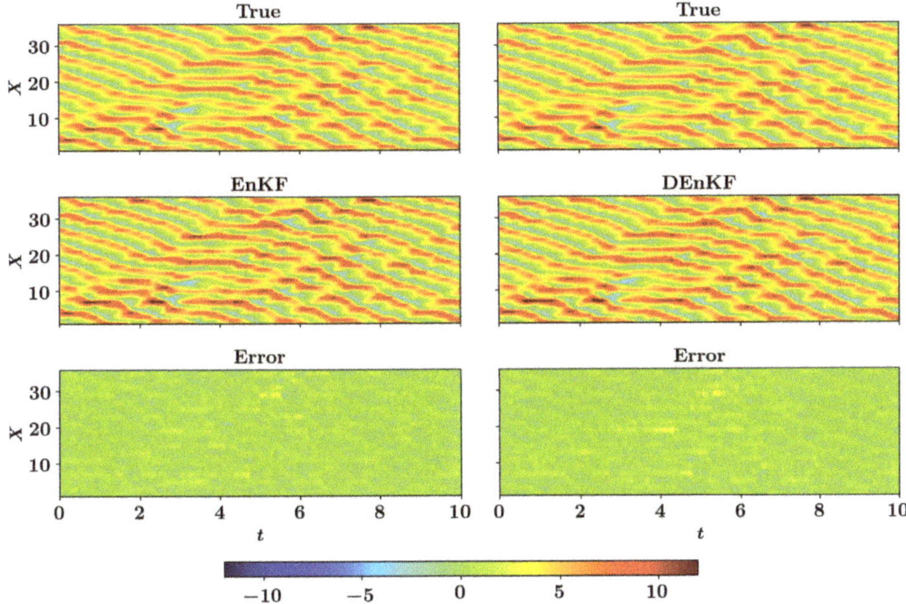

Figure 10. Full state trajectory of the multiscale Lorenz 96 model with no parameterizations in the forecast model. The EnKF algorithm uses the inflation factor $\lambda = 1.04$ and $N = 50$ and the DEnKF uses the inflation factor $\lambda = 1.05$ and $N = 45$. The observation data for both EnKF an DEnKF algorithm is obtained by adding measurement noise to the exact solution of the two-level Lorenz 96 system.

9.3. Kuramato Sivashinsky

In this section, we describe the Kuramoto–Sivashinsky (K-S) equation derived by Kuramoto [36], which is used as a turbulence model for different flows. The one-dimensional K-S equation can be written as

$$\frac{\partial u}{\partial t} = -\nu \frac{\partial^4 u}{\partial x^4} - \frac{\partial^2 u}{\partial x^2} - u \frac{\partial u}{\partial x}, \tag{65}$$

where ν is the viscosity coefficient. The K-S equation is characterized by the second-order unstable diffusion term responsible for an instability at large scales, the fourth-order stabilizing viscous term that provides damping at small scales, and a quadratic nonlinear term which transfers energy between large and small scales. We use the computational domain extending from 0 to L, i.e., $x \in [0, L]$ and time $t \in [0, \infty]$. We impose the Dirichlet and Neumann boundary conditions as given below

$$u(0, t) = u(L, t) = 0, \tag{66}$$

$$\frac{\partial u}{\partial x}\Big|_{x=0} = \frac{\partial u}{\partial x}\Big|_{x=L} = 0. \tag{67}$$

We spatially discretize the domain with the grid size $\Delta x = L/(n-1)$, where n is the degrees of freedom. We set $L = 50$ and $n = 129$ for our numerical experiments. The state of the system at discretized grid is denoted as $u_i = u((i-1)\Delta x)$ for $i = 1, \ldots, n$. Using the second-order finite difference discretization, the discretized K-S equation can be written as

$$\frac{du_i}{dt} = -\nu \frac{u_{i+2} - 4u_{i+1} + 6u_i - 4u_{i-1} + u_{i-2}}{\Delta x^4} - \frac{u_{i+1} - 2u_i + u_{i-1}}{\Delta x^2} - \frac{1}{2}\frac{u_{i+1}^2 - u_{i-1}^2}{2\Delta x}. \tag{68}$$

The first term on the right hand side is computed by utilizing ghost nodes and the Neumann boundary condition is assigned for ghost points. We impose $u_0 = u_2$ and $u_{n+1} = u_{n-1}$ using the second-order discretization for Equation (67) at boundary points u_1 and u_n, respectively.

We use the fourth-order Runge–Kutta scheme for time integration with a time step $\Delta t = 0.25$. To generate an initial condition for the forward run we start with an equilibrium condition at time $t = -50$ and integrate up to time $t = 0$. The equilibrium condition for the model is $u_i = 0.1$ for $j \in \{1, \ldots, n\}$. Once the true initial condition is generated, we run the forward solver up to time $t = 50$. We test the prediction capability of sequential data assimilation algorithms for forecast up to $t = 50$. The K-S equation exhibits different levels of chaotic behavior depending on the value of viscosity coefficient v. The chaos depends upon the bifurcation parameter $\tilde{L} = L/2\pi\sqrt{v}$. We utilize $v = 1/2$ which represent the less chaotic behaviour. The observations for twin experiments are obtained by adding some noise to the true state of the system to account for experimental uncertainties and measurement errors. The observations are also sparse in time, meaning that the time interval between two observations can be different from the time step of the forecast model. For our twin experiments, we assume that observations are recorded at every 10th time step of the model for $v = 1/2$. Therefore, the time difference between two observations is $\delta t = 2.5$ for the K-S equation. We present results for the EnKF algorithm with three sets of observations. The first set of observations is very sparse with only 12.5% of the full state of the system. The first set utilizes observations for states $[u_8, u_{16}, \ldots, u_{128}] \in R^{16}$. In a second set of observations we employ observations at $[u_4, u_8, \ldots, u_{128}] \in R^{32}$ for the assimilation. The third set of observations consists of 50% of the full state of the system, i.e., observations at states $[u_2, u_4, \ldots, u_{128}] \in R^{64}$ for the assimilation. We apply $\sigma_o^2 = 1.0 \times 10^{-2}$ and $\sigma_i^2 = 1.0 \times 10^{-2}$ as the variance of observation noise and initial condition uncertainty, respectively.

In Figure 11, we present the time evolution of selected states for three different number of observations included in the assimilation of the EnKF algorithm. There is an excellent agreement between true and assimilated states u_{51} and u_{101}, for which observations are not present. We also provide the full state trajectory of the K-S equation in Figure 12. The results obtained clearly indicate that the EnKF algorithm is able to determine the correct state trajectory even when the observation data are very sparse, i.e., $m = 16$. With an increase in the number of observations, the prediction of the full state trajectory gets smoother, and almost the exact state is recovered with 50% observations.

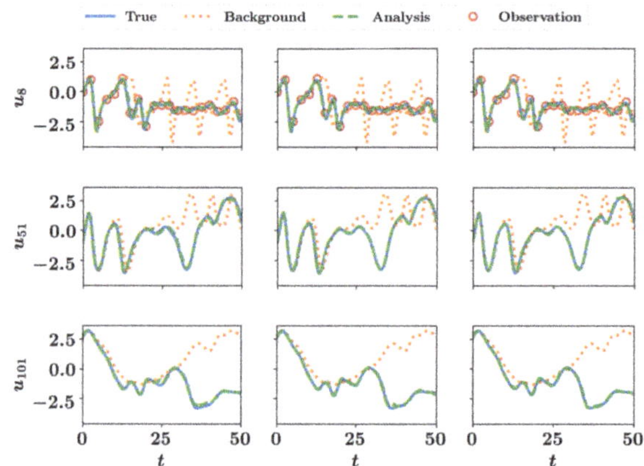

Figure 11. Selected trajectories of the Kuramoto–Sivashinsky model ($v = 1/2$) with the analysis performed by the ensemble Kalman filter (EnKF) using observations from $m = 16$ (**left**), $m = 32$ (**middle**), and $m = 64$ (**right**) state variables at every 10 time steps.

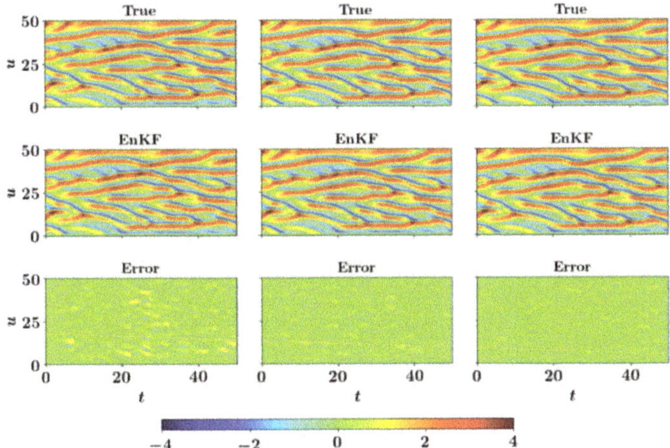

Figure 12. Full state trajectory of the Kuramoto–Sivashinsky model ($\nu = 1/2$) with the analysis performed by the ensemble Kalman filter (EnKF) using observations from $m = 16$ (**left**), $m = 32$ (**middle**), and $m = 64$ (**right**) state variables at every 10 time steps.

9.4. Quasi-Geostrophic (QG) Ocean Circulation Model

We consider a simple single-layer QG model to illustrate the application of sequential data assimilation for two-dimensional flows. Specifically, we use the deterministic ensemble Kalman filter (DenKF) algorithm discussed in Section 8.1 to improve the prediction of the single-layer QG model. The wind-driven oceanic flows exhibit a vast range of spatio-temporal scales and modeling of these scales with all the relevant physics has always been challenging. The barotropic vorticity equation (BVE) with various dissipative and forcing terms is one of the most commonly used models for geostrophic flows [37,38]. The dimensionless vorticity-streamfunction formulation for the BVE [39] with forcing and dissipative terms can be written as

$$\frac{\partial \omega}{\partial t} + J(\omega, \psi) - \frac{1}{Ro}\frac{\partial \psi}{\partial x} = \frac{1}{Re}\nabla^2 \omega + \frac{1}{Ro}\sin(\pi y), \tag{69}$$

where ω is the vorticity, ψ is the streamfunction, ∇^2 is the standard two-dimensional Laplacian operator, Re is the Reynolds number, and Ro is the Rossby number. The kinematic relation between vorticity and streamfunction is given by the following Poisson equation

$$\nabla^2 \psi = -\omega. \tag{70}$$

The nonlinear convection term is given by the Jacobian as follows

$$J(\omega, \psi) = \frac{\partial \psi}{\partial y}\frac{\partial \omega}{\partial x} - \frac{\partial \psi}{\partial x}\frac{\partial \omega}{\partial y}. \tag{71}$$

The computational domain for the QG model is $(x, y) \in [0, 1] \times [-1, 1]$ and is discretized using 128×256 grid resolution. Therefore, the QG model has the dimension of about 3.2×10^4. We utilize the homogeneous Dirichlet boundary condition for the vorticity and streamfunction at all boundaries. The vorticity and streamfunction is initialized from quiescent state, i.e., $\omega_{t=0} = \psi|_{t=0} = 0$. The QG model is numerically solved by discretizing Equation (69) using second-order finite difference scheme. The nonlinear Jacobian term is discretized with the energy-conserving Arakawa [40] numerical scheme. A third-order total-variation-diminishing Runge–Kutta scheme is used for the temporal integration and a fast sine transform

Poisson solver is utilized to update streamfunction from the vorticity [41]. For the physical parameters, we use values of Re $= 100$ and Ro $= 1.75 \times 10^{-3}$.

The QG model is integrated with a constant time step of 5×10^{-5} from time $t = 0$ to $t = 0.25$ to generate the true initial condition at final time $t = 0.25$. Then the data assimilation is conducted from time $t = 0.25$ to $t = 0.4$ with observations getting assimilated at every tenth time step. The synthetic observations are generated by sampling vorticity field on 16 equidistant points in x and y directions respectively and then adding the Gaussian noise, i.e., $\mathbf{v}_k \sim \mathcal{N}(0, \mathbf{R}_k)$, where $\mathbf{R}_k = \sigma_b^2 \mathbf{I}$. We set the observation noise variance at $\sigma_b^2 = 5$. The typical vorticity and streamfunction field along with the locations of measurements are shown in Figure 13.

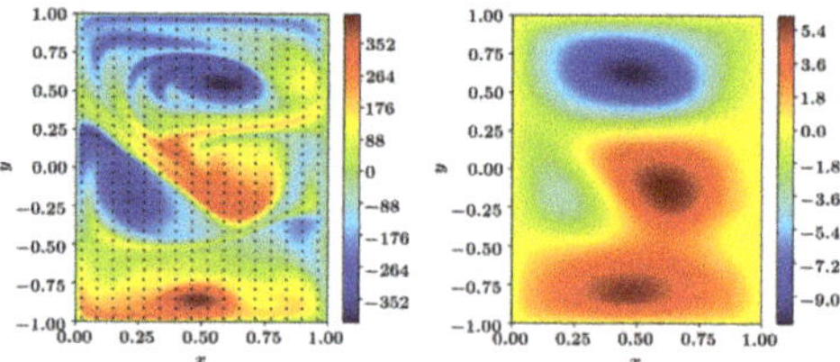

Figure 13. A Typical vorticity (left) and streamfunction (right) field for the single-layer QG model. The dots shows the locations of observations.

We employ 20 ensemble members for the DEnKF algorithm. The initialization of the ensemble members is an important step to get accurate prediction with any type of the EnKF algorithm. We initialize different ensemble members by randomly selecting the vorticity field snapshots between time $t = 0.24$ to $t = 0.25$. The other methods such as adding a random perturbation from the Gaussian distribution to the true initial condition can also be adopted. Figure 14 displays the vorticity field and the predicted vorticity field at three different time instances along with the difference (error) between the two. We can see that the true and analysis field are similar at all time instances and the magnitude of error is also small. We recall that we observe only around 2% of the system (i.e., observations at 16×32 locations). With more observations, the quality of the results can be further improved.

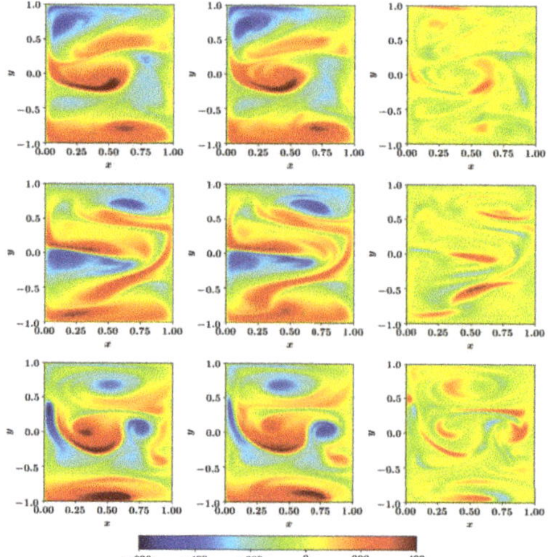

Figure 14. Snapshots of the true vorticity field (**left**), analysis estimate of the DEnKF algorithm (**middle**), and the difference (error) between the two fields (**right**) obtained for a particular run of the single-layer QG model. The snapshots of vorticity field are plotted at $t = 0.3$, 0.35, 0.4 (from top to bottom).

10. Concluding Remarks

In this tutorial paper, we provided a 101 introduction to common data assimilation techniques. In particular, we briefly covered the relevant mathematical foundation and the algorithmic steps for three dimensional variational (3DVAR), four dimensional variational (4DVAR), forward sensitivity method (FSM), and Kalman filtering approaches. Since it is considered as a first exposure to DA, we focused on the simplest implementations that anybody can easily follow. For example, to treat nonlinearity (e.g., in 3DVAR, 4DVAR, FSM and EKF), we only presented the first order Taylor expansions. We demonstrated the execution of the covered approaches with a series of Python modules that can be linked to each other easily. Again, we preferred to keep our codes as concise and simple as possible, even if it comes on the expense of computational efficiency. The Python codes used to generate this tutorial are publicly available through our GitHub repository https://github.com/Shady-Ahmed/PyDA.

Since it is introductory exploration, we should admit that we have bypassed a few important analyses and shortcut some key derivations. Interested readers are referred to well-established textbooks that offer in-depth discussions about various DA techniques [14,27,29,42–47]. Likewise, more advanced topics such as the particle filters [48], maximum likelihood ensemble filters [49,50], optimal sensor placement [51,52], higher-order analysis of variational methods [53], or hybrid methods [54–59] are omitted in our current presentation.

Author Contributions: Data curation, S.E.A., S.P., and O.S.; Supervision, O.S.; Writing—original draft, S.E.A., and S.P.; and Writing—review and editing, S.E.A., S.P., and O.S. All authors have read and agreed to the published version of the manuscript.

Funding: This material is based upon work supported by the U.S. Department of Energy, Office of Science, Office of Advanced Scientific Computing Research under Award Number DE-SC0019290. Omer San gratefully acknowledges their support. Disclaimer: This report was prepared as an account of work sponsored by an agency of the United States Government. Neither the United States Government nor any agency thereof, nor any of their employees, makes any warranty, express or implied, or assumes any legal liability or responsibility for the accuracy, completeness, or usefulness of any information, apparatus, product, or process disclosed, or represents that its use would not infringe privately owned rights. Reference herein to any specific commercial product, process, or service by trade name, trademark, manufacturer, or otherwise does not necessarily constitute or imply its endorsement, recommendation, or favoring by the United States Government or any agency thereof. The views and opinions of authors expressed herein do not necessarily state or reflect those of the United States Government or any agency thereof.

Acknowledgments: We thank Sivaramakrishnan Lakshmivarahan for his insightful comments as well as his archival NPTEL lectures [60] on dynamic data assimilation that greatly helped us in developing the PyDA module. Special thanks go to Ionel Michael Navon for providing his lecture notes on data assimilation.

Conflicts of Interest: The authors declare no conflict of interest.

References

1. Navon, I.M. Data assimilation for numerical weather prediction: A review. In *Data Assimilation for Atmospheric, Oceanic and Hydrologic Applications*; Springer: Berlin/Heidelberg, Germany, 2009; pp. 21–65.

2. Blum, J.; Le Dimet, F.X.; Navon, I.M. Data assimilation for geophysical fluids. *Handb. Numer. Anal.* **2009**, *14*, 385–441.

3. Le Dimet, F.X.; Navon, I.M.; Ştefănescu, R. Variational data assimilation: Optimization and optimal control. In *Data Assimilation for Atmospheric, Oceanic and Hydrologic Applications (Vol. III)*; Springer: Berlin/Heidelberg, Germany, 2017; pp. 1–53.

4. Attia, A.; Sandu, A. DATeS: A highly extensible data assimilation testing suite v1.0. *Geosci. Model Dev.* **2019**, *12*, 629–649. [CrossRef]

5. Lorenc, A.C. Analysis methods for numerical weather prediction. *Q. J. R. Meteorol. Soc.* **1986**, *112*, 1177–1194. [CrossRef]

6. Parrish, D.F.; Derber, J.C. The National Meteorological Center's spectral statistical-interpolation analysis system. *Mon. Weather Rev.* **1992**, *120*, 1747–1763. [CrossRef]

7. Courtier, P. Dual formulation of four-dimensional variational assimilation. *Q. J. R. Meteorol. Soc.* **1997**, *123*, 2449–2461. [CrossRef]

8. Rabier, F.; Järvinen, H.; Klinker, E.; Mahfouf, J.F.; Simmons, A. The ECMWF operational implementation of four-dimensional variational assimilation. I: Experimental results with simplified physics. *Q. J. R. Meteorol. Soc.* **2000**, *126*, 1143–1170. [CrossRef]

9. Elbern, H.; Schmidt, H.; Talagrand, O.; Ebel, A. 4D-variational data assimilation with an adjoint air quality model for emission analysis. *Environ. Model. Softw.* **2000**, *15*, 539–548. [CrossRef]

10. Courtier, P.; Thépaut, J.N.; Hollingsworth, A. A strategy for operational implementation of 4D-Var, using an incremental approach. *Q. J. R. Meteorol. Soc.* **1994**, *120*, 1367–1387. [CrossRef]

11. Lorenc, A.C.; Rawlins, F. Why does 4D-Var beat 3D-Var? *Q. J. R. Meteorol. Soc.* **2005**, *131*, 3247–3257. [CrossRef]

12. Gauthier, P.; Tanguay, M.; Laroche, S.; Pellerin, S.; Morneau, J. Extension of 3DVAR to 4DVAR: Implementation of 4DVAR at the Meteorological Service of Canada. *Mon. Weather Rev.* **2007**, *135*, 2339–2354. [CrossRef]

13. Lakshmivarahan, S.; Lewis, J.M. Forward sensitivity approach to dynamic data assimilation. *Adv. Meteorol.* **2010**, *2010*, 375615. [CrossRef]

14. Lakshmivarahan, S.; Lewis, J.M.; Jabrzemski, R. *Forecast Error Correction Using Dynamic Data Assimilation*; Springer: Cham, Switzerland, 2017.

15. Houtekamer, P.L.; Mitchell, H.L. Data assimilation using an ensemble Kalman filter technique. *Mon. Weather Rev.* **1998**, *126*, 796–811. [CrossRef]

16. Burgers, G.; Jan van Leeuwen, P.; Evensen, G. Analysis scheme in the ensemble Kalman filter. *Mon. Weather Rev.* **1998**, *126*, 1719–1724. [CrossRef]

17. Evensen, G. The ensemble Kalman filter: Theoretical formulation and practical implementation. *Ocean. Dyn.* **2003**, *53*, 343–367. [CrossRef]

18. Houtekamer, P.L.; Mitchell, H.L. A sequential ensemble Kalman filter for atmospheric data assimilation. *Mon. Weather Rev.* **2001**, *129*, 123–137. [CrossRef]

19. Houtekamer, P.L.; Mitchell, H.L. Ensemble kalman filtering. *Q. J. R. Meteorol. Soc.* **2005**, *131*, 3269–3289. [CrossRef]

20. Treebushny, D.; Madsen, H. A new reduced rank square root Kalman filter for data assimilation in mathematical models. In Proceedings of the International Conference on Computational Science, Melbourne, Australia, 2 June 2003; Springer: Berlin/Heidelberg, Germany, 2003; pp. 482–491.

21. Buehner, M.; Malanotte-Rizzoli, P. Reduced-rank Kalman filters applied to an idealized model of the wind-driven ocean circulation. *J. Geophys. Res. Ocean.* **2003**, *108*. [CrossRef]

22. Lakshmivarahan, S.; Stensrud, D.J. Ensemble Kalman filter. *IEEE Control. Syst. Mag.* **2009**, *29*, 34–46.

23. Apte, A.; Hairer, M.; Stuart, A.; Voss, J. Sampling the posterior: An approach to non-Gaussian data assimilation. *Phys. Nonlinear Phenom.* **2007**, *230*, 50–64. [CrossRef]

24. Bocquet, M.; Pires, C.A.; Wu, L. Beyond Gaussian statistical modeling in geophysical data assimilation. *Mon. Weather Rev.* **2010**, *138*, 2997–3023. [CrossRef]

25. Vetra-Carvalho, S.; Van Leeuwen, P.J.; Nerger, L.; Barth, A.; Altaf, M.U.; Brasseur, P.; Kirchgessner, P.; Beckers, J.M. State-of-the-art stochastic data assimilation methods for high-dimensional non-Gaussian problems. *Tellus Dyn. Meteorol. Oceanogr.* **2018**, *70*, 1–43. [CrossRef]

26. Attia, A.; Moosavi, A.; Sandu, A. Cluster sampling filters for non-Gaussian data assimilation. *Atmosphere* **2018**, *9*, 213. [CrossRef]

27. Lewis, J.M.; Lakshmivarahan, S.; Dhall, S. *Dynamic Data Assimilation: A Least Squares Approach*; Cambridge University Press: Cambridge, UK, 2006; Volume 104.

28. Evensen, G. Sequential data assimilation with a nonlinear quasi-geostrophic model using Monte Carlo methods to forecast error statistics. *J. Geophys. Res. Ocean.* **1994**, *99*, 10143–10162. [CrossRef]

29. Evensen, G. *Data Assimilation: The Ensemble Kalman Filter*; Springer: Berlin/Heidelberg, Germany, 2009.

30. Sakov, P.; Oke, P.R. A deterministic formulation of the ensemble Kalman filter: An alternative to ensemble square root filters. *Tellus Dyn. Meteorol. Oceanogr.* **2008**, *60*, 361–371. [CrossRef]

31. Whitaker, J.S.; Hamill, T.M. Ensemble data assimilation without perturbed observations. *Mon. Weather Rev.* **2002**, *130*, 1913–1924. [CrossRef]

32. Tippett, M.K.; Anderson, J.L.; Bishop, C.H.; Hamill, T.M.; Whitaker, J.S. Ensemble square root filters. *Mon. Weather Rev.* **2003**, *131*, 1485–1490. [CrossRef]

33. Lorenz, E.N. Predictability: A problem partly solved. In Proceedings of the Seminar on Predictability, Reading, UK, 9–11 September 1996; Volume 1.

34. Kerin, J.; Engler, H. On the Lorenz'96 Model and Some Generalizations. *arXiv* **2020**, arXiv:2005.07767.

35. Anderson, J.L.; Anderson, S.L. A Monte Carlo implementation of the nonlinear filtering problem to produce ensemble assimilations and forecasts. *Mon. Weather Rev.* **1999**, *127*, 2741–2758. [CrossRef]

36. Kuramoto, Y. Diffusion-induced chaos in reaction systems. *Prog. Theor. Phys. Suppl.* **1978**, *64*, 346–367. [CrossRef]

37. Majda, A.; Wang, X. *Nonlinear Dynamics and Statistical Theories for basic Geophysical Flows*; Cambridge University Press: New York, NY, USA, 2006.

38. Greatbatch, R.J.; Nadiga, B.T. Four-gyre circulation in a barotropic model with double-gyre wind forcing. *J. Phys. Oceanogr.* **2000**, *30*, 1461–1471. [CrossRef]

39. San, O.; Staples, A.E.; Wang, Z.; Iliescu, T. Approximate deconvolution large eddy simulation of a barotropic ocean circulation model. *Ocean. Model.* **2011**, *40*, 120–132. [CrossRef]

40. Arakawa, A. Computational design for long-term numerical integration of the equations of fluid motion: Two-dimensional incompressible flow. Part I. *J. Comput. Phys.* **1997**, *135*, 103–114. [CrossRef]

41. Press, W.H.; Flannery, B.P.; Teukolsky, S.A.; Vetterling, W.T. *Numerical Recipes*; Cambridge University Press: New York, NY, USA, 1989.

42. Cacuci, D.G.; Navon, I.M.; Ionescu-Bujor, M. *Computational Methods for Data Evaluation and Assimilation*; CRC Press: New York, NY, USA, 2013.

43. Kalnay, E. *Atmospheric Modeling, Data Assimilation and Predictability*; Cambridge University Press: New York, NY, USA, 2003.

44. Law, K.; Stuart, A.; Zygalakis, K. *Data Assimilation: A Mathematical Introduction*; Springer: Cham, Switzerland, 2015.

45. Asch, M.; Bocquet, M.; Nodet, M. *Data Assimilation: Methods, Algorithms, and Applications*; SIAM: Philadelphia, PA, USA, 2016.

46. Simon, D. *Optimal State Estimation: Kalman, H Infinity, and Nonlinear Approaches*; John Wiley & Sons: Hoboken, NJ, USA, 2006.

47. Labbe, R. Kalman and bayesian filters in Python. *Chap* **2014**, *7*, 246.

48. Van Leeuwen, P.J.; Künsch, H.R.; Nerger, L.; Potthast, R.; Reich, S. Particle filters for high-dimensional geoscience applications: A review. *Q. J. R. Meteorol. Soc.* **2019**, *145*, 2335–2365. [CrossRef] [PubMed]

49. Zupanski, M. Maximum likelihood ensemble filter: Theoretical aspects. *Mon. Weather Rev.* **2005**, *133*, 1710–1726. [CrossRef]

50. Zupanski, M.; Navon, I.M.; Zupanski, D. The Maximum Likelihood Ensemble Filter as a non-differentiable minimization algorithm. *Q. J. R. Meteorol. Soc.* **2008**, *134*, 1039–1050. [CrossRef]

51. Kang, W.; Xu, L. Optimal placement of mobile sensors for data assimilations. *Tellus Dyn. Meteorol. Oceanogr.* **2012**, *64*, 17133. [CrossRef]

52. Mons, V.; Chassaing, J.C.; Sagaut, P. Optimal sensor placement for variational data assimilation of unsteady flows past a rotationally oscillating cylinder. *J. Fluid Mech.* **2017**, *823*, 230–277. [CrossRef]

53. Le Dimet, F.X.; Navon, I.M.; Daescu, D.N. Second-order information in data assimilation. *Mon. Weather Rev.* **2002**, *130*, 629–648. [CrossRef]

54. Lorenc, A.C.; Bowler, N.E.; Clayton, A.M.; Pring, S.R.; Fairbairn, D. Comparison of hybrid-4DEnVar and hybrid-4DVar data assimilation methods for global NWP. *Mon. Weather Rev.* **2015**, *143*, 212–229. [CrossRef]

55. Desroziers, G.; Camino, J.T.; Berre, L. 4DEnVar: Link with 4D state formulation of variational assimilation and different possible implementations. *Q. J. R. Meteorol. Soc.* **2014**, *140*, 2097–2110. [CrossRef]

56. Wang, X.; Barker, D.M.; Snyder, C.; Hamill, T.M. A hybrid ETKF–3DVAR data assimilation scheme for the WRF model. Part I: Observing system simulation experiment. *Mon. Weather Rev.* **2008**, *136*, 5116–5131. [CrossRef]

57. Buehner, M.; Morneau, J.; Charette, C. Four-dimensional ensemble-variational data assimilation for global deterministic weather prediction. *Nonlinear Process. Geophys.* **2013**, *20*, 669–682. [CrossRef]

58. Kleist, D.T.; Ide, K. An OSSE-based evaluation of hybrid variational-ensemble data assimilation for the NCEP GFS. Part I: System description and 3D-hybrid results. *Mon. Weather Rev.* **2015**, *143*, 433–451. [CrossRef]

59. Kleist, D.T.; Ide, K. An OSSE-based evaluation of hybrid variational-ensemble data assimilation for the NCEP GFS. Part II: 4DEnVar and hybrid variants. *Mon. Weather Rev.* **2015**, *143*, 452–470. [CrossRef]
60. Lakshmivarahan, S. *Video Lectures on Dynamic Data Assimilation*; NPTEL Program; IIT Madras: Chennai, India, 2016. Available online: https://nptel.ac.in/courses/111/106/111106082/ (accessed on November 28, 2020).

Publisher's Note: MDPI stays neutral with regard to jurisdictional claims in published maps and institutional affiliations.

fluids

Reduced Order Models for the Quasi-Geostrophic Equations: A Brief Survey

Changhong Mou [1], Zhu Wang [2], David R. Wells [3], Xuping Xie [4] and Traian Iliescu [1,*]

[1] Department of Mathematics, Virginia Tech, Blacksburg, VA 24061, USA; cmou@vt.edu
[2] Department of Mathematics, University of South Carolina, Columbia, SC 29208, USA; wangzhu@math.sc.edu
[3] Department of Mathematics, University of North Carolina, Chapel Hill, NC 27516, USA; drwells@email.unc.edu
[4] Courant Institute of Mathematical Sciences, New York University, New York, NY 10012, USA; xxie@nyu.edu
*   Correspondence: iliescu@vt.edu

Abstract: Reduced order models (ROMs) are computational models whose dimension is significantly lower than those obtained through classical numerical discretizations (e.g., finite element, finite difference, finite volume, or spectral methods). Thus, ROMs have been used to accelerate numerical simulations of many query problems, e.g., uncertainty quantification, control, and shape optimization. Projection-based ROMs have been particularly successful in the numerical simulation of fluid flows. In this brief survey, we summarize some recent ROM developments for the quasi-geostrophic equations (QGE) (also known as the barotropic vorticity equations), which are a simplified model for geophysical flows in which rotation plays a central role, such as wind-driven ocean circulation in mid-latitude ocean basins. Since the QGE represent a practical compromise between efficient numerical simulations of ocean flows and accurate representations of large scale ocean dynamics, these equations have often been used in the testing of new numerical methods for ocean flows. ROMs have also been tested on the QGE for various settings in order to understand their potential in efficient numerical simulations of ocean flows. In this paper, we survey the ROMs developed for the QGE in order to understand their potential in efficient numerical simulations of more complex ocean flows: We explain how classical numerical methods for the QGE are used to generate the ROM basis functions, we outline the main steps in the construction of projection-based ROMs (with a particular focus on the under-resolved regime, when the closure problem needs to be addressed), we illustrate the ROMs in the numerical simulation of the QGE for various settings, and we present several potential future research avenues in the ROM exploration of the QGE and more complex models of geophysical flows.

Keywords: reduced order models; quasi-geostrophic equations; closure models

Citation: Mou, C.; Wang, Z.; Wells, D.R.; Xie, X.; Iliescu, T. Reduced Order Models for the Quasi-Geostrophic Equations: A Brief Survey. *Fluids* **2021**, *6*, 16. https://doi.org/10.3390/fluids6010016

Received: 01 December 2020
Accepted: 24 December 2020
Published: 31 December 2020

1. Introduction

1.1. Reduced Order Models (ROMs)

Reduced order modeling aims at answering the following question:

$$\boxed{\text{For a given system, what is the model with the minimum number of degrees of freedom?}} \tag{1}$$

The resulting models, called reduced order models (ROMs), can decrease the computational cost of traditional full order models (FOMs) (i.e., models obtained through classical numerical discretizations, such as finite element, finite difference, finite volume, or spectral methods) by orders of magnitude without a significant decrease in numerical accuracy. Thus, ROMs can be used in the efficient numerical simulation of problems that require numerous runs, e.g., uncertainty quantification, control, and shape optimization.

ROMs come in different flavors. Projection ROMs have been used in the numerical simulation of both nonlinear [1–3] and linear [4] systems. In particular, projection ROMs

169

have been successful in the numerical simulation of complex fluid flows [2,5–7]. In this survey, we exclusively consider projection ROMs that answer question (1) as follows:

> To construct the ROM, use numerical or experimental data to find the "best" basis.   (2)

Once the "best" basis is found, the ROM is constructed by using projection methods. In Galerkin projection ROMs, the trial and test spaces are the same; in Petrov-Galerkin projection ROMs, the trial and test spaces are different. In this paper, we focus on Galerkin projection ROMs.

Specifically, to approximate the dynamics of a flow variable u of a given system

$$\dot{u} = f(u), \tag{3}$$

the ROM strategy proceeds as follows:

Algorithm 1 ROM Strategy

1: Use numerical or experimental data to choose modes $\{\varphi_1, \ldots, \varphi_R\}$, which represent the recurrent spatial structures in the flow.

2: Choose the dominant modes $\{\varphi_1, \ldots, \varphi_r\}$, $r \leq R$, as basis functions for the ROM.

3: Use a Galerkin truncation $u_r(x, t) = \sum_{j=1}^{r} a_j(t)\, \varphi_j(x)$.

4: Replace u with u_r in (3).

5: Use a Galerkin projection of the PDE obtained in step (4) onto the ROM space $\mathbf{X}^r := \text{span}\{\varphi_1, \ldots, \varphi_r\}$ to obtain the ROM:

$$\dot{a} = F(a), \tag{4}$$

where $a(t) = (a_i(t))_{i=1,\ldots,r}$ is the vector of coefficients in the Galerkin truncation in step (3) and F comprises the ROM operators.

6: In an offline stage, compute the ROM operators (e.g., vectors, matrices, and tensors), which are preassembled from the ROM basis.

7: In an online stage, repeatedly use the ROM (4) for various parameter settings and/or longer time intervals.

At this point, several remarks are in place.

ROMs are Galerkin methods with a data-driven basis: First, we note that the general form of projection ROMs (outlined in Algorithm 1) is strikingly similar to the general form of classical Galerkin methods used in a finite element, spectral, or spectral element context. Conceptually, the main difference between ROMs and classical Galerkin discretizations is the way the basis is constructed: In classical Galerkin methods, the basis is universal, i.e., it is the same for all the problems. For example, for finite elements, the basis functions are piecewise polynomial functions on a given mesh. In projection ROMs, however, the basis is a *data-driven basis*, i.e., a basis constructed from problem data. Thus, the ROM basis is adapted to the specific problem (see steps (1)–(2) in Algorithm 1): Once the problem changes, the ROM basis changes accordingly.

While the choice of basis is the main conceptual difference between ROMs and classical Galerkin methods, this choice can make a tremendous difference in the computational cost: For example, for a two-dimensional flow past a circular cylinder at a Reynolds number $Re = 1000$, a finite element discretization requires $\mathcal{O}(10^5)$ degrees of freedom, whereas a

ROM requires $\mathcal{O}(10)$ degrees of freedom [8,9]. Thus, for this particular test case, *the ROM dimension is four orders of magnitude lower than the FOM dimension.*

Recurrent, Dominant, Coherent Spatial Structures: ROMs do not work well for all problems. ROMs are numerical methods and, like any other numerical method, ROMs work well for certain classes of problems and not so well for other classes of problems. One class of problems for which ROMs have been particularly successful is flows that display *recurrent, dominant, coherent structures.* A classical example in this class is the two-dimensional flow past a circular cylinder, which has become the workhorse of ROMs for fluid flows [5,6,9]. The flow past a circular cylinder displays coherent spatial structures (the von Karman vortex street) that continuously recur in time. One can show that a few such structures have significantly higher kinetic energy content than the remaining structures, and, therefore, are expected to dominate the dynamics of the underlying system. Indeed, as mentioned above, for the two-dimensional flow past a circular cylinder, the dimension of the ROM constructed with these dominating structures can be four orders of magnitude lower than the FOM dimension. Thus, for this test problem, ROMs work extremely well. Other problems that display recurrent, dominant, coherent structures, for which ROMs work well, include: (i) lid driven cavity flow [10]; (ii) flow past a backward facing step [11]; (iii) flow in a constrained channel [12,13]; and (iv) flow in the boundary layer of a pipe [2].

We emphasize, however, that there are classes of problems for which ROMs do not work well. Homogeneous flows are one such example. Indeed, for homogeneous flows, it was proved in [2,14] that one of the most popular ROM techniques yields a ROM basis that is identical to the Fourier basis. Thus, in this case, the resulting ROM is nothing but a spectral method, which does not reduce the FOM dimension.

The take-home message is that *ROMs are appropriate for problems that display recurrent, coherent, dominant spatial structures. However, for problems that do not display these types of spatial structures (e.g., homogeneous flows), ROMs are not appropriate since they cannot reduce the FOM computational cost.*

1.2. ROMs for the Quasi-Geostrophic Equations

ROMs are an excellent fit for the numerical investigation of ocean flows. Indeed, large-scale ocean circulation includes large-scale coherent structures (gyres) that recur in time and permanent gyres (e.g., the Sargasso Sea) that have a relatively high kinetic energy content. Thus, as pointed out above, ROMs could enable an efficient and relatively accurate numerical simulation of large scale ocean circulation, decreasing the FOM computational cost by orders of magnitude and making possible efficient ensemble calculation and uncertainty quantification for climate modeling and weather prediction.

However, generating FOM data to build the ROM basis can be a daunting task. Specifically, using an accurate mathematical model (e.g., the Boussinesq equations), including all the relevant flow variables, and using realistic parameters, could require enormous computational resources on state-of-the-art computational platforms, both in terms of CPU time and memory. Thus, various *simplified mathematical models* for the large scale ocean circulation have been proposed over the years [15–17]. These simplified models are constructed by using *asymptotic expansions* with respect to both the time scales and the length scales. The *rotation* and *stratification* are the two main effects that are used to construct simplified models for geophysical flows.

One of the most popular simplified models for large scale ocean circulation is the *quasi-geostrophic equations (QGE)* (also known as the barotropic vorticity equations), which were proposed in the late 1940s by Jule Charney [18]. The QGE are a simplified model for geophysical flows in which rotation plays a central role, such as wind-driven ocean circulation in mid-latitude ocean basins. Specifically, the QGE ensure a near-geostrophic balance, i.e., the pressure gradient almost balances the Coriolis force (which is due to rotation). The one-layer QGE do not include stratification effects, but the N-layer or continuously stratified QGE model stratification.

The computational cost of numerical simulations of large scale ocean flows is significantly lower for the QGE than for the full-fledged Boussinesq equations. Since the QGE represent a practical compromise between the efficient numerical simulations of ocean flows and the accurate representation of large scale ocean dynamics, these equations have been often used in the testing of new numerical methods for ocean flows. Thus, to understand the potential of using ROMs for the efficient numerical simulation of ocean flows, ROMs have been tested on the QGE for various parameter settings. Of course, once the ROMs are calibrated for the simplified (yet relevant) setting of the QGE, they should be extended to more realistic mathematical models, such as the Boussinesq equations. In this brief survey, we summarize some of the ROM developments for the QGE.

The rest of the paper is organized as follows: In Section 2, we present and discuss the QGE. In Section 3, we summarize the main types of numerical discretizations used to generate the FOM data for the ROM construction. In Section 4, we present the Galerkin ROM approach for the QGE. In Section 5, we illustrate numerically the QGE reduced order modeling for one test case. Finally, in Section 6, we present conclusions and outline open problems in the reduced order modeling of the QGE.

2. Quasi-Geostrophic Equations (QGE)

The QGE describe the motion of stratified, rotating flows, and have been used extensively for modeling mid-latitude oceanic and atmospheric circulations. In 1950, a single-layer quasi-geostrophic model was used for modeling the atmospheric dynamics in the first successful numerical weather prediction performed on the ENIAC digital computer [18], which led to "enormous scientific advance", in Richardson's words [15,19,20]. Since then, the QGE have been widely investigated and applied in weather prediction and climate modeling.

The QGE can be derived from the primitive equations, that is, the incompressible Navier–Stokes equations under the Boussinesq approximation in a rotating framework [21–24]. The equations in Cartesian coordinates on a plane Ω tangent to the sphere read:

$$\frac{Du}{Dt} - f_c v = -\frac{1}{\rho}\frac{\partial p}{\partial x} + \frac{\partial}{\partial x}\left(\mathcal{A}\frac{\partial u}{\partial x}\right) + \frac{\partial}{\partial y}\left(\mathcal{A}\frac{\partial u}{\partial y}\right) + \frac{\partial}{\partial z}\left(\nu_E\frac{\partial u}{\partial z}\right), \tag{5}$$

$$\frac{Dv}{Dt} + f_c u = -\frac{1}{\rho}\frac{\partial p}{\partial y} + \frac{\partial}{\partial x}\left(\mathcal{A}\frac{\partial v}{\partial x}\right) + \frac{\partial}{\partial y}\left(\mathcal{A}\frac{\partial v}{\partial y}\right) + \frac{\partial}{\partial z}\left(\nu_E\frac{\partial v}{\partial z}\right), \tag{6}$$

$$0 = -\frac{1}{\rho}\frac{\partial p}{\partial z} - g, \tag{7}$$

$$0 = \frac{\partial u}{\partial x} + \frac{\partial v}{\partial y} + \frac{\partial w}{\partial z}, \tag{8}$$

$$\frac{D\rho}{Dt} = \frac{\partial}{\partial x}\left(\mathcal{A}\frac{\partial \rho}{\partial x}\right) + \frac{\partial}{\partial y}\left(\mathcal{A}\frac{\partial \rho}{\partial y}\right) + \frac{\partial}{\partial z}\left(\kappa_E\frac{\partial \rho}{\partial z}\right), \tag{9}$$

where u, v, and w are velocity components in the x, y, and z directions, $\frac{D}{Dt} = \frac{\partial}{\partial t} + u\frac{\partial}{\partial x} + v\frac{\partial}{\partial y} + w\frac{\partial}{\partial z}$ is the material derivative, ρ is density, p is the pressure, f_c is the Coriolis force, and eddy viscosity and diffusivity coefficients $\mathcal{A}$, ν_E, and κ_E are either constant or functions of flow variables and grid parameters. The dimensionless Rossby number Ro is defined as $\mathrm{Ro} = \frac{U}{f_c L}$, in which U and L represent the velocity and length scale of the geophysical flows. The Rossby number essentially characterizes the strength of inertia compared to the Coriolis and pressure forces. Another dimensionless number is the Ekman number, which is defined as $\mathrm{Ek} = \frac{\nu_E}{\Omega H^2}$, with H the vertical extent of the flow. Since Ro is the ratio of the respective scales $\frac{U^2}{L}$ and $f_c U$ of the first two terms in (5) and (6), and since Ek measures the ratio of viscous forces to Coriolis forces, when both Ro and Ek are much smaller than 1 (e.g.,

$Ro = 0.0036$ in Section 5), the Coriolis term dominates the left hand sides of momentum Equations (5) and (6), and the equations can be simplified, yielding the geostrophic balance:

$$-f_c v = -\frac{1}{\rho}\frac{\partial p}{\partial x}, \tag{10}$$

$$f_c u = -\frac{1}{\rho}\frac{\partial p}{\partial y}. \tag{11}$$

The resulting system reaches an equilibrium state in which the pressure gradient balances perfectly with the Coriolis force. When the Rossby and Ekman numbers are still small, but not nearly zero, the flow only achieves a near-geostrophic balance. Considering the beta-plane approximation $f_c = f_0 + \beta y$ and ignoring the stratification effect, one can obtain the single layer QGE by regular perturbation analysis [17,25,26]. The resulting equations are usually put in the following streamfunction-potential vorticity two-dimensional formulation:

$$\frac{\partial q}{\partial t} + J(q, \psi) = \mathrm{Re}^{-1}\Delta q + F_e, \tag{12a}$$

$$q = -\mathrm{Ro}\,\Delta\psi + y, \tag{12b}$$

where $\mathrm{Ro} = \frac{U}{\beta L^2}$ is the redefined Rossby number [27–29], $\mathrm{Re} = \frac{UL}{A}$ is the Reynolds number, ψ is the streamfunction, q is the potential vorticity, $J(q, \psi) = \frac{\partial q}{\partial x}\frac{\partial \psi}{\partial y} - \frac{\partial q}{\partial y}\frac{\partial \psi}{\partial x}$ is the Jacobian, F_e is the external forcing, and βy measures the beta-plane effect from the Coriolis force due to rotation. By eliminating the potential vorticity, we can obtain the pure streamfunction formulation:

$$-\frac{\partial}{\partial t}(\Delta\psi) + \mathrm{Re}^{-1}\Delta^2\psi - J(\Delta\psi, \psi) - \mathrm{Ro}^{-1}\frac{\partial\psi}{\partial x} = \mathrm{Ro}^{-1}F_e. \tag{13}$$

Equations (12) and (13) are supplemented by boundary conditions, such as $\psi = \frac{\partial\psi}{\partial n} = 0$ on $\partial\Omega$. More details regarding the parameters and nondimensionalization of the QGE are given in, e.g., [8,30–33]. Note that the velocity can be recovered from the streamfunction according to the following formula:

$$v = \left(\frac{\partial\psi}{\partial y}, -\frac{\partial\psi}{\partial x}\right). \tag{14}$$

One can also introduce the vorticity $\omega = \mathrm{Ro}^{-1}(q - y)$ and recast the QGE (12) in the following streamfunction-vorticity formulation:

$$\frac{\partial\omega}{\partial t} + J(\omega, \psi) - \mathrm{Ro}^{-1}\frac{\partial\psi}{\partial x} = \mathrm{Re}^{-1}\Delta\omega + \mathrm{Ro}^{-1}F_e, \tag{15a}$$

$$\omega = -\Delta\psi. \tag{15b}$$

This form is close to the streamfunction-vorticity formulation of the two-dimensional Navier–Stokes equations, but it has an additional convection term $\mathrm{Ro}^{-1}\frac{\partial\psi}{\partial x_1}$ and the forcing term is scaled by Ro^{-1} due to the rotation effect of the Earth. Such rotation effect can significantly change the behavior of QGE and yields a strong boundary layer in the solution, as shown in Figure 1: When Ro is unphysically large (i.e., close to 1) we have larger, circular gyres with lower kinetic energy but when Ro is decreased the gyres both increase in energy (due to increased forcing), which can be seen in the higher vorticity magnitudes, and move westward (due to the convection term).

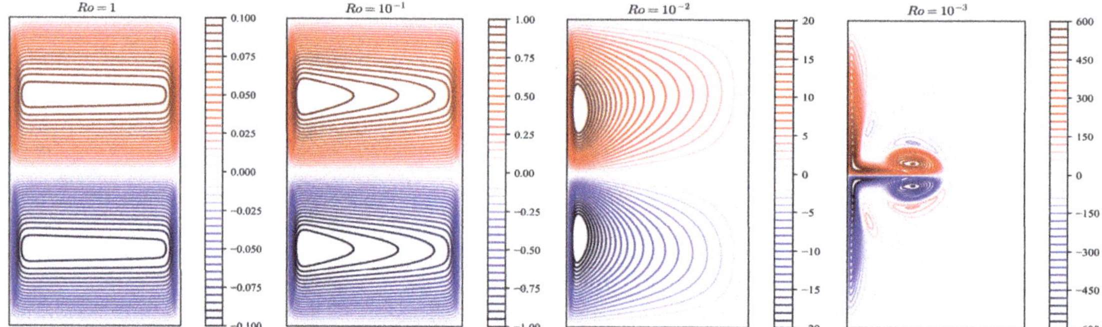

Figure 1. Solutions (vorticities) of the QGE subject to different Rossby numbers on the rectangular domain $[0,1] \times [0,2]$ when $\mathrm{Re} = 100$ and $F_e = \sin(\pi(y-1))$ at $t = 0.1$. From left to right: $\mathrm{Ro} = 1, 0.1, 0.01$, and 0.001. It is seen that decreasing the Rossby number yields a sharper western boundary layer.

When the fluid of interest is homogeneous, that is, no stratification is considered, we have the single layer QGE model. This is what we mainly focus on in this paper. However, to better approximate a continuously stratified fluid, a multi-layer model can be developed that assumes that the fluid consists of stacked isopyncal layers, the variation in the thickness of each layer is small compared to its mean thickness, and adjacent layer equations are coupled through the quasi-geostrophic potential vorticity [17]. Numerical investigations of multi-layer QGE have been made, for instance, in [34–36].

3. Full Order Model (FOM)

To generate FOM numerical data to construct the ROM basis, the QGE (12) need to be discretized both in space and in time. Popular QGE spatial discretizations include finite difference (FDM), finite volume (FVM), finite element (FEM), and pseudospectral methods. Essentially all discretizations use the method of lines (e.g., Runge-Kutta methods or other standard ODE solvers) to discretize in time. In this section, we survey each of these spatial discretizations for the QGE and, where available, comment on the existing numerical analysis results.

The primary intent of this paper is to survey the state-of-the-art for reduced order modeling of the QGE. While FOMs are required by ROMs, this section is not intended to be an exhaustive survey of the literature on the subject and instead only highlights the major trends.

3.1. Finite Difference Methods for the QGE

It is straightforward to apply the FDM to a geophysical flow model on rectangular grids. These were the first methods used [18,37] to simulate geophysical flows. In particular, the Arakawa grids were introduced by Arakawa and Lamb [38] to conserve energy and enstrophy at the grid level by effectively locating state variables across the mesh (i.e., a staggered-grid representation instead of nodal or cell-centered). See, e.g., [15,39] for detailed discussions. Among this class of grids, the C-grid places scalar quantities at the cell centers, while specifying the normal velocity components at the cell edges (which is essentially the classic MAC scheme [40]). Because of its excellent representation of the inertial-gravity waves, it has been widely used in geophysical flow simulations, for instance, for solving QGE in [33] and is the standard solver in the Modular Ocean Model version 6 [41]. Staggered-grid grid approximations like the C-grid can be thought of as either finite difference or finite volume schemes since the various velocity fluxes are explicitly solved for at cell faces rather than being reconstructed first from cell-centered values—this is the fundamental property that gives, e.g., the MAC scheme exactly zero divergence at cell centers (when calculated with standard second-order difference operators). Such schemes can also be extended to work with various turbulence modelling strategies [33,42–44].

3.2. Finite Volume Methods for the QGE

Like the staggered-grid finite difference schemes, the principal advantage of the FVM is preservation of the essential conservative quantities for the governing equations of geophysical fluid flows while additionally dealing with unstructured grids (i.e., complex geometries) more easily. This avoids the need for discretizing boundaries with staircasing, which results in inaccurate modelling of coastal phenomena like Kelvin waves [45]). This combination of properties makes the FVM the most common method for large-scale ocean simulations, such as those performed with [41] or [46].

Methods that use C-grid like discretizations (i.e., storing normal velocities on cell faces and mass or pressure in cell centers) on arbitrarily structured meshes must additionally introduce corrective measures to deal with the reconstruction of the tangential velocity (which is required by the discretization of the Coriolis force) [47–49]. These generalized C-grid methods are applicable to a wide class of meshes including latitude-longitude grids, Delaunay triangulations, Centroidal Voronoi tessellation (CVT), and spherical CVT. A different approach to overcome this issue was considered in [50], where the non-staggered Z-grid scheme [51] was used for the QGE model.

3.3. Pseudospectral and Spectral Methods for the QGE

Like finite difference methods, pseudospectral methods (due to their immediate applicability to hypercube geometries) have been used in a variety of different ways in QGE solvers. Some QGE solvers, like the one used in [52], use a pseudospectral discretization to compute turbulence statistics. Alternatively, some finite difference methods use a pseudospectral interpretation of solution grid values to do fast Laplace solves with a multidimensional discrete sine transformation [32,53] or resolve stability problems from nonlinearities via dealiasing [54,55].

Furthermore, pseudospectral methods have been used for the spatial discretization of the QGE [56,57]. The FOM results used in this paper to construct the ROM basis in Section 5 were also generated with a pseudospectral method. By *pseudospectral* we mean that spatial derivatives in (12) are evaluated by performing a discrete sine transform, wave number multiplication, and an another discrete sine transform in which dealiasing (the 3/2 s rule from [58]) is used in the nonlinear term of (12) for stability. The FOM solver exploits the homogeneous boundary conditions to ignore even-numbered Fourier modes (i.e., the RODFT00 transformation in [59]). Since this method permits very fast evaluation of spatial derivatives we essentially treat them as a black box operation as part of an explicit ODE solver for evolving the Fourier coefficients in time. The largest stable timestep is found by using the power method for computing the principal eigenvalue of the linearized discretization of (12). Numerical experiments imply that setting a strict error tolerance on the error caused by the ODE solver requires smaller timesteps than the one required for ODE stability, which validates the choice of explicit methods for the relevant range of Reynolds numbers and grid resolutions. See Section 5.4 for additional details on the numerical experiments used in this manuscript.

3.4. Finite Element Methods for the QGE

The FEM is particularly appealing because it combines advantages of multiple methods. It can easily handle adaptive mesh refinement and complex geometries (like the FVMs), but also can create higher-order schemes (like pseudospectral methods) at the same time, like the discretization used in [30,60], which is shown in Figure 2 (see also [61–63]). The first FE approximation of the QGE, to the best of our knowledge, was a scheme based on the mixed formulation developed in [64]. The conservation properties and stability of the FE discretization were proved as well as the suboptimal convergence of the FE method. The performance of FEM on simulating a multilayer QGE of ocean circulations has been compared to the FDM in [65]. Since the vorticity-streamfunction formulation (15) of the QGE results in a second-order PDE, for a conforming finite element discretization, a C^0 element can be utilized. Considering the finite element spaces $W_\psi \subset H_0^1(\Omega) \cap W^{1,4}(\Omega)$,

$W_\omega \subset H^1(\Omega)$ (see, e.g., [66] for the definition of these finite element spaces), the finite element discretization reads: Find $\psi_h \in W_\psi$ and $\omega_h \in W_\omega$ satisfying

$$\begin{cases} \left(\frac{\partial \omega_h}{\partial t}, \phi_h\right) + (J(\omega_h, \psi_h), \phi_h) = -\mathrm{Re}^{-1}(\nabla \omega_h, \nabla \phi_h) + \mathrm{Ro}^{-1}\left(\frac{\partial \psi_h}{\partial x}, \phi_h\right) + \mathrm{Ro}^{-1}(F_e, \phi_h) & \forall \phi_h \in W_\psi \\ (\omega_h, v_h) = (\nabla \psi_h, \nabla v_h) & \forall v_h \in W_\omega \end{cases}. \tag{16}$$

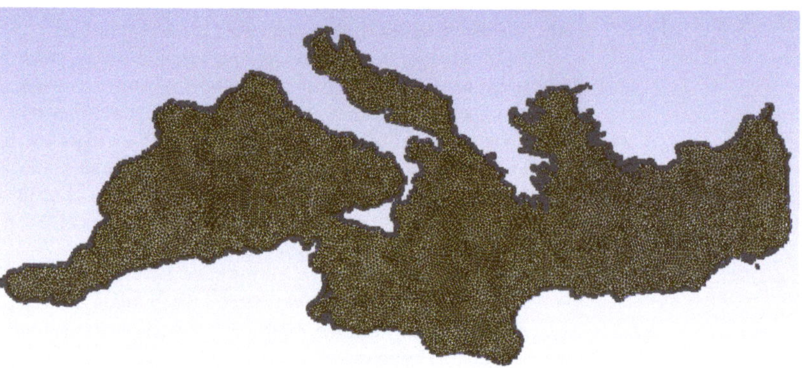

Figure 2. Triangulation of the Mediterranean Sea suitable for simulations with finite element methods which was used in [30,60] (see also [61–63]).

In [34,67], Medjo considered this formulation and proved bounds for the time discretization error. Cascon et al. [68] proved both a priori and a posteriori error estimates for the FE discretization of the linear Stommel-Munk model, which is a simplified version of the QGE obtained by dropping the nonlinear term.

The streamfunction formulation (13) of QGE is a fourth-order PDE, which naturally necessitates C^1 elements for a conforming finite element discretization. Considering the finite element space $W \subset H_0^2(\Omega)$, the finite element discretization reads: Find $\psi_h \in W$ [66] satisfying

$$\left(\frac{\partial}{\partial t}\nabla\psi_h, \nabla\phi_h\right) + \mathrm{Re}^{-1}(\Delta\psi_h, \Delta\phi_h) + (J(\psi_h, \Delta\psi_h), \phi_h) - \mathrm{Ro}^{-1}\left(\frac{\partial\psi_h}{\partial x}, \phi_h\right) = \mathrm{Ro}^{-1}(F_e, \phi_h), \forall \phi_h \in W. \tag{17}$$

To our knowledge, the first optimal error convergence results for the finite element approximation of the QGE (12) were proved for the streamfunction formulation (17) using Argyris elements in [30]. Several numerical tests, commonly employed in the geophysical literature, showed the accuracy of the finite element discretization and illustrated the theoretical estimates. Other recent developments of FEM for QGE include discontinuous Galerkin formulation using C^0 elements [62] and B-splines [69–73]. In particular, an adaptive refinement algorithm for B-splines finite element approximation was presented in [71] for the streamfunction formulation.

4. Reduced Order Models (ROMs)

ROMs for the QGE have been developed for decades (see, e.g., [8,32,61,74–83]). Most ROMs have been constructed by using a classical Galerkin projection framework, but data-driven modeling (e.g., machine learning) has also been recently used [84,85]. In Section 4.1, we outline the standard Galerkin ROM construction. In Section 4.2, we explain the importance of considering under-resolved regimes when developing ROMs for realistic, chaotic flows. Furthermore, we present several ROM closure strategies, which are generally needed when ROMs are used in an under-resolved regime. The flowchart of the ROMs presented in this section is illustrated in Figure 3.

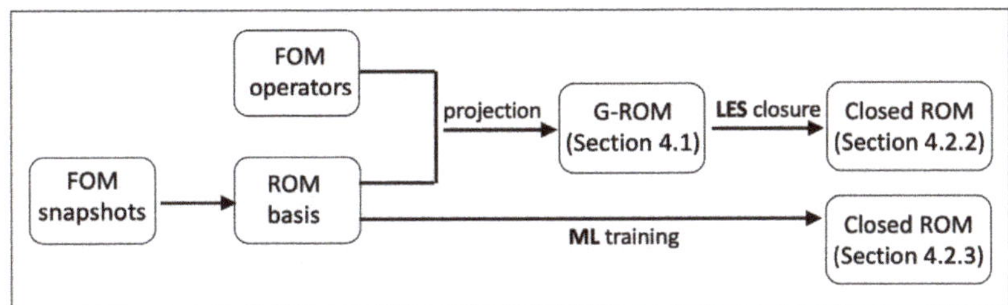

Figure 3. Framework of the ROMs presented in Section 4.

4.1. Galerkin Reduced Order Model (G-ROM)

To construct the standard Galerkin ROM, we start by generating the ROM basis. To this end, we use the proper orthogonal decomposition (POD) [2,6], which is also known as empirical orthogonal functions (EOF) and principal component analysis (PCA). We emphasize, however, that other ROM bases could be used, such as principal interaction patterns (PIPs) and optimal persistence patterns (OPPs) [74] (see also [1,3,5,7,75,86,87] for alternative strategies).

The POD starts by collecting the snapshots $\{\omega_h^1, \ldots, \omega_h^M\}$, which are numerical approximations of the vorticity in the QGE (15) at M different time instances. We consider relatively accurate snapshots. If inaccurate snapshots are used to construct the POD basis, the resulting ROM can be inaccurate (see, e.g., [88]). For clarity of presentation, in this paper we use the finite element discretization, but other numerical discretizations could be used. The POD seeks a low-dimensional basis that approximates the snapshots optimally with respect to a certain norm. In this presentation, we use the L^2 norm and the L^2 inner product:

$$\left(\omega_1, \omega_2\right) = \int_\Omega \omega_1(\mathbf{x})\, \omega_2(\mathbf{x})\, d\mathbf{x}. \tag{18}$$

We note that, although the L^2 norm and the L^2 inner product are the most popular choices in reduced order modeling, other norms and inner products could also be used (see, e.g., [89]). To construct POD basis functions that approximate the snapshots optimally with respect to the L^2 norm, we solve the following minimization problem [89]:

$$\min_{\widetilde{\varphi}_1, \cdots, \widetilde{\varphi}_N} \sum_{j=1}^{M} \left\| \omega_h^j - \sum_{i=1}^{N} \left(\omega_h^j, \widetilde{\varphi}_i\right) \widetilde{\varphi}_i \right\|_{L^2}^2 \tag{19}$$

$$\text{s.t. } \left(\widetilde{\varphi}_l, \widetilde{\varphi}_m\right) = \delta_{lm} \quad \text{for } 1 \leq l, m, \leq N,$$

where δ_{lm} is the Kronecker delta. The solution of the minimization problem (19) is equivalent to the solution of the eigenvalue problem

$$Y^T M_h Y \widetilde{\varphi}_j = \lambda_j \widetilde{\varphi}_j, \quad j = 1, \ldots, N, \tag{20}$$

where Y denotes the snapshot matrix, whose columns correspond to the finite element coefficients of the snapshots, M_h denotes the finite element mass matrix, and N is the dimension of the finite element space. The eigenvalues are real and non-negative, so they can be ordered as follows: $\lambda_1 \geq \lambda_2 \geq \ldots \geq \lambda_R \geq \lambda_{R+1} = \ldots = \lambda_N = 0$, where R is the rank of the snapshot matrix. It can be shown [89] that these eigenvalues determine how well the corresponding POD modes represent the given vorticity snapshots: the lower the eigenvalue index, the more important the corresponding POD mode. Thus, we choose the POD vorticity basis functions $\{\varphi_j\}_{j=1}^r$ from the eigenfunctions in (20) that correspond to the first $r \leq R$ largest eigenvalues and define the ROM vorticity space as $X^r := \text{span}\{\varphi_1, \ldots, \varphi_r\}$.

To determine the POD streamfunction basis functions, we use the POD vorticity basis functions and follow the approach in [8,32]. Specifically, we define the POD streamfunction basis functions as the normalized functions $\{\phi_j\}_{j=1}^r$, which are chosen such that the satisfy the following Poisson problem with homogeneous Dirichlet boundary conditions:

$$-\Delta\phi_j = \varphi_j, \quad j = 1,\ldots,r. \tag{21}$$

Next, we define the ROM approximations of the vorticity and streamfunction as follows:

$$\omega_r(\boldsymbol{x},t) \;=\; \sum_{j=1}^r a_j(t)\,\varphi_j(\boldsymbol{x})\,, \tag{22}$$

$$\psi_r(\boldsymbol{x},t) \;=\; \sum_{j=1}^r a_j(t)\,\phi_j(\boldsymbol{x})\,, \tag{23}$$

where $\{a_j(t)\}_{j=1}^r$ are the sought time-varying ROM coefficients. We note that we made two important choices in our approach: (i) We enforced the coupling between the POD vorticity and streamfunction basis functions in (21); and (ii) We used the same ROM coefficients in the ROM vorticity approximation (22) and in the ROM streamfunction approximation (23). The motivation for making these two choices is efficiency. Indeed, we only need to construct a ROM for the vorticity; once the coefficients a_j are determined from (15a), Equation (15b) is automatically satisfied. (Of course, one could use a different approach and construct two different ROM bases and two different ROM approximations for the vorticity and streamfunction, but that would increase the ROM computational cost.) To construct a ROM for the vorticity, we replace the vorticity ω by ω_r in the QGE (15a), and then we use a Galerkin projection onto X^r. Thus, we obtain the Galerkin ROM (G-ROM) for the QGE: $\forall\, i = 1,\ldots,r$,

$$\left(\frac{\partial\omega_r}{\partial t},\varphi_i\right) + (J(\omega_r,\psi_r),\varphi_i) - \mathrm{Ro}^{-1}\left(\frac{\partial\psi_r}{\partial x},\varphi_i\right) + \mathrm{Re}^{-1}\left(\nabla\omega_r,\nabla\varphi_i\right) = \mathrm{Ro}^{-1}\left(F_e,\varphi_i\right). \tag{24}$$

The G-ROM (24) yields the following autonomous dynamical system for the vector of time coefficients, $\mathbf{a}(t) = (a_i(t))_{i=1,\ldots,r}$:

$$\dot{\mathbf{a}} = \mathbf{b} + \mathbf{A}\,\mathbf{a} + \mathbf{a}^\top\mathbf{B}\,\mathbf{a}, \tag{25}$$

where $\mathbf{b}$, $\mathbf{A}$, and $\mathbf{B}$ are an $r \times 1$ vector, an $r \times r$ matrix, and an $r \times r \times r$ tensor, which correspond to the constant, linear, and quadratic terms in the numerical discretization of the QGE (15), respectively. The r-dimensional system (25) can be written componentwise as follows: For all $i = 1,\ldots,r$,

$$\dot{a}_i(t) = b_i + \sum_{m=1}^r A_{im}a_m(t) + \sum_{m=1}^r\sum_{n=1}^r B_{imn}\,a_m(t)\,a_n(t), \tag{26}$$

where

$$b_i = \mathrm{Ro}^{-1}\left(F_e,\varphi_i\right), \tag{27}$$

$$A_{im} = \mathrm{Ro}^{-1}\left(\frac{\partial\phi_m}{\partial x},\varphi_i\right) - \mathrm{Re}^{-1}\left(\nabla\varphi_m,\nabla\varphi_i\right), \tag{28}$$

$$B_{imn} = -\left(J(\varphi_m,\phi_n),\varphi_i\right). \tag{29}$$

The G-ROM (25) has been investigated in the numerical simulation of the QGE (15) (see, e.g., [8,32,80,83]), where it was shown that it can decrease the FOM computational cost by orders of magnitude. However, the numerical simulations in [8,32] have also shown that a low-dimensional G-ROM is not able to produce accurate approximations of the streamfunction and the velocity fields. The G-ROM's numerical inaccuracy in [8,32] is due to the lack of a closure model [8,9,90], which we discuss in Section 4.2.

4.2. ROM Closure Models

In this section, we survey the *ROM closure models* developed for the QGE (12). First, we define closure modeling and we explain why it is needed when ROMs are used in the under-resolved regime (Section 4.2.1). Then, we present the two main types of ROM closure modeling for the QGE that are in current use: large eddy simulation (LES) ROM closure models (Section 4.2.2) and machine learning (ML) ROM closure models (Section 4.2.3). While LES and ML ROM closures are both data-driven modeling approaches, they are different in the way they use data to develop a closure model: The LES approach is based on ROM spatial filtering and least squares methods, whereas the ML approach is based on machine learning techniques.

4.2.1. Under-Resolved ROMs Require Closure Models

The concept of under-resolved simulations is central in classical CFD. *Under-resolved simulations are those simulations in which the number of degrees of freedom* (e.g., the number of mesh points or basis functions) *is not enough to capture the dynamics of the underlying system.* For example, in turbulent flow simulations the available number of mesh points in a finite element or finite volume discretization, or the number of basis functions in a spectral discretization are not enough to resolve all the lengthscales in the turbulent flow, down to the Kolmogorov scale [91–93]. The numerical simulations at these inherently coarse resolutions are called under-resolved simulations.

In under-resolved simulations of turbulent flows, standard discretizations yield inaccurate results, which are not acceptable in practical engineering settings, e.g., large relative errors, inaccurate quantities of interest (e.g., lift and drag), and inaccurate flow features (e.g., vortex shedding frequency for the flow past a cylinder). In these cases, the classical computational models (e.g., the Navier–Stokes equations) are generally supplemented with correction terms that model the effect of the neglected scales (e.g., the scales smaller than the given coarse mesh size). These correction terms are generally called *closure models* [91–93].

The concept of under-resolved simulations is also relevant to reduced order modeling: *Under-resolved ROM simulations are those simulations in which the ROM dimension is not enough to capture the dynamics of the underlying system.* But how exactly do we determine whether a ROM simulation is resolved or under-resolved? Next, we present several potential answers to this question. Some of these answers are *a priori* criteria (i.e., can be used before the ROM simulation), some are *a posteriori* criteria (i.e., can be used only after the ROM simulation).

Kolmogorov n-width: The Kolmogorov n-width is an *a priori* criterion to determine whether the ROM simulation is resolved or under-resolved. Given the solution manifold $\mathcal{M}$ of the underlying system's dynamics, the Kolmogorov n-width [94] provides a way to quantify the best n-dimensional trial subspace $\mathcal{X}^n$:

$$d_n(\mathcal{M}) := \inf_{\mathcal{X}^n} \sup_{\omega \in \mathcal{M}} \inf_{g \in \mathcal{X}^n} \|\omega - g\|.$$

Of course, calculating the Kolmogorov n-width for general systems can be challenging. There are, however, cases when the relative size of the Kolmogorov n-width is known. For example, it is known that, for computational problems dominated by diffusion, the Kolmogorov n-width decays fast, while for those dominated by convection, it decays slowly [95]. As a result, in order to obtain an accurate approximation of the solution manifold, the dimension of the ROM trial space is expected to be much higher in the convection-dominated case than in the diffusion-dominated case. Thus, for convection-dominated systems, if we use a very high-dimensional (i.e., of the same order as the Kolmogorov n-width) ROM, we obtain a resolved ROM simulation. If, however, we use a low-dimensional (i.e., much lower than the Kolmogorov n-width) ROM, we obtain an under-resolved ROM simulation.

Eigenvalue decay rate: The eigenvalue decay rate is an *a priori* criterion to determine whether the ROM simulation is resolved or under-resolved. The eigenvalues $\lambda_1, \dots, \lambda_R$ in the eigenvalue problem (20) (used to construct the ROM basis) represent the energy content of the corresponding ROM modes [2,89]. Thus, the ratio

$$\frac{\sum_{i=1}^{r} \lambda_i}{\sum_{i=1}^{R} \lambda_i} \tag{30}$$

defines the relative energy content of the first r ROM basis functions with respect to the total energy of the system (see, e.g., page 16 in [89]). We emphasize that the concept of "energy" in this context is used in a generic sense. For example, when the snapshots are FOM approximations of a the velocity field in a fluid flow, the energy in (30) is the kinetic energy; when the snapshots are FOM approximations of the vorticity field in the QGE, the energy in (30) is the enstrophy. We can define the resolved regime as the regime in which the ROM dimension r is large enough to ensure that the relative energy ratio (30) is larger than a certain threshold (e.g., 90%). Thus, we expect a low-dimensional ROM to be in the resolved regime when the eigenvalues have a fast decay, and in the under-resolved regime when the eigenvalues have a slow decay.

To illustrate this point, in Figure 4 we plot the scaled eigenvalues λ_k/λ_1, $k = 1, \dots, 150$ for two flow settings: the 2D flow past a cylinder at Re = 1000 and the QGE with Re = 450 and Ro = 0.0036 (the latter will be used in the numerical investigation in Section 5). This plot shows that the eigenvalues decay much faster for the flow past a cylinder case than for the QGE case.

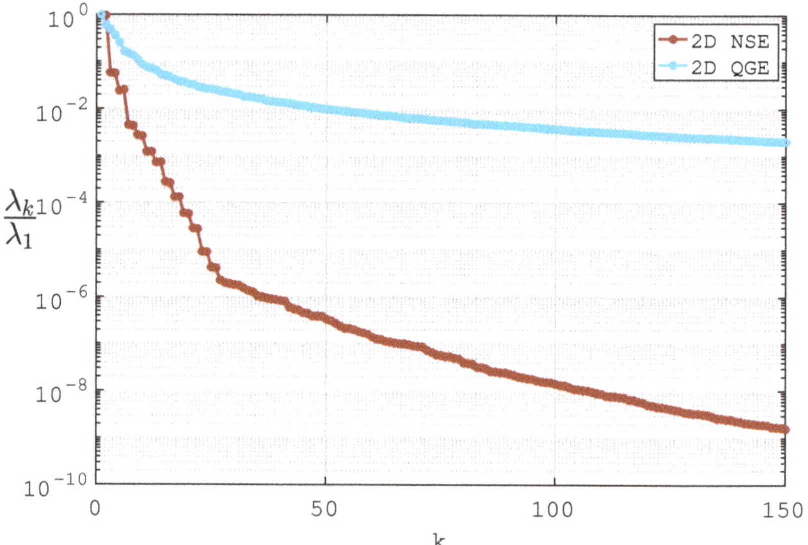

Figure 4. Scaled eigenvalues $\frac{\lambda_k}{\lambda_1}$ for the 2D flow past a circular cylinder with Re = 1000 and the QGE with Re = 450 and Ro = 0.0036 (see Section 5 for details).

Indeed, the results in Table 1 show that, in order to achieve a 90% relative energy ratio in (30), we need to use only 2 ROM modes for the flow past a cylinder, and 77 ROM modes for the QGE case. Thus, if we use only a handful of ROM modes to ensure a low computational cost, we expect the resulting low-dimensional ROM to accurately capture the dynamics of the flow past a cylinder, but not the dynamics of the QGE. In this case, we perform a resolved ROM simulation of the flow past a cylinder, and an under-resolved ROM simulation for the QGE.

Table 1. Number of ROM modes needed to achieve a given relative energy content (30) for the 2D flow past a circular cylinder with Re = 1000 and the QGE with Re = 450 and Ro = 0.0036 (see Section 5 for details).

Relative Energy Content	90%	95%	99%
2D flow past a cylinder	2	4	6
QGE	77	152	380

ROM Lengthscale: The ROM lengthscale is an *a priori* criterion to determine whether the ROM simulation is resolved or under-resolved. In principle, the ROM lengthscale criterion follows the same algorithm as the standard CFD lengthscale criterion: Start with a lengthscale that is large enough to capture the relevant dynamics, and then choose the input discretization parameters such that phenomena occurring at the chosen lengthscale can be approximated. Choosing the discretization parameters is, however, fundamentally different in classical CFD and ROMs: In classical CFD, the *spatial meshsize* (e.g., for finite difference or finite element methods) or the cutoff wavenumber in a Fourier truncation (e.g., for spectral methods) clearly determines what lengthscale can be approximated. For ROMs, however, *there is no straightforward definition of a lengthscale based on the ROM discretization parameters*, i.e., the ROM dimension (r), the ROM basis ($\{\varphi_1, \ldots, \varphi_r\}$), and the ROM eigenvalues ($\{\lambda_1, \ldots, \lambda_r\}$). To our knowledge, only very few ROM lengthscale definitions based on the ROM discretization parameters have been proposed. In [96], a ROM lengthscale was defined for the 3D flow past a circular cylinder at $Re = 1000$ (see also [2] for related work). This lengthscale was then used in [96] to build ROM closure models.

Trial and error: The trial and error approach is an *a posteriori* criterion to determine whether the ROM simulation is resolved or under-resolved. Specifically, a few ROM simulations are run in the offline stage in order to determine the ROM discretization parameters that yield accurate results, which are acceptable in practical engineering settings, e.g., small relative errors, accurate quantities of interest (e.g., lift and drag), and accurate flow features (e.g., vortex shedding frequency for the flow past a cylinder).

In Section 5, we show that under-resolved ROM simulations of the QGE can yield inaccurate results. To increase the accuracy of these under-resolved ROM simulations, the standard G-ROM (25) is generally supplemented with a closure model:

$$\dot{\mathbf{a}} = \mathbf{b} + \mathbf{A}\,\mathbf{a} + \mathbf{a}^\top \mathbf{B}\,\mathbf{a} + \boldsymbol{\tau}^{ROM}, \tag{31}$$

where $\boldsymbol{\tau}^{ROM}$ is the closure model that needs to be determined.

There are two main types of ROM closure modeling approaches, i.e., approaches to modeling the term $\boldsymbol{\tau}^{ROM}$ in (31) in the offline stage:

- **Black box** ROM closure models: These models consider the true closure model $\boldsymbol{\tau}^{FOM}$ as a black box, i.e., the specific form of $\boldsymbol{\tau}^{FOM}$ is not determined. Instead, one first postulates a model form for $\boldsymbol{\tau}^{FOM}$, i.e., $\boldsymbol{\tau}^{FOM} \approx \boldsymbol{\tau}^{ROM}$, and then determines the parameters of the model form $\boldsymbol{\tau}^{ROM}$, either by using available data or physical insight.

- **Mathematical** ROM closure models: These models use *filtering/averaging* (e.g., with respect to space, time, or initial conditions) to determine the specific form of the true ROM closure term $\boldsymbol{\tau}^{FOM}$. As in the black box ROM closure models, one postulates a model form for $\boldsymbol{\tau}^{FOM}$, i.e., $\boldsymbol{\tau}^{FOM} \approx \boldsymbol{\tau}^{ROM}$. However, the mathematical ROM closure modeling utilizes *data for the specific form* of $\boldsymbol{\tau}^{FOM}$ to determine the ROM closure model $\boldsymbol{\tau}^{ROM}$.

In this paper, we do not survey the ROM closure models. Instead, we only focus on ROM closure modeling for the QGE. Specifically, in Sections 4.2.2 and 4.2.3, we present two different ROM closure modeling strategies for the QGE. We note, however, that there are alternative ROM closure modeling strategies for the QGE, e.g., the stochastic mode reduction strategy developed by Majda and his collaborators (see [76] and references therein).

4.2.2. Large Eddy Simulation ROM Closure Models

The large eddy simulation (LES) ROM closure modeling is inspired from classical LES of turbulent flows [91–93]. The LES-ROM closure models come in two flavors: black box and mathematical.

The black box LES-ROM closure models developed for the QGE use physical insight to postulate a model form for the closure term. Specifically, they postulate that the ROM closure term has to be dissipative. In [80], a linear damping term (i.e., a third-order Laplace operator) is used as a ROM closure model. In [32], a nonlinear damping term (i.e., a simplified Smagorinsky model [97]) is used as a ROM closure model. A significant improvement to the Smagorinsky ROM closure model used in [32] is the dynamic subgrid-scale ROM closure model which was first proposed in [96] and later adapted to the QGE in [79].

The mathematical LES-ROM closure models developed for the QGE are an elegant approach to closure modeling. These ROM closure models are built in three steps: In the first step, the QGE are *spatially filtered* to obtain the large structures which can be approximated at the given coarse resolution. In this first step, an exact formula for the ROM closure term τ^{FOM} is also obtained. In the second step, a specific model form is postulated for the LES-ROM closure model, i.e., $\tau^{FOM} \approx \tau^{ROM}$ in (31). Finally, in the third step, FOM data is used to find the parameters in the general form τ^{ROM} that yield the closest (in a least squares sense) approximation to true ROM closure term τ^{FOM}. One example of mathematical LES-ROMs is the recently developed *data-driven variational multiscale* ROM (DD-VMS-ROM) [8,9,90], which is centered around the variational multiscale framework. There are two versions of the DD-VMS-ROM: a two-scale model [8,9] and an improved three-scale model [90]. For clarity of presentation, we present the two-scale DD-VMS-ROM. To construct the DD-VMS-ROM, the ROM projection from the ROM space $X^R = \mathrm{span}\{\varphi_1,\ldots,\varphi_R\}$ to the subspace $X^r = \mathrm{span}\{\varphi_1,\ldots,\varphi_r\}$, $r \leq R$ is used as a spatial filter. The filtered QGE yield an exact formula for the ROM closure term, τ^{FOM}. Next, a specific model form is prescribed for the exact closure term

$$\tau^{FOM} \approx \tau^{ROM} := \widetilde{A}\,a + a^\top \widetilde{B}\,a \tag{32}$$

and the entries of the LES-ROM closure operators $\widetilde{A}$ and $\widetilde{B}$ are found by solving the following *least squares problem*:

$$\min_{\widetilde{A},\widetilde{B}} \sum_{j=1}^{M} \left\| \tau^{FOM}(t_j) - \left(\widetilde{A}\,a^{FOM}(t_j) + (a^{FOM}(t_j))^\top \widetilde{B}\,a^{FOM}(t_j) \right) \right\|^2, \tag{33}$$

where a^{FOM} are computed from the FOM data. Finally, the LES-ROM closure operators obtained in (33) are used to build the DD-VMS-ROM:

$$\dot{a} = b + \left(A + \widetilde{A} \right) a + a^\top \left(B + \widetilde{B} \right) a. \tag{34}$$

In Section 5, we investigate the DD-VMS-ROM in the under-resolved simulation of the QGE.

4.2.3. Machine Learning ROM Closure Models

Machine learning (ML) methods have recently started to make an impact in reduced order modeling of fluid flows. The ML methods most frequently used to build ROMs include multilayer perceptron (MLP) [85,98,99], convolution neural networks (CNN) [100], recurrent neural networks (RNN) [84,101,102], and variational autoencoder (VAE) [100,103]. Some of these ML-ROMs are *nonintrusive*, i.e., they use the FOM codes as black boxes, only to generate output data from different inputs. For these nonintrusive ML-ROMs, no prior information about the underlying governing equations is required to construct the model. These models fully rely on data combined with ML methods to discover the ROM dynamics and can be written as follows:

$$\dot{a} = \widehat{F}(a, \theta), \tag{35}$$

where a is the vector of ROM coefficients and θ is the vector of learnable parameters in the ML model $\widehat{F}$. The nonintrusive ML-ROMs are fundamentally different from classical intrusive modeling strategies, such as the Galerkin method used to generate the G-ROM (25), which need access to the underlying governing equations in order to construct the ROM.

Only few ML-ROMs have been developed for the QGE. For example, an extreme learning machine concept with neural networks was introduced for the ROM closure of the QGE in [85]. Furthermore, a nonintrusive reduced order modeling framework embedded with a long short-term memory (LSTM) network was developed for quasi-geostrophic turbulence to improve the time series prediction of ROMs in [84]. The LSTM-ROM for QGE [84] was constructed (trained) in two steps:

1. The ROM coefficients in a given time window $\{a^{FOM,(n-k)}, a^{FOM,(n-k+1)}, \ldots, a^{FOM,(n)}\}$ were extracted from the high-resolution FOM data by projecting the snapshots onto the ROM modes.

2. The LSTM neural network was used to construct an ML-ROM that mapped the old ROM coefficients $\{a^{FOM,(n-k)}, a^{FOM,(n-k+1)}, \ldots, a^{FOM,(n)}\}$ to the ROM coefficients at the new time step $a^{FOM,(n+1)}$.

The resulting model was then used in the testing stage to predict the ROM coefficients at new time instances.

Recently, *hybrid* ROMs that combine classical Galerkin modeling with machine learning have started to become popular. For example, a hybrid ROM closure was proposed in [104] for the QGE. This hybrid ROM combined classical Galerkin projection methods with neural network closures to perform near real-time prediction of mesoscale ocean flows. The numerical investigation in [104] showed that the hybrid ROM was more accurate than both the classical G-ROM and a pure ML-ROM (i.e., a ROM built entirely from data by using machine learning).

5. Numerical Results

In this section, we present an illustration of the projection ROMs constructed in Section 4 in the numerical simulation of the QGE described in Section 2. First, we describe the details of the computational setting that we use in our ROM numerical investigation: the regimes (Section 5.1), the test problem (Section 5.2), the criteria (Section 5.3), and the generation of the FOM data used to construct the ROM basis (Section 5.4). After we clarify these details, we perform a numerical investigation of the ROM accuracy and ROM efficiency (Section 5.5).

5.1. Regimes

In our numerical illustration, we use four regimes: (i) a *reconstructive* regime, which is an easier test case, in which the ROM is validated on the same time interval as the time interval used to train the ROM; (ii) a *predictive* regime, which is a harder test case, in which the ROM is trained on a short time interval and validated on a longer time interval; (iii) a *resolved* regime, in which the number of ROM basis functions is enough to represent the system's dynamics; and (iv) an *under-resolved* regime, in which the number of ROM basis functions is not enough to represent the system's dynamics. These four regimes illustrate different features of the ROMs.

The reconstructive regime is the first step in a ROM investigation. At the very least, the proposed ROM needs to provide an efficient and accurate approximation of the FOM data used to train it (i.e., the FOM results used to construct the ROM basis). The predictive regime is a harder test in the ROM investigation. In order to be practical, the proposed ROM needs to be able to approximate the FOM results on time intervals and parameter ranges that are wider than those used to train the ROM. (For clarity, in this section, we consider only a longer time interval, but wider parameter ranges could also be considered.) Of course, the proposed ROM generally has a harder time approximating data that it has not seen in the training process, but the ROM needs to perform well in the predictive regime in order to be deemed successful in practice.

The resolved regime is an easier test in the ROM investigation. Since the ROM uses a relatively large number of ROM basis functions, which is enough to capture the underlying system's dynamics, a straightforward, standard G-ROM is expected to perform well in the resolved regime. The under-resolved regime is a much harder test in the ROM investigation. In the under-resolved regime, the proposed ROM needs to use a relatively small (i.e., not enough to capture the system's dynamics) number of ROM basis functions and somehow still be able to approximate the FOM data. In classical computational fluid dynamics (CFD), the under-resolved regime is one of the most important tests for the practicality of the proposed numerical method. Indeed, many realistic CFD applications are turbulent and chaotic, and standard resolved discretizations (e.g., direct numerical simulation (DNS)) are simply not possible, since they require an unrealistic number of degrees of freedom. The under-resolved regime is relatively much less investigated in the ROM world. We believe, however, that to develop ROMs that can be used in the numerical simulation of *realistic, chaotic* geophysical flows, the proposed ROMs need to be investigated in the under-resolved regime.

Since the reconstructive and predictive regimes, on the one hand, and the resolved and under-resolved regimes, on the other hand, serve different purposes, we consider four regime pairs in the ROM numerical investigation in Section 5.5: First, we consider the resolved and reconstructive, and the resolved and predictive regimes. The goal here is to investigate the reconstructive and, more importantly, the predictive capabilities of the standard G-ROM in the relatively simple resolved regime. Second, we consider the under-resolved and reconstructive, and the under-resolved and predictive regimes. The goal here is different. We want to investigate the reconstructive and predictive capabilities of the standard G-ROM in the challenging under-resolved regime. We expect that, when only a few ROM basis functions are used to build it, the standard G-ROM will perform poorly in the under-resolved regime. Thus, to address the G-ROM's potential inaccuracies, we also consider the LES-ROM proposed in Section 4.2.2, i.e., the DD-VMS-ROM. In the under-resolved regime, we expect the LES-ROM to be more accurate than the standard G-ROM.

5.2. Test Problem Setup

In our ROM numerical investigation in a QGE setting, we need to make several choices. Specifically, in the QGE (15), we need to choose the spatial domain, the time interval, the forcing (F_e), the Reynolds number (Re), and the Rossby number (Ro). We emphasize that these choices are important: Some choices yield a relatively easy test problem, i.e., a problem in which a standard ROM built with relatively few ROM basis functions can generate an accurate and efficient approximation. Other choices, however, yield a challenging test problem, in which standard low-dimensional ROMs produce inaccurate results.

In our numerical investigation, we choose parameters that yield a *challenging test problem*, which has been used in numerous studies (see, e.g., [8,27–29,31–33,79,85,105]) as a simplified model for more realistic ocean dynamics. Specifically, we choose the simple spatial domain $[0,1] \times [0,2]$, the relatively long time interval $[0,100]$, and a symmetric double-gyre wind forcing given by $F_e = \sin(\pi (y - 1))$, which yields a four-gyre circulation in the time mean. We also choose the same Reynolds number and Rossby number as those used in [8,28,32,33], i.e., Re = 450 and Ro = 0.0036.

We emphasize that this four-gyre QGE test problem represents a significant challenge for FOM simulations with standard numerical methods. Indeed, as shown in [27], although a double-gyre wind forcing is used, the long term time-average yields a *four-gyre* pattern (see Figure 5). On realistic coarse meshes, classical numerical methods (e.g., finite element and finite volume methods) generally produce inaccurate approximations to this test problem. In particular, standard numerical discretizations fail to recover the correct four-gyre pattern (see, e.g., [32,33]). One of the main reasons for the challenging character of the four-gyre test problem is the relatively low Rossby number used (i.e., $Ro = 0.0036$). Indeed, as shown in Figure 1, a relatively small Rossby number yields a sharp western boundary layer, which makes the test problem challenging for FOM simulations (see, e.g., [32,33]). While the Reynolds number used (i.e., $Re = 450$) is not large by turbulence modeling standards, it turns out that it yields a convection-dominated regime that is challenging for FOM simulations. Overall, these parameter choices together with the chosen spatial domain, time interval, and forcing function, yield a challenging FOM test problem. This is clearly illustrated in the plot of the FOM kinetic energy in Figure 6, which suggests that this is a *chaotic* system, with non-periodic time evolution.

Given the non-periodic, chaotic evolution of the this four-gyre test problem, we expect it to represent a challenging test not only for FOM simulations, but also for ROM simulations. This expectation is supported by projecting the FOM data on the ROM basis functions to obtain the true ROM coefficients, which the proposed ROMs need to approximate. These true ROM coefficients, which are plotted in Figure 7, have a non-periodic, chaotic evolution, which is challenging to capture by standard ROMs. In Section 5.5, we will show that this four-gyre test problem does indeed represent a challenging test for ROMs.

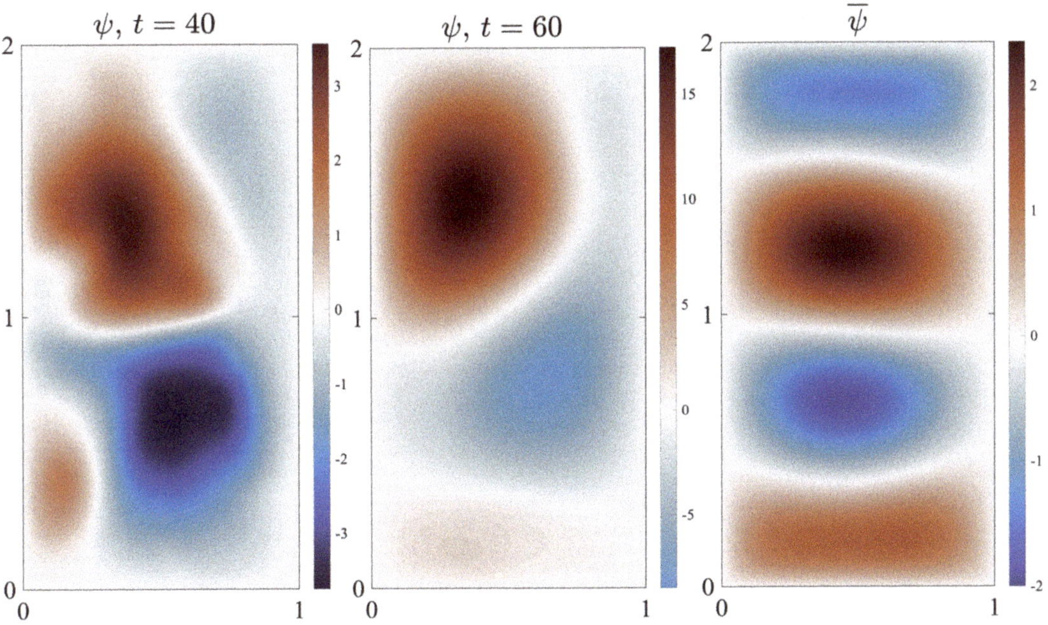

Figure 5. FOM streamfunction contour plots at $t = 40$ (**left**), $t = 60$ (**middle**), and time-averaged (**right**).

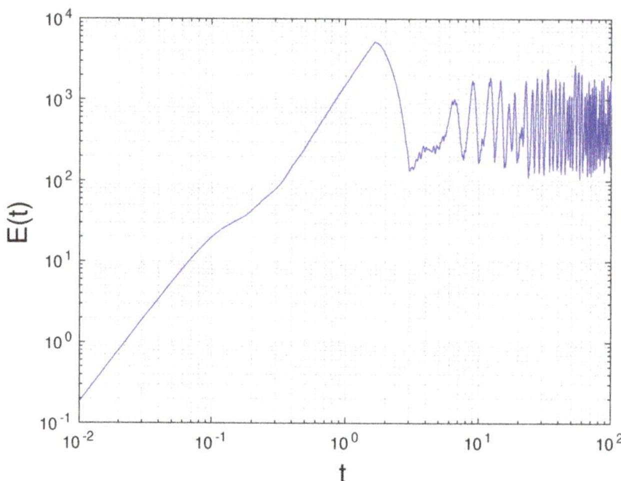

Figure 6. Time evolution of the kinetic energy of the FOM.

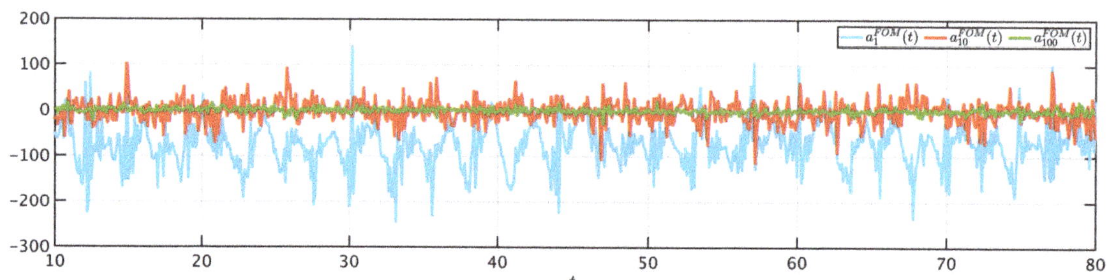

Figure 7. Time evolution of $a_1^{FOM}(t), a_{10}^{FOM}(t), a_{100}^{FOM}(t)$.

Remark (QGE vs. 2D Flow Past a Cylinder). *The 2D flow past a cylinder at low Reynolds numbers has become one of the most popular test problems in the ROM world. The reason is that the time evolution of the true ROM coefficients is periodic and a few ROM modes are required to capture the system's dynamics. By comparison, the QGE test problem used in our numerical investigation is a significantly harder test problem: Its true ROM coefficients display a non-periodic, chaotic time evolution and relatively many ROM modes are required to capture the system's dynamics. This statement is supported by the plot in Figure 4 and the results in Table 1: The plot shows that the eigenvalues decay much faster for the flow past a cylinder test case than for the QGE test case. The results in Table 1 show that, in order to achieve a 90% relative energy content (which is defined in (30)), the flow past a cylinder test case requires only 2 ROM modes, whereas the QGE test case requires 77 ROM modes.*

5.3. Criteria

To investigate the ROMs, we use the following three criteria: (i) the relative L^2 norm of the time-averaged streamfunction errors between ψ^{FOM} and ψ^{ROM}:

$$\left\| \frac{1}{M}\sum_{j=1}^{M} \psi^{FOM}(t_j) - \frac{1}{M}\sum_{j=1}^{M} \psi^{ROM}(t_j) \right\|_{L^2}^2 \Bigg/ \left\| \frac{1}{M}\sum_{j=1}^{M} \psi^{FOM}(t_j) \right\|_{L^2}^2. \tag{36}$$

(ii) The ROM's ability to recover the four-gyre pattern of the time-average of the FOM streamfunction in Figure 5. (iii) The ROM computational cost. The first two criteria quantify the ROM numerical accuracy, whereas the third criterion quantifies the ROM efficiency. We note that the first two criteria utilize time-averages. The reason for using time-averages is that, in the numerical investigation of chaotic systems (such as the four-gyre test problem), pointwise in time quantities are less robust (e.g., prone to phase errors) and can yield deceiving results.

To define the resolved and under-resolved ROM regimes, we use two of the four criteria outlined in Section 4.2.1: (i) the trial and error criterion; and (ii) the eigenvalue decay rate criterion. Specifically, when we use the trial and error criterion in our numerical investigation, we call the ROM regime resolved if its relative L^2 norm of the error (36) is $\mathcal{O}(10^{-1})$, and under-resolved otherwise. We note that an $\mathcal{O}(10^{-1})$ relative error is large by engineering standards. However, our numerical investigation will show that even this large threshold requires high-dimensional ROMs. When we use the eigenvalue decay rate criterion in our numerical investigation, we call the ROM regime resolved if its relative kinetic energy content (which was defined in (30)) is above 90%, and under-resolved otherwise.

5.4. FOM Snapshot Generation

To generate the FOM data (i.e., snapshots) that is used to construct the ROM basis functions, we utilize fine resolution spatial and temporal discretizations. Specifically, for the FOM spatial discretization, we use a pseudospectral method with a 257×513 spatial resolution [8]. For the FOM time discretization, we use an explicit Runge-Kutta method (Tanaka-Yamashita, an order 7 method with an embedded order 6 method for error control), and an error tolerance of 10^{-8} in time with adaptive time refinement and coarsening [8] in addition to an eigenvalue-based time step restriction for ensuring numerical stability. These spatial and temporal discretizations yield numerical results that are similar to the fine resolution numerical results obtained in [32,33].

To collect FOM snapshots, we first need to decide what time interval we utilize. To this end, in Figure 6, we plot the time evolution of the kinetic energy, $E(t)$. Figure 6 (see also Figure 1 in [32]) shows that the flow starts with a short transient interval (approximately $[0, 10]$), after which it converges to a *statistically steady state*. We emphasize that, although the flow is statistically steady, it still displays a complex, chaotic behavior. To illustrate this, in Figure 5, we display the instantaneous contour plot for the streamfunction field at $t = 40$ and $t = 60$. While $t = 40$ and $t = 60$ are well within the statistically steady state regime, the flow displays a non-periodic, complex time evolution, with a high degree of variability. Furthermore, in Figure 7, we plot the time evolution of the true ROM coefficients $a_1^{FOM}(t), a_{10}^{FOM}(t)$, and $a_{100}^{FOM}(t)$, which are obtained by projecting the FOM vorticity data onto the ROM bases, φ_1, φ_{10}, and φ_{100}, respectively:

$$a_i^{FOM}(t) = \left(\omega^{FOM}(t), \varphi_i \right), \tag{37}$$

where $\omega^{FOM}(t)$ is the FOM vorticity at time t. The true ROM coefficients display a non-periodic, chaotic behavior within the time interval $[10, 80]$. Thus, the numerical approximation of this statistically steady regime remains challenging for the ROMs that we investigate in this section.

In our numerical investigation, we follow [8,32,33] and collect 701 FOM snapshots in the time interval $[T_{min}, T_{max}] = [10, 80]$ at equidistant time intervals. Collecting a large number of snapshots ensures that the FOM data used to train the ROM is rich enough to capture the relevant dynamics. Next, we use the algorithm outlined in Section 4 and the FOM snapshots to construct the ROM basis. In Figure 8, we plot selected ROM streamfunction basis functions. We observe that, as the ROM basis index increases, the spatial structures displayed by the ROM basis functions become smaller and smaller. This is consistent with the idea that the ROM modes are arranged in decreasing importance (dominance) order: The first ROM mode is the most dominant, the second ROM mode is the second most dominant, and so on.

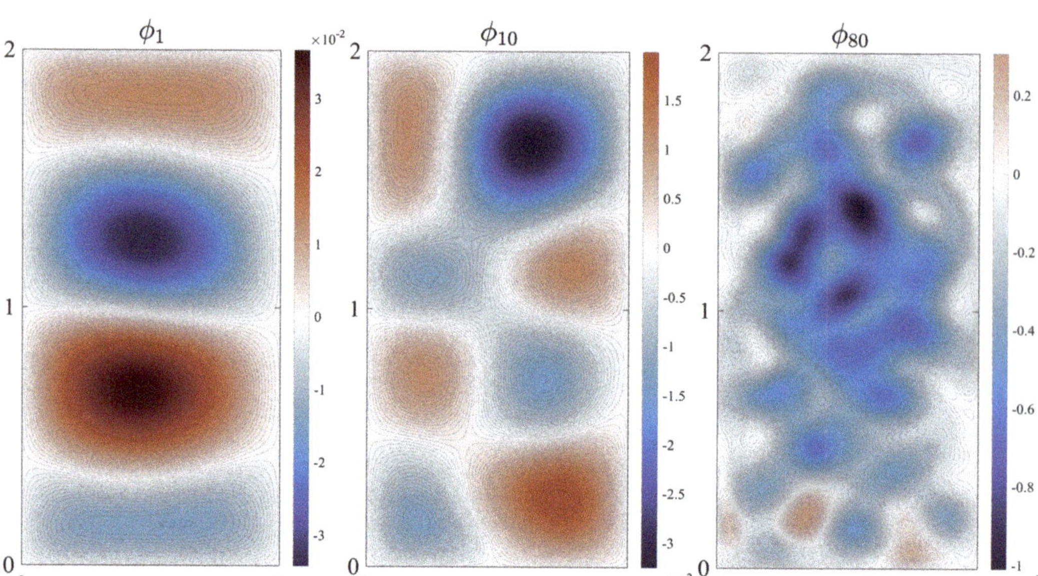

Figure 8. Streamfunction basis functions: ϕ_1, ϕ_{10}, and ϕ_{80}.

5.5. ROM Numerical Investigation

In this section, we perform a numerical investigation of the ROM accuracy and efficiency.

To investigate the ROM accuracy, we consider the four regimes discussed in Section 5.1. First, we consider the resolved regime, both in the reconstructive (Section 5.5.1) and predictive (Section 5.5.2) settings. In these two regimes, we investigate only the standard G-ROM (25), since in the resolved case there is no need for ROM closure. The goal of these two sections is to use the two criteria presented in Section 5.3 (i.e., the relative L^2 error and the relative energy content) to determine the minimum ROM dimension (r) that is necessary in the resolved regime. Next, we consider the under-resolved regime, both in the reconstructive (Section 5.5.3) and predictive (Section 5.5.4) settings. In these two regimes, we investigate the standard G-ROM (25) and one LES-ROM, i.e., the DD-VMS-ROM presented in Section 4.2.2. The goal of these two sections is to determine whether the LES-ROM can significantly increase the standard G-ROM accuracy in the under-resolved regime.

To investigate the ROM efficiency, in Section 5.5.5 we discuss the computational cost of the standard G-ROM and the LES-ROM.

For both the G-ROM and the LES-ROM, we use the same time discretization on the time interval $[10, 80]$: the RK4 method with a uniform step size $\Delta t = 10^{-3}$.

5.5.1. Resolved, Reconstructive Regime

In this section, we consider the resolved, reconstructive regime.

In Table 2, we list the relative L^2 errors (36) of the time-averaged streamfunction and the relative energy content (30) for G-ROM with several r values: $r = 10, 20, 40,$ and 80. As expected, as the G-ROM dimension (r) increases, the relative errors converge to 0 and the relative energy content increases. We emphasize, however, that one needs a relatively large r value to attain what we defined as a resolved regime: *To attain an $\mathcal{O}(10^{-1})$ relative error and 90% relative energy content, one needs to take $r = \mathcal{O}(10^2)$.*

Table 2. Resolved, reconstructive regime. Relative L^2 errors (36) of the time-averaged streamfunction and relative energy content (30) for G-ROM with different r values.

r	10	20	40	80	120
Relative error	2.009×10^2	7.377×10^0	4.595×10^{-1}	2.999×10^{-1}	1.493×10^{-1}
Relative energy content	65.24%	75.25%	83.65%	90.33%	93.48%

In Figure 9, for $r = 10, 40$, and 120, we plot the time-average of the streamfunction ψ over the time interval $[10, 80]$ for the FOM and G-ROM. We note that we use the same scale for the FOM and the G-ROM with large r values (i.e., $r = 40$ and $r = 120$). However, for the G-ROM with a low r value (i.e., $r = 10$), we use a different scale, since the magnitude of these G-ROM results is much larger than the rest. The plots in Figure 9 show that the G-ROM with low r values (i.e., $r = 10$ and $r = 40$) fails to recover the FOM four-gyre pattern. The G-ROM with $r = 120$ captures the FOM four-gyre pattern, but even in this case the magnitude of the time-averaged streamfunction is only marginally accurate. Thus, the plots in Figure 9 support the results in Table 2: *To recover the FOM four-gyre pattern, one needs to take $r = \mathcal{O}(10^2)$.*

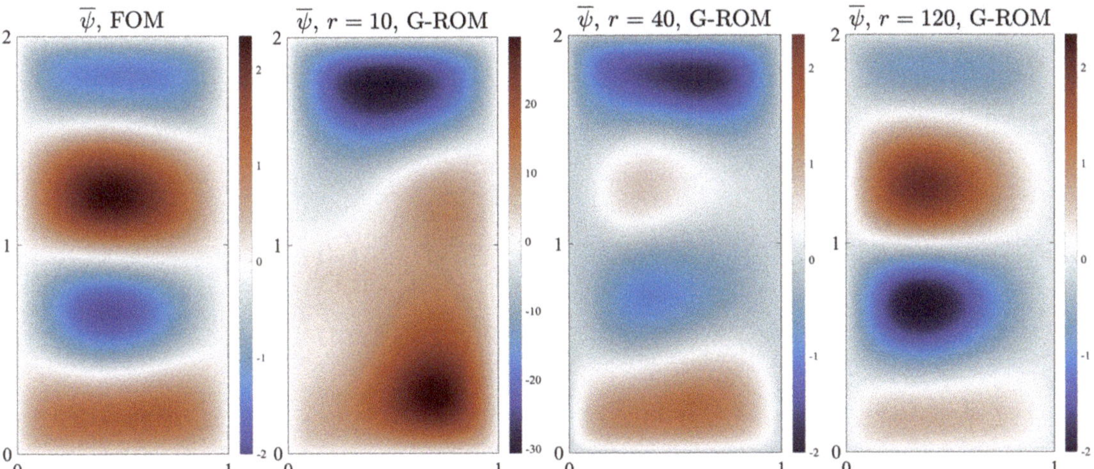

Figure 9. Resolved, reconstructive regime. Time-averaged streamfunction, $\overline{\psi}$, for FOM and G-ROM with $r = 10, 40$, and 120.

5.5.2. Resolved, Predictive Regime

In this section, we consider the resolved, predictive regime. To construct the G-ROM basis functions, we use data (snapshots) from the time interval $[10, 45]$ and test the G-ROM on a longer time interval (i.e., $[10, 80]$) to test the predictive capabilities of the G-ROM.

In Table 3, we list the relative L^2 errors (36) of the time-averaged streamfunction and the relative energy content (30) for G-ROM with several r values: $r = 10, 20, 40$, and 80. We note that, as the G-ROM dimension (r) increases, the errors converge to 0 and the relative energy content increases. As expected, the errors in the predictive regime are worse than the errors in the reconstructive regime in Section 5.5.1. Furthermore, as in the reconstructive regime, one needs a relatively large r value to attain what we defined as a resolved regime: *To attain an $\mathcal{O}(10^{-1})$ error and 90% relative energy content, one needs to take $r = \mathcal{O}(10^2)$.*

Table 3. Resolved, predictive regime. Relative L^2 errors (36) of the time-averaged streamfunction and relative energy content (30) for Galerkin ROM (G-ROM) with different r values.

r	10	20	40	80	120
Relative error	2.030×10^2	1.015×10^1	5.115×10^{-1}	3.892×10^{-1}	2.619×10^{-1}
Relative energy content	66.03%	76.38%	85.17%	92.23%	95.41%

In Figure 10, for $r = 10, 40$, and 120, we plot the time-average of the streamfunction ψ over the time interval $[10, 80]$ for the FOM and G-ROM. We note that we use the same scale for the FOM and the G-ROM with large r values (i.e., $r = 40$ and $r = 120$). For the G-ROM with a low r value (i.e., $r = 10$), we use a different scale, since the magnitude of these G-ROM results is much larger than the rest. The plots in Figure 10 show that, as in the reconstructive regime in Section 5.5.1, the G-ROM with low r values (i.e., $r = 10$ and $r = 40$) fails to recover the FOM four-gyre pattern. The G-ROM with $r = 120$ captures the FOM four-gyre pattern, but even in this case the magnitude of the time-averaged streamfunction is only marginally accurate. Thus, the plots in Figure 10 support the results in Table 3: *To recover the FOM four-gyre pattern, one needs to take* $r = \mathcal{O}(10^2)$.

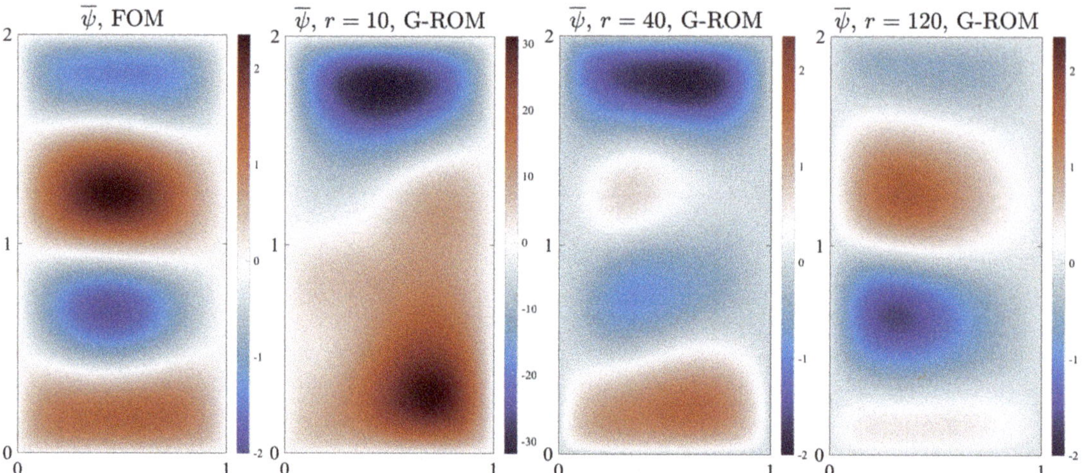

Figure 10. Resolved, predictive regime. Time-averaged streamfunction, $\overline{\psi}$, for FOM and G-ROM with $r = 10, 40$, and 120.

5.5.3. Under-Resolved, Reconstructive Regime

In this section, we consider the under-resolved, reconstructive regime. Since we use the under-resolved regime, we investigate the standard G-ROM and an LES-ROM. Specifically, we investigate the improved, three-scale version [90] of the DD-VMS-ROM (34).

In Table 4, we list the relative L^2 errors (36) of the time-averaged streamfunction of the G-ROM and LES-ROM for several r values: $r = 10, 15$, and 20. For all the r values considered, *the LES-ROM is orders of magnitude more accurate than the G-ROM*. More importantly, for $r = 20$, *the LES-ROM is almost one order of magnitude more accurate than the G-ROM with $r = 120$*, which was used in Table 2.

In Figure 11, for $r = 10$, we plot the time-average of the streamfunction ψ over the time interval $[10, 80]$ for the FOM, G-ROM, and LES-ROM. We note that we use the same scale for the FOM and the LES-ROM. For the G-ROM, however, we use a different scale, since the magnitude of the G-ROM results is much larger than the rest. The plots in Figure 11 show that the G-ROM fails to recover the FOM four-gyre pattern. On the other hand, the LES-ROM successfully captures the four-gyre pattern and its correct magnitude. In fact, the LES-ROM with $r = 10$ is even more accurate than the resolved G-ROM with $r = 120$

in Figure 9. Thus, the plots in Figure 11 support the results in Table 4: *The LES-ROM is dramatically more accurate than the G-ROM.*

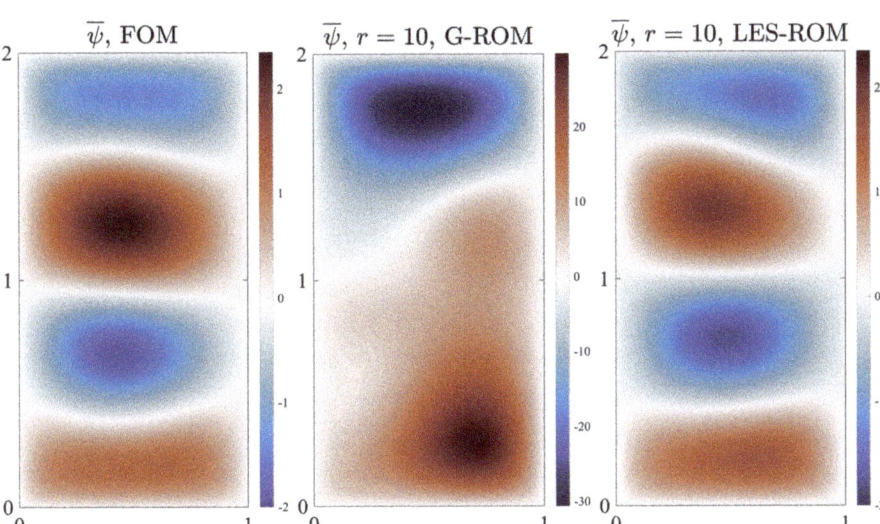

Figure 11. Under-resolved, reconstructive regime. Time-averaged streamfunction, $\overline{\psi}$, for FOM, G-ROM, and large eddy simulation (LES)-ROM with $r = 10$.

Table 4. Under-resolved, reconstructive regime. Relative L^2 errors (36) of the time-averaged streamfunction for G-ROM and LES-ROM for different r values.

r	G-ROM	LES-ROM
10	2.009×10^2	1.074×10^{-1}
15	5.569×10^1	6.780×10^{-2}
20	7.377×10^0	2.784×10^{-2}

5.5.4. Under-Resolved, Predictive Regime

In this section, we consider the under-resolved, predictive regime for the G-ROM and LES-ROM. To construct the G-ROM and LES-ROM basis functions, we use data (snapshots) from the time interval $[10, 45]$ and test the G-ROM and LES-ROM on a longer time interval (i.e., $[10, 80]$) to test the predictive capabilities of the G-ROM and LES-ROM.

In Table 5, we list the relative L^2 errors (36) of the time-averaged streamfunction of the G-ROM and LES-ROM for several r values: $r = 10, 15$, and 20. For all the r values considered, *the LES-ROM is orders of magnitude more accurate than the G-ROM.* Most importantly, for $r = 20$, *the LES-ROM is more accurate than the G-ROM with $r = 120$*, which was used in Table 3.

In Figure 12, for $r = 10$, we plot the time-average of the streamfunction ψ over the time interval $[10, 80]$ for the FOM, G-ROM, and LES-ROM. We note that we use the same scale for the FOM and the LES-ROM. For the G-ROM, however, we use a different scale, since the magnitude of the G-ROM results is much larger than the rest. The plots in Figure 12 show that the G-ROM fails to recover the FOM four-gyre pattern. On the other hand, the LES-ROM successfully captures the four-gyre pattern and its correct magnitude. Thus, the plots in Figure 12 support the results in Table 5: *The LES-ROM is significantly more accurate than the G-ROM.*

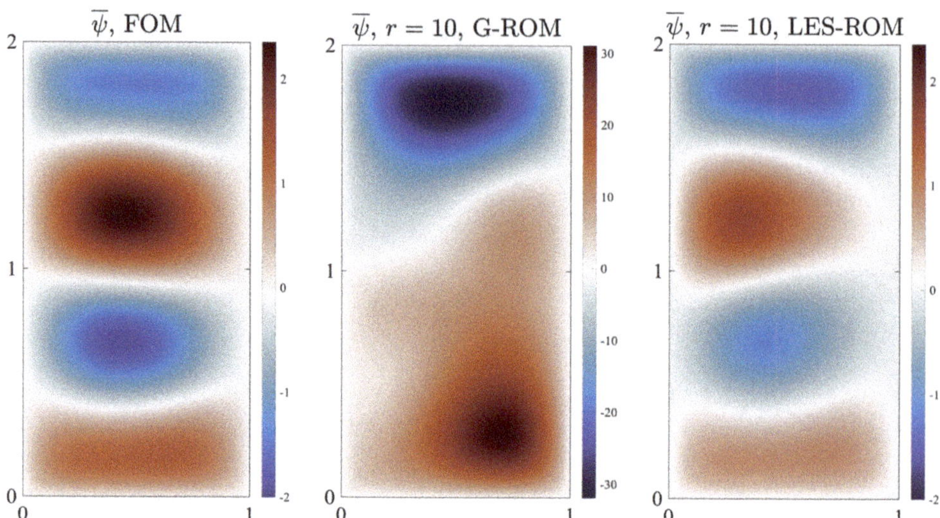

Figure 12. Under-resolved, predictive regime. Time-averaged streamfunction, $\overline{\psi}$, for FOM, G-ROM, and LES-ROM with $r = 10$.

Table 5. Under-resolved, predictive regime. Relative L^2 errors (36) of the time-averaged streamfunction for G-ROM and LES-ROM for different r values.

r	G-ROM	LES-ROM
10	2.030×10^2	1.622×10^{-1}
15	2.880×10^2	2.385×10^{-1}
20	1.015×10^1	1.266×10^{-1}

5.5.5. Computational Cost

The ROM computational cost has two components: (i) the computational cost of the *offline stage*, i.e., when the ROM operators are assembled; and (ii) the computational cost of the *online stage*, i.e., when the ROM is actually used in practical computations. While the offline computational cost can be high, it is often offset in the online stage, when the ROM is used for numerous runs.

In Table 6, we list the CPU time for the FOM, G-ROM, and LES-ROM in the online stage. We note that the CPU time of the G-ROM is similar to the CPU time of the LES-ROM. We emphasize that *both the G-ROM and the LES-ROM CPU times are orders of magnitude lower than the FOM CPU time.* Furthermore, the G-ROM CPU time increases significantly as r increases.

Table 6. CPU time for FOM, G-ROM, and LES-ROM in the online stage.

FOM CPU time			2.19×10^5 s		

G-ROM CPU time	$r = 10$	$r = 20$	$r = 40$	$r = 80$	$r = 120$
	2.69×10^0 s	4.80×10^0 s	4.58×10^1 s	1.32×10^2 s	6.45×10^2 s

LES-ROM CPU time	$r = 10$	$r = 15$	$r = 20$
	3.22×10^0 s	3.85×10^0 s	5.07×10^0 s

5.5.6. Summary

The results in our numerical investigation yield the following conclusions:

1. For our test problem, the resolved regime requires ROMs that have a *large dimension* (i.e., $r = \mathcal{O}(10^2)$) in both the reconstructive and the predictive regimes.
2. In the realistic, under-resolved regime, *the LES-ROM is orders of magnitude more accurate than the G-ROM* in both the reconstructive and the predictive regimes.
3. The LES-ROM in the under-resolved regime (i.e., with $r = 20$) is *significantly more accurate* and *dramatically more efficient* than the G-ROM in the resolved regime (i.e., with $r = 120$).

6. Conclusions and Outlook

The quasi-geostrophic equations (QGE) (also known as the barotropic vorticity equations) are a simplified mathematical model for large scale wind-driven ocean circulation. Since the QGE computational cost is significantly lower than the computational cost of full fledged mathematical models of ocean flows, the QGE have often been used to test new numerical methods for geophysical flows, such as reduced order models (ROMs).

In this brief survey, we summarized projection-based ROMs developed for the QGE in order to understand ROMs' potential in efficient numerical simulations of ocean flows. Specifically, in Section 2, we briefly explained how the QGE are derived from the primitive equations by using simple scaling arguments. We also outlined the various QGE formulations currently used, and we illustrated the importance of the Rossby number, which quantifies the rotation effects in the QGE. In Section 3, we surveyed the main numerical methods used in the spatial discretization of the QGE: finite difference, finite volume, pseudospectral and spectral, and finite element methods. In Section 4, we presented the main steps in the construction of the standard Galerkin ROM (G-ROM). Specifically, we showed how the full order model (FOM) simulations generate data (snapshots) that is used to build the ROM basis, which is then utilized in a Galerkin projection framework to construct the G-ROM. We also emphasized the importance of appropriate treatment of the under-resolved regime, i.e., when the number of ROM modes is not enough to capture the relevant QGE dynamics. The ROM under-resolved regime is often encountered in realistic geophysical settings dominated by convection, when the Kolmogorov n-width is large. One of the main approaches for tacking the ROM under-resolved regime is ROM closure modeling, i.e., modeling the effect of the discarded ROM modes. We reviewed two types of ROM closure models for the QGE: large eddy simulation (LES) ROM closure models (which are based on spatial filtering and data driven modeling), and machine learning (ML) ROM closure models. Finally, in Section 5, we showed how ROMs are used in the numerical simulation of the QGE. To this end, we considered a QGE test problem in which long-term time averaging yields a four-gyre pattern. We showed that, if enough ROM modes were used (i.e., in the resolved regime), the standard G-ROM yielded accurate results at a low computational cost. If, however, only a few ROM modes were used (i.e., in the under-resolved regime), the standard G-ROM yielded inaccurate results, whereas the LES-ROM yielded accurate results at a low computational cost.

ROMs have a significant potential in efficient and relatively accurate numerical simulations of geophysical flows that display recurrent dominant spatial structures. This brief survey aimed at showcasing the ROMs' potential in simplified settings, i.e., for QGE simulations. We emphasize, however, that the ultimate goal is to use ROMs in realistic many query atmospheric and oceanic applications, e.g., uncertainty quantification and data assimilation. While the first steps have been made (see, e.g., [106–112]), there are significant challenges that still need to be addressed. Next, we present several potential future research avenues in the ROM exploration of the QGE and more complex models of geophysical flows.

To develop ROMs for geophysical flows, *realistic* computational settings need to be considered. For example, realistic parameters (e.g., the Reynolds number, *Re*), and realistic complex geometries need to be investigated. Since realistic oceanic and atmospheric flows display an enormous range of spatial and temporal scales, new ROMs need to be constructed for under-resolved regimes in which the ROM closure problem becomes central, just as in FOM. Thus, novel robust, stable, accurate, and efficient ROM closure models for realistic geophysical flows need to be built. But how should these ROM closure models be developed? By using physical insight (as in classical FOMs), data (as currently done in many research areas), or both? Furthermore, in addition to the rotation effects modeled by the QGE, *stratification* should also be investigated. In the simplified QGE setting, stratification could be included by considering the multilayer QGE or the continuously stratified QGE. More realistic western boundary layers should also be investigated. Of course, all these problems are compounded when mathematical models that are more accurate than the QGE are considered, such as the Boussinesq equations. Finally, mathematical support for these new ROMs needs to be provided. The first steps in this direction have been made (see, e.g., [61,75]), but much more remains to be done.

Author Contributions: Conceptualization, C.M., Z.W., D.R.W., X.X. and T.I.; methodology, C.M., Z.W., D.R.W., X.X. and T.I.; software, C.M., Z.W. and D.R.W.; validation, C.M., Z.W. and D.R.W.; formal analysis, C.M., Z.W. and D.R.W.; investigation, C.M., Z.W., D.R.W., X.X. and T.I.; resources, C.M., Z.W., D.R.W., X.X. and T.I.; data curation, C.M., Z.W. and D.R.W.; writing—original draft preparation, C.M., Z.W., D.R.W., X.X. and T.I.; writing—review and editing, C.M., Z.W., D.R.W., X.X. and T.I.; visualization, C.M., Z.W. and D.R.W.; supervision, T.I.; project administration, T.I.; funding acquisition, Z.W., D.R.W. and T.I. All authors have read and agreed to the published version of the manuscript.

Funding: This research was funded by National Science Foundation, DMS-2012253, DMS-1953113, DMS-1913073, OAC-1450327, and by U.S. Department of Energy, DE-SC0020270.

Institutional Review Board Statement: Not applicable.

Informed Consent Statement: Not applicable.

Conflicts of Interest: The authors declare no conflict of interest.

References

1. Hesthaven, J.S.; Rozza, G.; Stamm, B. *Certified Reduced Basis Methods for Parametrized Partial Differential Equations*; Springer: Berlin/Heidelberg, Germany, 2015.
2. Holmes, P.; Lumley, J.L.; Berkooz, G. *Turbulence, Coherent Structures, Dynamical Systems and Symmetry*; Cambridge University Press: Cambridge, UK, 1996.
3. Quarteroni, A.; Manzoni, A.; Negri, F. *Reduced Basis Methods for Partial Differential Equations: An Introduction*; Springer: Berlin/Heidelberg, Germany, 2015; Volume 92.
4. Benner, P.; Gugercin, S.; Willcox, K. A survey of projection-based model reduction methods for parametric dynamical systems. *SIAM Rev.* **2015**, *57*, 483–531. [CrossRef]
5. Brunton, S.L.; Kutz, J.N. *Data-Driven Science and Engineering: Machine Learning, Dynamical Systems, and Control*; Cambridge University Press: Cambridge, UK, 2019.
6. Noack, B.R.; Morzynski, M.; Tadmor, G. *Reduced-Order Modelling for Flow Control*; Springer: Berlin/Heidelberg, Germany, 2011; Volume 528.
7. Taira, K.; Hemati, M.S.; Brunton, S.L.; Sun, Y.; Duraisamy, K.; Bagheri, S.; Dawson, S.T.M.; Yeh, C.A. Modal analysis of fluid flows: Applications and outlook. *AIAA J.* **2020**, *58*, 998–1022. [CrossRef]
8. Mou, C.; Liu, H.; Wells, D.R.; Iliescu, T. Data-Driven Correction Reduced Order Models for the Quasi-Geostrophic Equations: A Numerical Investigation. *Int. J. Comput. Fluid Dyn.* **2020**, *34*, 147–159. [CrossRef]
9. Xie, X.; Mohebujjaman, M.; Rebholz, L.G.; Iliescu, T. Data-Driven Filtered Reduced Order Modeling of Fluid Flows. *SIAM J. Sci. Comput.* **2018**, *40*, B834–B857. [CrossRef]
10. Star, S.K.; Stabile, G.; Belloni, F.; Rozza, G.; Degroote, J. Extension and comparison of techniques to enforce boundary conditions in finite volume POD-Galerkin reduced order models for fluid dynamic problems. *arXiv* **2019**, arXiv:1912.00825.
11. Couplet, M.; Sagaut, P.; Basdevant, C. Intermodal energy transfers in a proper orthogonal decomposition-Galerkin representation of a turbulent separated flow. *J. Fluid Mech.* **2003**, *491*, 275. [CrossRef]
12. Hess, M.W.; Quaini, A.; Rozza, G. Reduced basis model order reduction for Navier–Stokes equations in domains with walls of varying curvature. *Int. J. Comput. Fluid Dyn.* **2020**, *34*, 119–126. [CrossRef]
13. Pitton, G.; Quaini, A.; Rozza, G. Computational reduction strategies for the detection of steady bifurcations in incompressible fluid-dynamics: Applications to Coanda effect in cardiology. *J. Comput. Phys.* **2017**, *344*, 534–557. [CrossRef]

14. Skitka, J.; Marston, J.B.; Fox-Kemper, B. Reduced-Order Quasilinear Model of Ocean Boundary-Layer Turbulence. *J. Phys. Oceanogr.* **2020**, *50*, 537–558. [CrossRef]

15. Cushman-Roisin, B.; Beckers, J.M. *Introduction to Geophysical Fluid Dynamics: Physical and Numerical Aspects*; Academic Press: Cambridge, MA, USA, 2011.

16. Majda, A.J.; Wang, X. *Nonlinear Dynamics and Statistical Theories for Basic Geophysical Flows*; Cambridge University Press: Cambridge, UK, 2006; p. xii+551.

17. Vallis, G.K. *Atmospheric and Oceanic Fluid Dynamics: Fundamentals and Large-Scale Circulation*; Cambridge University Press: Cambridge, UK, 2006.

18. Charney, J.G.; Fjörtoft, R.; Neumann, J.V. Numerical Integration of the Barotropic Vorticity Equation. *Tellus* **1950**, *2*, 237–254. [CrossRef]

19. Majda, A.J.; Shefter, M.G. Nonlinear instability of elementary stratified flows at large Richardson number. *Chaos Interdiscip. J. Nonlinear Sci.* **2000**, *10*, 3–27. [CrossRef] [PubMed]

20. Majda, A.J.; Shefter, M.G. Elementary stratified flows with instability at large Richardson number. *J. Fluid Mech.* **1998**, *376*, 319–350. [CrossRef]

21. Majda, A.J.; Embid, P. Averaging over fast gravity waves for geophysical flows with unbalanced initial data. *Theor. Comput. Fluid. Dyn.* **1998**, *11*, 155–169. [CrossRef]

22. Embid, P.F. Averaging over fast gravity waves for geophysical flows with arbitary. *Commun. Partial. Differ. Equ.* **1996**, *21*, 619–658. [CrossRef]

23. Embid, P.F.; Majda, A.J. Low Froude number limiting dynamics for stably stratified flow with small or finite Rossby numbers. *Geophys. Astro. Fluid.* **1998**, *87*, 1–50. [CrossRef]

24. Majda, A.J.; Grote, M.J. Model dynamics and vertical collapse in decaying strongly stratified flows. *Phys. Fluids.* **1997**, *9*, 2932–2940. [CrossRef]

25. Dijkstra, H.A. *Dynamical Oceanography*; Springer: Berlin, Germany, 2008; p. xvi+407.

26. Pedlosky, J. *Geophysical Fluid Dynamics*, 2nd ed.; Springer: Berlin, Germany, 1992.

27. Greatbatch, R.J.; Nadiga, B.T. Four-gyre circulation in a barotropic model with double-gyre wind forcing. *J. Phys. Oceanogr.* **2000**, *30*, 1461–1471. [CrossRef]

28. Holm, D.D.; Nadiga, B.T. Modeling mesoscale turbulence in the barotropic double-gyre circulation. *J. Phys. Oceanogr.* **2003**, *33*, 2355–2365. [CrossRef]

29. Nadiga, B.T.; Margolin, L.G. Dispersive-dissipative eddy parameterization in a barotropic model. *J. Phys. Oceanogr.* **2001**, *31*, 2525–2531. [CrossRef]

30. Foster, E.L.; Iliescu, T.; Wang, Z. A finite element discretization of the streamfunction formulation of the stationary quasi-geostrophic equations of the ocean. *Comput. Methods Appl. Mech. Eng.* **2013**, *261*, 105–117. [CrossRef]

31. Monteiro, I.O.; Manica, C.C.; Rebholz, L.G. Numerical study of a regularized barotropic vorticity model of geophysical flow. *Numer. Methods Partial. Differ. Equ.* **2015**, *31*, 1492–1514. [CrossRef]

32. San, O.; Iliescu, T. A stabilized proper orthogonal decomposition reduced-order model for large scale quasigeostrophic ocean circulation. *Adv. Comput. Math.* **2015**, *41*, 1289–1319. [CrossRef]

33. San, O.; Staples, A.E.; Wang, Z.; Iliescu, T. Approximate deconvolution large eddy simulation of a barotropic ocean circulation model. *Ocean Model.* **2011**, *40*, 120–132. [CrossRef]

34. Medjo, T.T. Numerical Simulations of a Two-Layer Quasi-Geostrophic Equation of the Ocean. *SIAM J. Numer. Anal.* **2000**, *37*, 2005–2022. [CrossRef]

35. Medjo, T.T. Multi-layer quasi-geostrophic equations of the ocean with delays. *Discret. Contin. Dyn. Syst. Ser. B* **2008**, *10*, 171.

36. Shevchenko, I.; Berloff, P. Multi-layer quasi-geostrophic ocean dynamics in eddy-resolving regimes. *Ocean Model.* **2015**, *94*, 1–14. [CrossRef]

37. Phillips, N.A. The general circulation of the atmosphere: A numerical experiment. *Q. J. R. Meteorol. Soc.* **1956**, *82*, 123–164. [CrossRef]

38. Arakawa, A.; Lamb, V.R. Computational Design of the Basic Dynamical Processes of the UCLA General Circulation Model. In *General Circulation Models of the Atmosphere*; Elsevier: Amsterdam, The Netherlands, 1977; Volume 17, pp. 173–265. [CrossRef]

39. Collins, S.N.; James, R.S.; Ray, P.; Chen, K.; Lassman, A.; Brownlee, J. Grids in numerical weather and climate models. In *Climate Change and Regional/Local Responses*; IntechOpen: Rijeka, Croatia, 2013. [CrossRef]

40. Harlow, F.H.; Welch, J.E. Numerical calculation of time-dependent viscous incompressible flow of fluid with free surface. *Phys. Fluids* **1965**, *8*, 2182–2189. [CrossRef]

41. Modular Ocean Model (MOM)—Geophysical Fluid Dynamics Laboratory. Available online: https://www.gfdl.noaa.gov/mom-ocean-model/# (accessed on 9 November 2020).

42. Maulik, R.; San, O. Dynamic modeling of the horizontal eddy viscosity coefficient for quasigeostrophic ocean circulation problems. *J. Ocean Eng. Sci.* **2016**, *1*, 300–324. [CrossRef]

43. Maulik, R.; San, O. A novel dynamic framework for subgrid scale parametrization of mesoscale eddies in quasigeostrophic turbulent flows. *Comput. Math. Appl.* **2017**, *74*, 420–445. [CrossRef]

44. San, O.; Staples, A.E.; Iliescu, T. Approximate Deconvolution Large Eddy Simulation of a Stratified Two-Layer Quasigeostrophic Ocean Model. *Ocean Model.* **2013**, *63*, 1–20. [CrossRef]

45. Griffiths, S.D. Kelvin wave propagation along straight boundaries in C-grid finite-difference models. *J. Comput. Phys.* **2013**, *255*, 639–659. [CrossRef]

46. Campin, J.M.; Heimbach, P.; Losch, M.; Forget, G.; Adcroft, A.; Menemenlis, D.; Hill, C.; Jahn, O.; Scott, J.; Mazloff, M.; et al. MITgcm/MITgcm: Mid 2020 Version. 2020. Available online: https://zenodo.org/record/3967889/export/xd#.X-08 7RYRVPY(accessed on 9 November 2020).

47. Thuburn, J.; Ringler, T.D.; Skamarock, W.C.; Klemp, J.B. Numerical representation of geostrophic modes on arbitrarily structured C-grids. *J. Comput. Phys.* **2009**, *228*, 8321–8335. [CrossRef]

48. Ringler, T.D.; Thuburn, J.; Klemp, J.B.; Skamarock, W.C. A unified approach to energy conservation and potential vorticity dynamics for arbitrarily-structured C-grids. *J. Comput. Phys.* **2010**, *229*, 3065–3090. [CrossRef]

49. Chen, Q.; Ringler, T.D.; Gunzburger, M. A co-volume scheme for the rotating shallow water equations on conforming non-orthogonal grids. *J. Comput. Phys.* **2013**, *240*, 174–197. [CrossRef]

50. Chen, Q.; Ju, L. Conservative finite-volume schemes for the quasi-geostrophic equation on coastal-conforming unstructured primal–dual meshes. *Q. J. R. Meteorol. Soc.* **2018**, *144*, 1106–1122. [CrossRef]

51. Randall, D.A. Geostrophic adjustment and the finite-difference shallow-water equations. *Mon. Weather Rev.* **1994**, *122*, 1371–1377. [CrossRef]

52. Nadiga, B. Nonlinear evolution of a baroclinic wave and imbalanced dissipation. *J. Fluid Mech.* **2014**, *756*, 965–1006. [CrossRef]

53. San, O.; Staples, A.E. An efficient coarse grid projection method for quasigeostrophic models of large-scale ocean circulation. *Int. J. Multiscale Comput. Eng.* **2013**, *11*, 463–495. [CrossRef]

54. Phillips, N.A. An example of non-linear computational instability. *Atmos. Sea Motion* **1959**, *501*, 504.

55. Orszag, S.A. On the elimination of aliasing in finite-difference schemes by filtering high-wavenumber components. *J. Atmos. Sci.* **1971**, *28*, 1074. [CrossRef]

56. Abernathey, R.; Rocha, C.B.; Poulin, F.; Jansen, M. pyqg: v0.1.4. 2015. Available online: https://zenodo.org/record/32539#.X-09 JhYRVPY (accessed on 9 November 2020).

57. Hogg, A.M.C.; Dewar, W.K.; Killworth, P.D.; Blundell, J.R. A Quasi-Geostrophic Coupled Model (Q-GCM). *Mon. Weather Rev.* **2003**, *131*, 2261–2278. [CrossRef]

58. Canuto, C.; Hussaini, M.Y.; Quarteroni, A.; Zang, T.A. *Spectral Methods: Fundamentals in Single Domains*; Scientific Computation; Springer: Berlin, Germany, 2006.

59. Frigo, M.; Johnson, S.G. The Design and Implementation of FFTW3. *Proc. IEEE* **2005**, *93*, 216–231. [CrossRef]

60. Foster, E.L. Finite Elements for the Quasi-Geostrophic Equations of the Ocean. Ph.D. Thesis, Virginia Tech, Blacksburg, VA, USA, 2013.

61. Galán del Sastre, P. Estudio Numérico Del Atractor en Ecuaciones de Navier-Stokes Aplicadas a Modelos de Circulación Del océano. Ph.D. Thesis, Universidad Complutense de Madrid, Madrid, Spain, 2004.

62. Kim, T.Y.; Park, E.J.; Shin, D.W. A C0-discontinuous Galerkin method for the stationary quasi-geostrophic equations of the ocean. *Comput. Meth. Appl. Mech. Engrg.* **2016**, *300*, 225–244. [CrossRef]

63. Shin, D.W.; Kang, Y.; Park, E.J. C0-discontinuous Galerkin methods for a wind-driven ocean circulation model: Two-grid algorithm. *Comput. Meth. Appl. Mech. Engrg.* **2018**, *328*, 321–339. [CrossRef]

64. Fix, G. Finite element models for ocean circulation problems. *SIAM J. Appl. Math.* **1975**, *29*, 371–387. [CrossRef]

65. LeProvost, C.; Bernier, C.; Blayo, E. A comparison of two numerical methods for integrating a quasi-geostrophic multilayer model of ocean circulations: finite element and finite difference methods. *J. Comput. Phys.* **1994**, *110*, 341–359. [CrossRef]

66. Temam, R. *Navier–Stokes Equations: Theory and Numerical Analysis*; American Mathematical Society: Providence, RI, USA, 2001; Volume 2.

67. Medjo, T.T. Mixed Formulation of the Two-Layer Quasi-Geostrophic Equations of the Ocean. *Numer. Methods Partial. Differ. Equ. Int. J.* **1999**, *15*, 489–502. [CrossRef]

68. Cascon, J.M.; Garcia, G.C.; Rodriguez, R. A Priori and A Posteriori Error Analysis for a Large-Scale Ocean Circulation Finite Element Model. *Comput. Methods Appl. Mech. Eng.* **2003**, *192*, 5305–5327. [CrossRef]

69. Kim, T.Y.; Iliescu, T.; Fried, E. B-spline based finite-element method for the stationary quasi-geostrophic equations of the ocean. *Comput. Methods Appl. Mech. Eng.* **2015**, *286*, 168–191. [CrossRef]

70. Jiang, W.; Kim, T.Y. Spline-based finite-element method for the stationary quasi-geostrophic equations on arbitrary shaped coastal boundaries. *Comp. Meth. Appl. Mech. Eng.* **2016**, *299*, 144–160. [CrossRef]

71. Al Balushi, I.; Jiang, W.; Tsogtgerel, G.; Kim, T.Y. Adaptivity of a B-spline based finite-element method for modeling wind-driven ocean circulation. *Comp. Meth. Appl. Mech. Eng.* **2018**, *332*, 1–24. [CrossRef]

72. Kim, D.; Kim, T.Y.; Park, E.J.; Shin, D.w. Error estimates of B-spline based finite-element methods for the stationary quasi-geostrophic equations of the ocean. *Comp. Meth. Appl. Mech. Eng.* **2018**, *335*, 255–272. [CrossRef]

73. Rotundo, N.; Kim, T.Y.; Jiang, W.; Heltai, L.; Fried, E. Error analysis of a B-spline based finite-element method for modeling wind-driven ocean circulation. *J. Sci. Comput.* **2016**, *69*, 430–459. [CrossRef]

74. Crommelin, D.T.; Majda, A.J. Strategies for model reduction: comparing different optimal bases. *J. Atmos. Sci.* **2004**, *61*, 2206–2217. [CrossRef]

75. Galán del Sastre, P.; Bermejo, R. Error estimates of proper orthogonal decomposition eigenvectors and Galerkin projection for a general dynamical system arising in fluid models. *Numer. Math.* **2008**, *110*, 49–81. [CrossRef]

76. Franzke, C.; Majda, A.J.; Vanden-Eijnden, E. Low-order stochastic mode reduction for a realistic barotropic model climate. *J. Atmos. Sci.* **2005**, *62*, 1722–1745. [CrossRef]
77. Kondrashov, D.; Berloff, P. Stochastic modeling of decadal variability in ocean gyres. *Geophys. Res. Lett.* **2015**, *42*, 1543–1553. [CrossRef]
78. Kondrashov, D.; Chekroun, M.D.; Berloff, P. Multiscale Stuart-Landau emulators: Application to wind-driven ocean gyres. *Fluids* **2018**, *3*, 21. [CrossRef]
79. Rahman, S.M.; Ahmed, S.E.; San, O. A dynamic closure modeling framework for model order reduction of geophysical flows. *Phys. Fluids* **2019**, *31*, 046602. [CrossRef]
80. Selten, F.M. An efficient description of the dynamics of barotropic flow. *J. Atmos. Sci.* **1995**, *52*, 915–936. [CrossRef]
81. Selten, F.M. Baroclinic empirical orthogonal functions as basis functions in an atmospheric model. *J. Atmos. Sci.* **1997**, *54*, 2099–2114. [CrossRef]
82. Selten, F.M. A statistical closure of a low-order barotropic model. *J. Atmos. Sci.* **1997**, *54*, 1085–1093. [CrossRef]
83. Strazzullo, M.; Ballarin, F.; Mosetti, R.; Rozza, G. Model Reduction for Parametrized Optimal Control Problems in Environmental Marine Sciences and Engineering. *SIAM J. Sci. Comput.* **2018**, *40*, B1055–B1079. [CrossRef]
84. Rahman, S.M.; Pawar, S.; San, O.; Rasheed, A.; Iliescu, T. A nonintrusive reduced order modeling framework for quasigeostrophic turbulence. *Phys. Rev. E* **2019**, *100*, 053306. [CrossRef]
85. San, O.; Maulik, R. Extreme learning machine for reduced order modeling of turbulent geophysical flows. *Phys. Rev. E* **2018**, *97*, 042322. [CrossRef]
86. Xie, X.; Nolan, P.J.; Ross, S.D.; Mou, C.; Iliescu, T. Lagrangian Data-Driven Reduced Order Modeling Using Finite Time Lyapunov Exponents. *Fluids* **2020**, *5*, 189. [CrossRef]
87. Perotto, S.; Reali, A.; Rusconi, P.; Veneziani, A. HIGAMod: A Hierarchical IsoGeometric Approach for MODel reduction in curved pipes. *Comput. Fluids* **2017**, *142*, 21–29. [CrossRef]
88. Caiazzo, A.; Iliescu, T.; John, V.; Schyschlowa, S. A numerical investigation of velocity-pressure reduced order models for incompressible flows. *J. Comput. Phys.* **2014**, *259*, 598–616. [CrossRef]
89. Volkwein, S. *Proper Orthogonal Decomposition: Theory and Reduced-Order Modelling*; Lecture Notes; University of Konstanz: Konstanz, Germany, 2013.
90. Mou, C.; Koc, B.; San, O.; Rebholz, L.G.; Iliescu, T. Data-Driven Variational Multiscale Reduced Order Models. *Comput. Methods Appl. Mech. Engrg.* **2021**, *373*, 113470. [CrossRef]
91. Berselli, L.C.; Iliescu, T.; Layton, W.J. *Mathematics of Large Eddy Simulation of Turbulent Flows*; Scientific Computation; Springer: Berlin, Germany, 2006; p. xviii+348.
92. Pope, S. *Turbulent Flows*; Cambridge University Press: Cambridge, UK, 2000; p. xxxiv+771.
93. Sagaut, P. *Large Eddy Simulation for Incompressible Flows*, 3rd ed.; Scientific Computation; Springer: Berlin, Germany, 2006; p. xxx+556.
94. Pinkus, A. *N-Widths in Approximation Theory*; Springer: Berlin, Germany, 2012; Volume 7.
95. Ohlberger, M.; Rave, S. Reduced basis methods: Success, limitations and future challenges. *arXiv* **2015**, arXiv:1511.02021.
96. Wang, Z.; Akhtar, I.; Borggaard, J.; Iliescu, T. Proper orthogonal decomposition closure models for turbulent flows: A numerical comparison. *Comput. Meth. Appl. Mech. Eng.* **2012**, *237–240*, 10–26. [CrossRef]
97. Smagorinsky, J.S. General circulation experiments with the primitive equations. *Mon. Weather Rev.* **1963**, *91*, 99–164. [CrossRef]
98. Maulik, R.; San, O.; Rasheed, A.; Vedula, P. Subgrid modelling for two-dimensional turbulence using neural networks. *J. Fluid Mech.* **2019**, *858*, 122–144. [CrossRef]
99. Pawar, S.; San, O.; Rasheed, A. Deep Learning Based Sub-Grid Scale Closure for LES of Kraichnan Turbulence. APS 2019, G17-007. Available online: https://ui.adsabs.harvard.edu/abs/2019APS..DFDG17007P/abstract (accessed on 9 November 2020).
100. Lee, K.; Carlberg, K.T. Model reduction of dynamical systems on nonlinear manifolds using deep convolutional autoencoders. *J. Comput. Phys.* **2020**, *404*, 108973. [CrossRef]
101. Ahmed, S.E.; San, O.; Rasheed, A.; Iliescu, T. A long short-term memory embedding for hybrid uplifted reduced order models. *Phys. D Nonlinear Phenom.* **2020**, *409*, 132471. [CrossRef]
102. Ahmed, S.; Rahman, S.M.; San, O.; Rasheed, A. LSTM based nonintrusive ROM of convective flows. APS 2019, L10-003. Available online: https://ui.adsabs.harvard.edu/abs/2019APS..DFDL10003A/abstract (accessed on 9 November 2020).
103. Parish, E.J. *Machine Learning Closure Modeling for Reduced-Order Models of Dynamical Systems*; Technical Report; Sandia National Lab. (SNL-CA): Livermore, CA, USA, 2019.
104. Rahman, S.M.; San, O.; Rasheed, A. A hybrid approach for model order reduction of barotropic quasi-geostrophic turbulence. *Fluids* **2018**, *3*, 86. [CrossRef]
105. Cummins, P.F. Inertial gyres in decaying and forced geostrophic turbulence. *J. Mar. Res.* **1992**, *50*, 545–566. [CrossRef]
106. Daescu, D.N.; Navon, I.M. A dual-weighted approach to order reduction in 4DVAR data assimilation. *Mon. Weather Rev.* **2008**, *136*, 1026–1041. [CrossRef]
107. Kaercher, M.; Boyaval, S.; Grepl, M.A.; Veroy, K. Reduced basis approximation and a posteriori error bounds for 4D-Var data assimilation. *Optim. Eng.* **2018**, *19*, 663–695. [CrossRef]
108. Maday, Y.; Patera, A.T.; Penn, J.D.; Yano, M. A parameterized-background data-weak approach to variational data assimilation: formulation, analysis, and application to acoustics. *Int. J. Num. Meth. Engng.* **2015**, *102*, 933–965. [CrossRef]

109. Popov, A.A.; Mou, C.; Iliescu, T.; Sandu, A. A multifidelity ensemble Kalman filter with reduced order control variates. *arXiv* **2020**, arxiv:2007.00793.
110. Ştefănescu, R.; Sandu, A.; Navon, I.M. POD/DEIM reduced-order strategies for efficient four dimensional variational data assimilation. *J. Comput. Phys.* **2015**, *295*, 569–595. [CrossRef]
111. Xiao, D.; Du, J.; Fang, F.; Pain, C.C.; Li, J. Parameterised non-intrusive reduced order methods for ensemble Kalman filter data assimilation. *Comput. Fluids* **2018**, *177*, 69–77. [CrossRef]
112. Zerfas, C.; Rebholz, L.G.; Schneier, M.; Iliescu, T. Continuous data assimilation reduced order models of fluid flow. *Comput. Meth. Appl. Mech. Eng.* **2019**, *357*, 112596. [CrossRef]

Article

A Swing of Beauty: Pendulums, Fluids, Forces, and Computers

Michael Mongelli [1] and Nicholas A. Battista [2],*

[1] Department of Computer Science, 2000 Pennington Road, The College of New Jersey, Ewing Township, NJ 08628, USA; mongelm1@tcnj.edu

[2] Department of Mathematics and Statistics, 2000 Pennington Road, The College of New Jersey, Ewing Township, NJ 08628, USA

* Correspondence: battistn@tcnj.edu; Tel.: +1-609-771-2445

Received: 27 January 2020; Accepted: 5 April 2020; Published: 12 April 2020

Abstract: While pendulums have been around for millennia and have even managed to swing their way into undergraduate curricula, they still offer a breadth of complex dynamics to which some has still yet to have been untapped. To probe into the dynamics, we developed a computational fluid dynamics (CFD) model of a pendulum using the open-source fluid-structure interaction (FSI) software, *IB2d*. Beyond analyzing the angular displacements, speeds, and forces attained from the FSI model alone, we compared its dynamics to the canonical damped pendulum ordinary differential equation (ODE) model that is familiar to students. We only observed qualitative agreement after a few oscillation cycles, suggesting that there is enhanced fluid drag during our setup's initial swing, not captured by the ODE's linearly-proportional-velocity damping term, which arises from the Stokes Drag Law. Moreover, we were also able to investigate what otherwise could not have been explored using the ODE model, that is, the fluid's response to a swinging pendulum—the system's underlying fluid dynamics.

Keywords: fluid dynamics education; damped pendulums; fluid drag; fluid-structure interaction; computational fluid dynamics

---

1. Introduction

Historically, pendulums have been used for a multitude of purposes. From seismometers used almost two thousand years ago [1,2], to devices that increase efficiency for societal capacity, such as reciprocating saws and pumps [3,4], to keeping time [5,6], to even medieval torture devices [7], applications of pendulums are far and wide. Edgar Allan Poe even wrote a short story about one, *The Pit and the Pendulum* [8]. The esteemed polymath Galileo Galilei dreamt of the first pendulum clock in 1637, but it wasn't constructed until 1656 when Dutch physicist Christiaan Huygens drew out the plans, thus designing it. He enlisted clockmaker Salomon Coster to build it. The introduction of a pendulum clock increased time keeping accuracy from 15 minutes to a quarter of minute [5]— pendulums changed history!

It is no surprise that the study of pendulums swings its way into many foundational courses in science, mathematics, and engineering. Students are introduced to them in courses such as mechanics, dynamics, or differential equations, where they are first exposed to the idea of a *simple* gravity pendulum. A simple gravity pendulum is an idealized pendulum—a weight (*bob*) is attached to a massless string, which is tethered to a fixed pivot point, and is allowed to swing freely, without friction [9]. It will continue to swing forever. Realistic? Not unless one lives in a vacuum, but ultimately a good place to begin a student's exploration of simple harmonic motion.

If $\theta(t)$ represents the angular displacement (in radians) from the vertical at time t (see Figure 1a), the ordinary differential equation (ODE) describing such a simple pendulum system takes the form:

$$I\frac{d^2\theta}{dt^2} + mgL\sin\theta = 0, \tag{1}$$

where I, m, and L are the pendulum bob's moment of inertia and overall mass and length, respectively, and g is the gravitational acceleration. Since the only external force acting on this pendulum is gravity, it will swing forever, with no loss in oscillatory amplitude, see Figure 1b for an example. Figure 1b shows simulated results for different pendulum cases, each with a circular bob of a different radius. Note that the ODE was solved numerically, as no small angle approximation was used [9]. For these cases of circular bobs, of radius R, attached to a massless string of length L, the moment of inertia is calculated to be:

$$I = m\frac{1}{2}R^2 + mL^2. \tag{2}$$

There are two things to note from Figure 1b. The first is that over time the oscillation amplitudes do not decay. The second is that although amplitudes of oscillation are not affected, the period of oscillation is affected by bobs of different radii. Larger bobs have larger periods, due to their moment of inertia being greater [9].

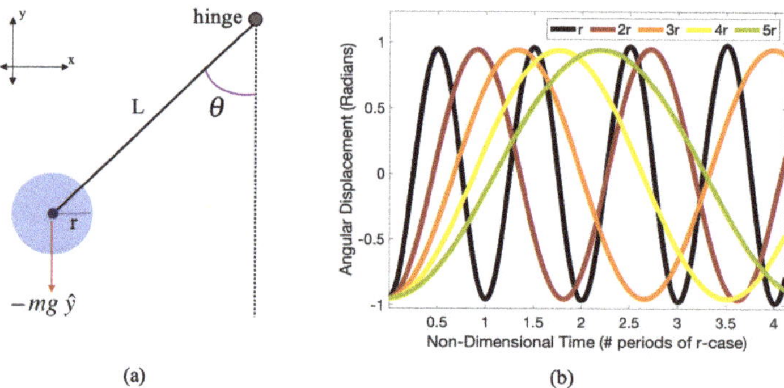

Figure 1. (**a**) A pendulum of length L with circular bob of radius, r, and mass, m. (**b**) Angular displacement (in radians) over time for various gravity pendulums of differing radii. The non-dimensional time is given in terms of the number of periods of the case with radius, r.

In a world (or classroom) like ours which does not exist in a vacuum, a table-sized pendulum demonstration would eventually lead to its angular oscillations reaching zero, i.e., standing still. This is due to the pendulum swinging in air—a fluid. The air resists the pendulum's motion, effectively pushing back on the pendulum. This is known as *fluid drag*.

The concept of fluid drag is probably familiar to you already. It is the reason why parachutes work. The equation governing a pendulum swinging in a fluid environment is given by

$$I\frac{d^2\theta}{dt^2} + b\frac{d\theta}{dt} + mgL\sin\theta = 0, \tag{3}$$

where the parameter b is deemed a *damping* parameter. This is a reduced order model of the system, as the contribution of the fluid onto the pendulum is entirely contained within the parameter, b. That is, this equation models how the fluid is believed to affect the swinging motion of the pendulum, while providing no mechanism to understand the underlying fluid's dynamics. Numerical simulation results from solving Equation (3) are presented in Figure 2.

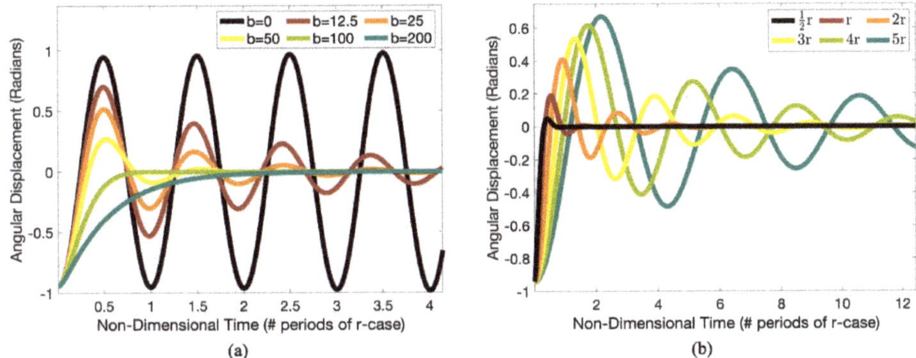

Figure 2. Angular displacements against non-dimensional time for damped physical pendulums in the case of (**a**) constant radius and varied damping, *b*, and (**b**) constant *b* and varied radii.

Figure 2a holds the radius constant at *r*, the same *r* from Figure 1b, while varying the damping coefficient, *b*. Compared to Figure 1b, angular oscillations decay in all cases when $b > 0$. The damping induced from $b > 0$ will eventually lead to its equilibrium—a stationary pendulum hanging straight down. However, the decay rate is dependent on *b*; larger values of *b* lead to a quicker decay of oscillation. Note that *b* has units of $\frac{\text{kg·m}^2}{\text{s}^2}$ and in realistic situations, $b > 0$. Moreover, depending on the value of *b*, the pendulum system will exhibit one of three behaviorial cases:

1. *Under-damped*: The pendulum will swing back and forth, although its amplitude of oscillation will steadily decline, until it asymptotically approaches its equilibrium.
2. *Critically-damped*: the pendulum returns to equilibrium as quickly as it can. If the damping parameter were made slightly more or slightly less, it would result in the pendulum returning slower to its equilibrium position.
3. *Over-damped*: the pendulum moves towards its equilibrium position slower than the critically-damped case. There is no oscillation.

The simulations shown in Figure 2b held the damping parameter fixed ($b = 150$ from Figure 2a) and varied the radius of the bob. Note that changing the radius *r* will vary the moment of inertia (see Equation (2)). This data suggests that as *r* increases for a given *b*, this would lead to more oscillatory behavior. That is, smaller *r* tends to make the pendulum system more damped. Intuitively this doesn't make much sense as is—a larger pendulum bob should feel more drag since it has a larger surface area. It would be like jumping out of an airplane with a parachute with a surface area of 10 m^2 and floating down slower (and more softly) than a parachute of 40 m^2. How could this be?

For the simulations in Figure 2b, we fixed the damping parameter *b* and then varied *r*. We did not consider that the damping parameter may depend on the radius (among a variety of other parameters), that is, the geometry of the pendulum bob. Furthermore, we have yet to motivate where the damping term in Equation (3) comes from. Let's settle that.

It comes from the seminal work of prolific physicist and mathematician Sir George Stokes, who even studied drag force using pendulums [10]! In particular, he derived a drag force equation, now known as *Stokes Law*, by investigating spheres moving through a fluid at low Reynolds numbers, i.e., situations in which either the fluid is moving extremely slowly or the fluid's viscosity is very high [11,12]. Stokes Law (for a sphere at low *Re*) is presented as the following:

$$F_D = 6\pi\mu rv, \tag{4}$$

where F_D is the fluid drag, μ is the fluid's dynamic viscosity and *r* and *v* are radius and speed of the sphere that is moving through the fluid. Note that the Reynolds number (*Re*) depends on two fluid parameters, i.e., its density, ρ, and dynamic viscosity, μ, as well as two parameters based on the physical system being investigated, i.e., a characteristic length and velocity scale, L and V, respectively. The *Re* is defined to be

$$Re = \frac{\rho L V}{\mu}.\tag{5}$$

Thus *Stokes Drag* describes that this damping frictional force acting on the sphere is proportional to its size, $F_D \sim r$, and speed, $F_D \sim v$. Careful to remember that this may only be true in low Reynolds number situations, where either v, r, or both may be small. Notice that the form that the damping term took in Equation (3) was similar, but used angular displacement ($\theta(t)$) and angular velocity ($\frac{d\theta}{dt}$). As suggested by numerical simulations presented above, this damping equation gives rise to exponential decay in angular displacement (Figure 2).

At higher Reynolds numbers, i.e., situations in which fluid viscosity is low or the speed or size of an object is large, the drag force takes a different form. For these situations, the drag law was discovered by none other than the infamous Lord Rayleigh (John William Strutt) using dimensional analysis [13]. For high Reynolds numbers settings, the fluid drag force takes the following form [14,15]:

$$F_D = \rho r^2 K v^2,\tag{6}$$

where ρ is the fluid density, r is the sphere's radius, v is the sphere's velocity, and K is a non-dimensional number that is based on the fluid flow's speed and direction as well as the object's shape, size, and orientation in respect to the flow, and the fluid's density and viscosity. In a nutshell, for a specific object, this constant K may significantly change if one or more of these parameters are varied.

This drag force is traditionally represented in the following generalized manner:

$$F_D = \frac{1}{2}\rho A C_D v^2,\tag{7}$$

where A is a cross-sectional area of an object in flow and C_D is a dimensionless number called the *drag coefficient*. In this representation C_D is analogous to K above.

Moreover, work in the latter half of the 20th century and early 21st century has shown that in particular situations correction terms must be included into the drag force equations [16,17]. Furthermore there are still unknown dynamics of pendulums involving small amplitude oscillations [18]. Although physical pendulums have been used for thousands of years and studied by students in foundational courses for over a century, they remain an active research area [19].

With the advent of new technologies, e.g., experimental measurement and flow visualization tools, researchers have further probed into the complex interactions of pendulums and the fluid environments they are immersed within [20–25]. In particular, Mathai et al. [25] investigated how fluid drag on pendulums may be enhanced due to dynamic interactions with their own vortex wake as they swing—something not quantified previously!

Mathai et al. [25] went on to note *Even with the wake history force included, the current model is still quite basic. In reality, the dynamics <of a pendulum> is highly nonlinear, with changes in direction, curvilinear trajectories and wide variations in instantaneous Re <Reynolds number>...Fully resolved direct numerical simulations...can provide better insights into the flow-induced forces.* That is exactly where our work on pendulums began, although there have been two previous studies using CFD models of pendulums [26,27].

In this paper we implemented a fluid-structure interaction (FSI) computational model of a swinging pendulum containing a spherical bob (a circular bob in two dimensions). In our FSI model, we varied the size of the circular pendulum bob, i.e., its radius, and its mass. We then analyzed the resulting data in terms of angular displacement, speed of the pendulum bob, and fluid forces acting on it, as well as compared the dynamics between our FSI model and the canonical reduced ODE model for a damped physical pendulum, Equation (3). Furthermore, we visualized (and qualitatively analyzed) the fluids motion in response to a swinging pendulum.

In addition, we provide instructor resources, such as slides and movies, in the Supplemental Materials (the items are listed in Appendix A), with the goal to streamline use of this work in educational settings. Moreover, we also offer the science community the first open-source pendulum models in a fluid-structure interaction framework. The models can be found at github.com/nickabattista/IB2d in the sub-directory:

$$\text{IB2d} \rightarrow \text{matIB2d} \rightarrow \text{Examples} \rightarrow \text{Examples\_Education} \rightarrow \text{Pendulum}.$$

Note that each example is of a point mass model bob and was selected for its computational speed in comparison to circular bobs (those of non-zero radius). Moreover, three versions are presented, each corresponding to a different grid resolution. The least spatially resolved case, 256×256, will be the fastest to run, but also the least accurate of the 3, while the 1024×1024 case has the highest spatial resolution, but will run the slowest. The default setting is to only save the pendulum (Lagrangian) data. To store the fluid (Eulerian) data, flags corresponding to the desired fluid quantities may be selected within the input2d file. We also provided the scripts used to solve Equations (1) and (3) that produce Figures 1 and 2.

2. Methods

To investigate the swinging motion of a two-dimensional pendulum bob immersed in a viscous, incompressible fluid, we used computational fluid dynamics (CFD). In our model, the bob starts at rest and begins to swing under gravitational acceleration acting on the mass of the pendulum. An immersed boundary (IB) framework was used to couple the pendulum's motion and the fluid it is immersed within. Scientists and engineers can use IB to study the interactions of an object and the fluid it is contained within, i.e., you can explore how the fluid affects the object and vice versa.

The immersed boundary method was developed by Charles Peskin, a mathematical physiologist at the Courant Institute of Mathematics [28–30]. Even though IB was invented in the 1970s, it is still extensively used for investigating fluid-structure interaction problems today. Many mathematicians, engineers, and scientists have since improved the original algorithm in attempts to increase its accuracy without significantly increasing the computational expense and time required [31–38]. IB is still a leading numerical framework for studying problems in FSI due to its robustness [39,40].

It has previously been applied to study problems ranging from cardiac fluid dynamics [41–44] to aquatic locomotion [45–49] to insect flight [50–52] to dating and relationships [39]. Additional details on the IB method can be found in Appendix B.

In the remainder of this section we will introduce our FSI pendulum's implementation into the *IB2d* framework, i.e., the computational geometry, geometrical and fluid parameters, and model assumptions.

2.1. Model Geometry

Figure 3 presents our pendulum model's computational geometry. In particular, Figure 3a illustrates the modeling idea—a bob (of radius r) is composed of a central point mass (of mass m) and outer neutrally-buoyant shell layer. It is tethered to a particular fixed location, a distance L away. The pendulum is immersed in a viscous, incompressible fluid of uniform density ρ, and dynamic viscosity, μ. Note that the fluid within and outside the pendulum bob is uniform in its properties. We define the pendulum's angular displacement, θ, to be the angle from the vertical dotted line. Gravity acts on the central mass point; as the rest of the pendulum bob's geometry is neutrally-buoyant, all acceleration of the bob is due to this single gravitational interaction. However, due to the structure properties of the pendulum bob, the neutrally-buoyant shell will undergo fluid drag due to the fluid's resistance in letting the pendulum bob move through it.

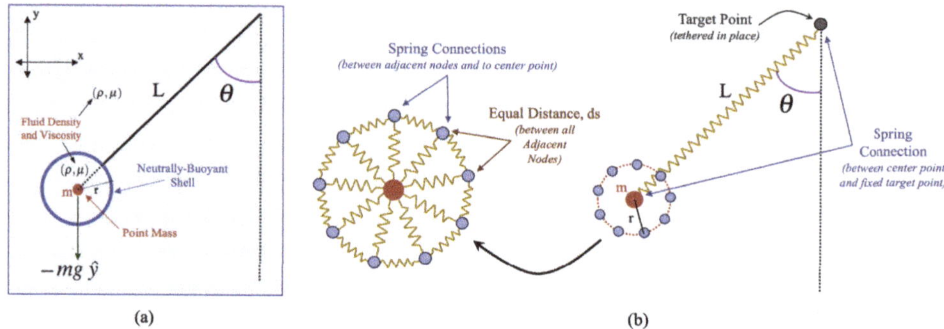

Figure 3. (**a**) Model of an immersed pendulum with circular bob of radius r in a viscous, incompressible fluid. The fluid has density and viscosity of ρ and μ, respectively. The pendulum has length L and the bob has mass m, concentrated at its center. (**b**) The computational geometry illustrating the fiber model construction of the discretized Lagrangian mesh.

As we wished to vary the pendulum bob's radius and mass of its central point, we considered the parameters listed in Table 1 for our FSI pendulum model. The explicit radii, r, and masses, m studied are $r \in \{0.001, 0.0025, 0.005, 0.0075, 0.01, 0.0125, 0.015, 0.0175, 0.02, 0.0225, 0.025\}$ m and $m \in \{2 \times 10^2, 5 \times 10^2, 1 \times 10^3, 2 \times 10^3, 5 \times 10^3, 1 \times 10^4\}$ kg, respectively. The initial angular displacement, θ_0, was $-\frac{\pi}{2} + \frac{\pi}{5} = -\frac{3\pi}{10}$ radians. We did not vary properties of the fluid, neither its density nor viscosity. Note that the kinematic viscosity in our simulations was $\nu = \frac{\mu}{\rho} = 10^{-5}$ m^2/s. Some common liquids with kinematic viscosities around 10^{-5} m^2/s are sulphuric acid at room temperature or a variety of oils (coconut, SAE Motor Oils, peanut, whale, etc.) at $\sim$100–130 degrees Fahrenheit [53]. Kinematic viscosity, ν, measures the fluid's internal resistance to flow under gravity.

Table 1. Table of geometric and fluid parameters used in our pendulum study.

Parameter	Description	Value
L	Pendulum Length	0.2 m
r	Pendulum Bob's Radius	$r \in [0.001, 0.025]$ m
m	Mass	$m \in [2 \times 10^2, 1 \times 10^4]$ kg
ρ	Fluid Density	1000 kg/m^3
μ	Fluid (dynamic) Viscosity	0.01 kg/(m · s)
g	Gravitational Acceleration	9.81 m/s^2
θ_0	Initial Angular Displacement	$-\frac{3\pi}{10}$ radians

2.2. Model Construction

Figure 3b provides a more detailed overview of the computational geometry. In particular it provides details regarding how the structure is modeled using *IB* fiber models, which are used to mimic desired material properties between discretized points, i.e., Lagrangian points, that compose the geometry [39]. The single mass point is tethered to the fixed point, a distance L away, via a *virtual spring*. The static hinge point is tethered in place using the *target point* model. Target points can be used to hold Lagrangian points nearly rigid. The individual mass point uses a *massive point* model that tethers the individual Lagrangian point to a *mass*, which dampens its movement [54]. Note that the target point and massive point models use spring-like mathematical formulations to achieve their desired effects, see [39].

The neutrally-buoyant shell is composed of equally-spaced Lagrangian points, to which are tethered to their neighboring points via virtual spring connections. Furthermore, each of these Lagrangian points are further tethered to the massive point in the center of the pendulum bob by a virtual spring. All of the virtual spring connections in the model use *stiff* springs with a particular spring resting length in order to keep the geometry from changing shape, i.e., trying to ensure that the Lagrangian points maintain a specific distance from other points. The number of Lagrangian points in a circular shell varies by the radius, see Table 2. Note that due to the coupled nature of the Lagrangian structure and fluid in the standard immersed boundary framework, it was more straightforward for us to approximately model rigid structures in this manner. Additional steps would have been necessary to solve the problem of each Lagrangian point only being allowed to move in a constrained way, due to the imposed rigidity, under forces from the fluid and other external forces (like gravity) pushing on it. Please see [46,55] for further information regarding immersed boundary formulations with rigid bodies.

Table 2. Table providing number of Lagrangian Points in the circular shell for a particular radius, r.

Radius (m)	0.001	0.0025	0.005	0.0075	0.01	0.0125	0.015	0.0175	0.02	0.0225	0.025
# Lag. Pts in Shell	12	32	64	96	128	160	194	226	258	290	320

Fiber models use a variety of different deformation force laws to model material properties. To model virtual (Hookean) springs, deformation forces were calculated as follows,

$$\mathbf{F}_{spr} = k_{spr} \left(1 - \frac{R_L(t)}{||\mathbf{X}_A(t) - \mathbf{X}_B(t)||} \right) \cdot \left(\begin{array}{c} x_A(t) - x_B(t) \\ y_A(t) - y_B(t) \end{array} \right) \tag{8}$$

where k_{spr} is the spring stiffness, $R_L(t)$ is the spring's resting lengths, and $\mathbf{X}_A = \langle x_A, y_A \rangle$ and $\mathbf{X}_B = \langle x_B, y_B \rangle$ are the Lagrangian nodes tethered by the spring. Note that in our model the resting lengths are time-independent, hence $R_L(t) = R_L$. As mentioned previously, the spring stiffnesses are large to ensure minimal stretching or compression of the computational geometry. The spring stiffness used to tether the massive point to the fixed hinge point and the pendulum bob points to both one another and the massive point are denoted by k_{spr_L} and k_{spr_B}, respectively.

In the simple case where a preferred position is enforced, boundary points are tethered to target points via springs. Its corresponding deformation force equation, which describes the force applied to the fluid by the boundary in Lagrangian coordinates is given by $\mathbf{F}_{target}$ and is explicitly written as,

$$\mathbf{F}_{target} = k_{target} \left(\mathbf{Y}_A(t) - \mathbf{X}_A(t) \right), \tag{9}$$

where k_{target} is the stiffness coefficient, and $\mathbf{Y}_A(t)$ is the prescribed Lagrangian position of the target point. In all simulations the hinge point was held nearly rigid by applying a force proportional to the distance between the location of the actual Lagrangian point and its preferred target position. Using a large value of k_{target} helps mitigate a small deviation between the actual and preferred position.

Artificial mass is modeled using the massive point approach of Kim et al. [54]. It is similar to target points. $\mathbf{Z}_A$ gives the Cartesian coordinates of the *massive* points, with associated mass density M_A. Note that such points do not interact with the fluid directly, similar to target points. $\mathbf{X}_A(t)$ give the Cartesian coordinates of a neutrally-buoyant Lagrangian boundary point, which do interact with the fluid. Similar to target points, if a Lagrangian point deviates from its massive point, a restoring force drives them back together, as shown in its mathematical description below

$$\mathbf{F}_{Mass} = k_{mass} (\mathbf{Z}_A(t) - \mathbf{X}_A(t)) \tag{10}$$

$$M_A \frac{\partial^2 \mathbf{Z}_A(t)}{\partial t^2} = -\mathbf{F}_{Mass} - M_A g \hat{y}, \tag{11}$$

where k_{mass} is a stiffness coefficient with $k_M \gg 1$, M_A is the mass of the massive point, g is the acceleration due to gravity in vertical direction, $\hat{y}$, and $\mathbf{Z}_A(t)$ is the position of the massive point to which the Lagrangian point, $\mathbf{X}_A(t)$, is tethered to at time t.

All numerical stiffness parameters are given in Table 3. The stiffnesses were selected to be as high as possible while also maintaining stability and fidelity of our numerical solver. Each pendulum simulated was of length 0.2 m and was immersed in a square computational domain of size $(L_x, L_y) = (1\text{ m}, 1\text{ m})$, with a grid resolution of 1024×1024, i.e., $dx = dy = \frac{L_x}{N_x} = \frac{L_y}{N_y} = 0.0009765625$ m. Points that compose the circular pendulum bob were evenly spaced apart at a distance of $ds = \frac{dx}{2}$. Note that this is the standard convention in the immersed boundary literature when choosing the Lagrangian point spacing, ds. It is used to avoid leaky boundaries [30]. Thus, fluid will not be allowed to flow in or out of the pendulum bob, unless due to numerical error. Moreover, note that the adjacent nodes along the circle were a distance r from the massive point at the center of the pendulum bob, which was tethered a distance of L from the fixed hinge target point. Each of the spring connections between specific points used a spring resting length equal to such corresponding distances in an effort to maintain the geometry while the pendulum was swinging. A time-step of $dt = 2.5 \times 10^{-5}$ s was used to march forward in time.

Table 3. Table of numerical temporal, spatial, and fiber model parameters used in our pendulum study.

Parameter	Description	Value
dt	time-step	2.5×10^{-5} s
$L_x \times L_y$	Grid Size	$1\text{ m} \times 1\text{ m}$
(N_x, N_y)	Grid Resolution	$(1024, 1024)$
$dx = dy$	Spatial Step	$L_x/N_x = L_y/N_y = 0.0009765625$ m
ds	Lagrangian Point Spacing	$\sim \frac{L_x}{2N_x}$
k_{spr_L}	Spring Stiffness Coefficient (Mass to Hinge)	1.25×10^8 kg $\cdot$ m/s^2
k_{spr_B}	Spring Stiffness Coefficient (Pendulum Bob)	2.5×10^8 kg $\cdot$ m/s^2
k_{target}	Target Point Stiffness Coefficient	5×10^7 kg $\cdot$ m/s^2
k_{mass}	Massive Point Stiffness Coefficient	2.5×10^6 kg $\cdot$ m/s^2

While running the simulations, we stored the following data every 0.005 s of simulation time:

1. Position of Lagrangian Points
2. Forces on Each Lagrangian Point (Horizontal/Vertical and Normal/Tangential Forces)
3. Fluid Velocity
4. Fluid Vorticity
5. Forces spread from the Lagrangian mesh onto the Eulerian grid

We then used the open-source software VisIt [56], created and maintained by Lawrence Livermore National Laboratory for visualization, see Figure 4, and the data analysis package software within *IB2d* [39] for data analysis. Figure 4 provides a visualization of some of the data produced for a pendulum of mass and radius of $m = 5 \times 10^2$ kg and $r = 0.0175$ m, respectively, at one snapshot in time. Section 3.4 explores the underlying fluid dynamics in further detail, including the time evolution over a pendulum's first swing, see Figure 19.

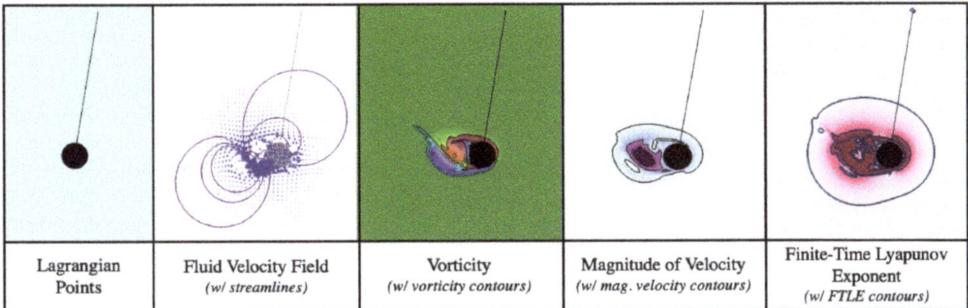

<table>
<tr><td>Lagrangian
Points</td><td>Fluid Velocity Field
(w/ streamlines)</td><td>Vorticity
(w/ vorticity contours)</td><td>Magnitude of Velocity
(w/ mag. velocity contours)</td><td>Finite-Time Lyapunov
Exponent
(w/ FTLE contours)</td></tr>
</table>

Figure 4. Snapshots of a single moment in time for a pendulum with mass, $m = 5 \times 10^2$ kg, and radius, $r = 0.0175$ m, providing some of the data stored during the time-step in which the simulation time reached $t = 0.70$ s, i.e., positions of Lagrangian points (pendulum), the velocity vector field, magnitude of velocity, and vorticity. Note that data from giving the force spread from the Lagrangian grid (pendulum) onto the Eulerian (fluid) grid is not shown. Lagrangian Coherent Structures (LCS) via finite time Lyapunov exponents (FTLE) are also illustrated, although they were computed during the post-processing stage, after the data was collected.

3. Results

Using an open source implementation of the immersed boundary method, IB2d, we modeled the motion of a pendulum with a circular bob immersed within a fluid undergoing gravity's influence. For this education focused paper, we explored angular displacement and speed of the pendulum bob as well as forces acting upon the pendulum bob to impede its motion. Upon doing so we quantified the decay in oscillation amplitude and speed damping. This was done for a variety of pendulum bob masses as well as radii (size). We also explored the effect that the motion of the pendulum bob has onto the fluid it was immersed. Lastly, we compared the reduced ODE model of a damped physical pendulum and our FSI model. We organized our results into the following five subsections:

1. Angular Displacement of the pendulum bob
2. Speed of the pendulum bob
3. Forces acting on the pendulum bob
4. Effect the pendulum bob has onto the fluid
5. Comparison between reduced ODE model and FSI model

3.1. Angular Displacement of the Pendulum Bob

As suggested from Section 1, since the pendulum is immersed within a fluid environment, its oscillation amplitude will decay over time. Figure 5 provides snapshots of multiple pendulums' angular displacement over time for pendulum bobs of differing radius and $m = 1 \times 10^3$ kg. Note that all simulations were run independently of each other and Lagrangian position data is being overlaid during post-processing.

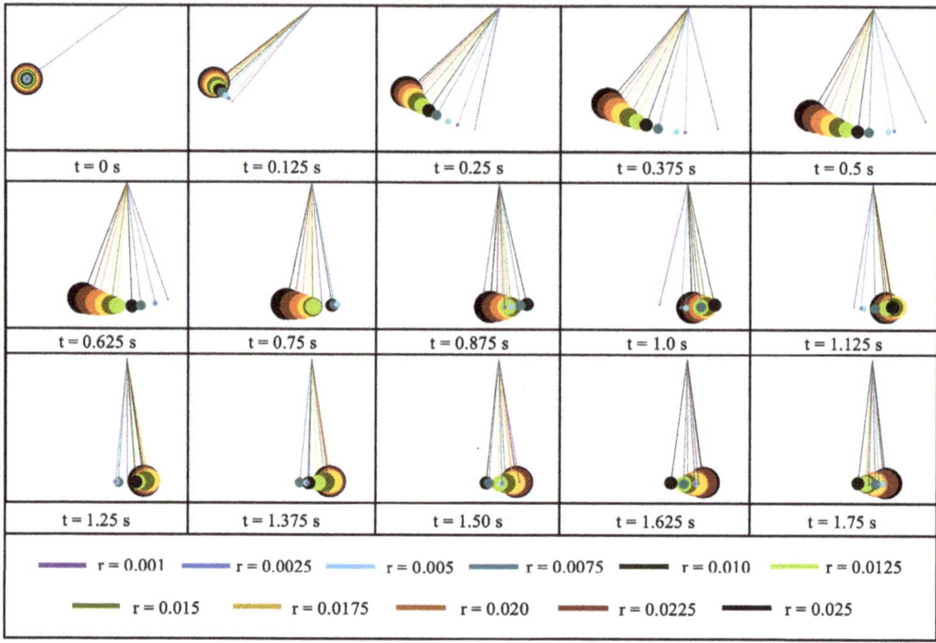

Figure 5. Snapshots of multiple pendulums' (of differing radius) angular displacement over time in the case of $m = 1 \times 10^3$ kg.

Moreover, both the size of the bob and the mass of the bob will affect its dynamics. Figure 6 illustrates that pendulums with the same size and shape bob may experience significantly different oscillation patterns due to different mass. In particular, depending on the mass, the pendulum could fall into any of the 3 damping cases (under-, over-, or critically-damped). See Figure A1 for the counterpart case where a specific mass is tested for a variety of radii. Consistent dynamics are observed.

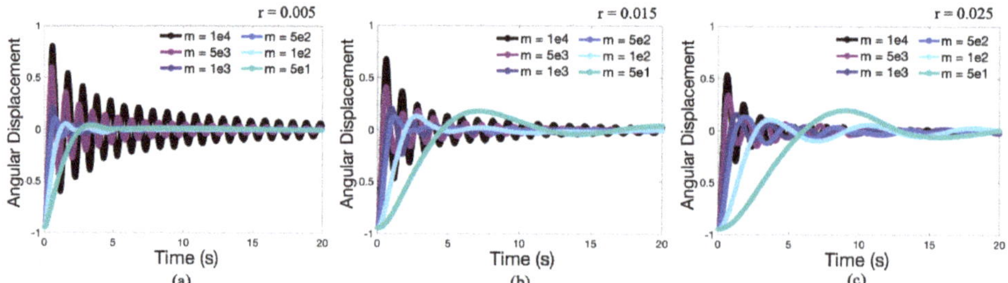

Figure 6. Depicting the angular displacement (radians) vs. time (s) for pendulums with the same radius but different masses. (a–c) give data for a specific radius, either $r = 0.005$ m, 0.015 m, or 0.025 m, respectively, for 4 orders of magnitudes in mass in each.

Next we calculated the maximum amplitude during each oscillation cycle for a variety of masses. The amplitude decays exponentially, see Figure 7. Figure 7 presents the displacement amplitude against peak number (number of half swings) for a variety of masses for $r = 0.005$ m. The data shows a linear relationship between the logarithm of the amplitude and the peak number, suggesting exponential decay, although lines seem a better fit starting at the first peak, rather than the initial displacement. Note that Figure 8 shows a steady increase in time between successive peaks for three different masses ($m = 2 \times 10^2, 1 \times 10^3$, and 5×10^3 kg) for a variety of radii. As mass increases, the time between peaks decreases. Moreover, generally as more swings occur the time become peaks becomes more consistent. We also note that the time between the start of each simulation and their first peak generally does not fit the data. Sections 3.2, 3.3 and 3.5 will also explore this observation in more detail.

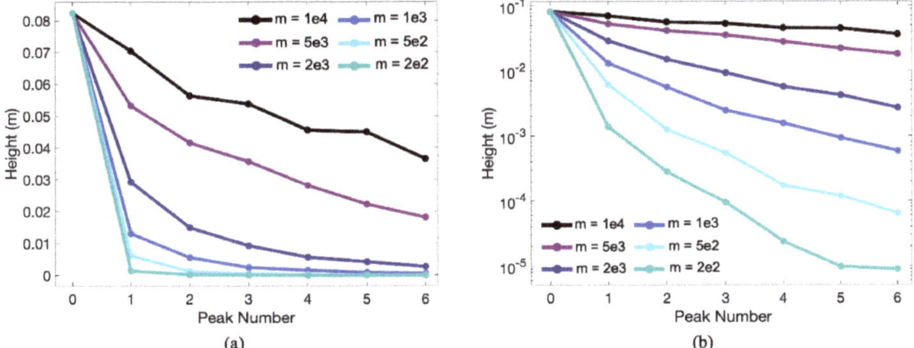

Figure 7. (**a**) Plot illustrating the decay of the height (m) that the pendulum bob reaches as the pendulum continues to swing for the case of $r = 0.005$ m for a spectrum of masses. The peak amplitude decays exponentially as illustrated by the linear relationship between the logarithm of the amplitude against peak number, as shown in (**b**).

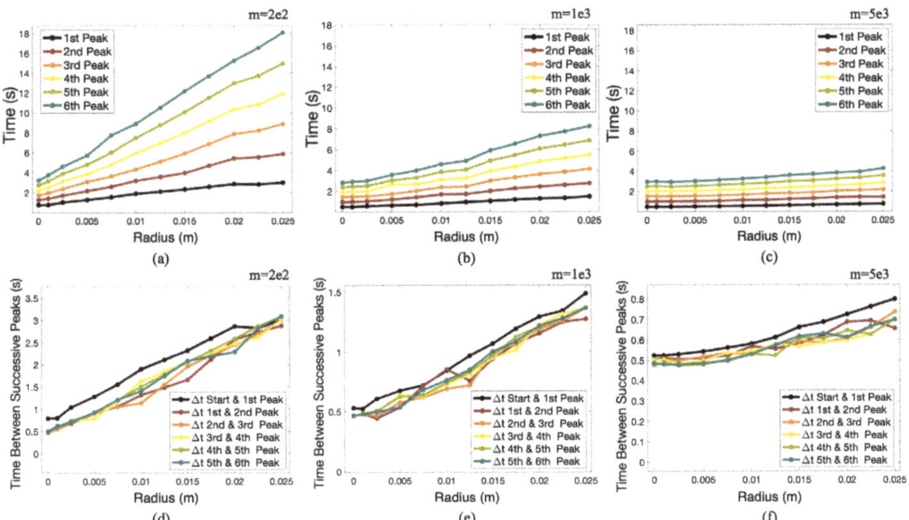

Figure 8. Plots illustrating the time of the peak in angular displacement against the pendulum bob's radius for its 1st through 6th peak (**a–c**) and the time difference between the peaks (**d–f**) against the pendulum bob's radius for three different masses: (**a,d**) $m = 2 \times 10^2$ kg, (**b,e**) $m = 1 \times 10^3$ kg, and (**c,f**) $m = 5 \times 10^3$ kg.

From this data, we computed the approximate damped period of oscillation, see Figure 9a,b. Figure 9b provides a colormap with contour lines of the period data from Figure 9a. As mass increases, the approximate period decreases, as suggested previously. Moreover, as radius increases, the period also increases. Note that when mass is high enough (e.g., the case of $m = 1 \times 10^4$ kg), the period does not significantly change between different radii; however, there is still exponential decay in the oscillatory amplitude (see Figure A2). Furthermore, for very small radii there appears to be a non-monotonic trend in period as a function of mass. The period was computed by averaging the time between every other peak over the first 5 full oscillations.

Next we explore how the speed of the pendulum bob is affected by variations in its mass and radius in a fluid environment.

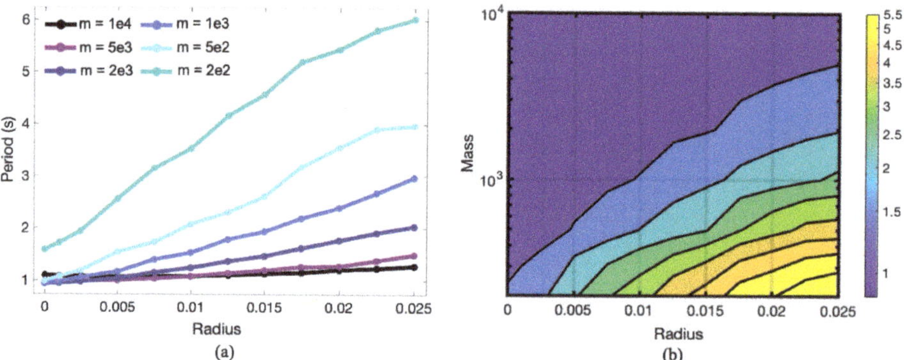

Figure 9. (a) The period given as a function of a pendulum bob's radius for a variety of masses. (b) A contour map showing the period as a function of both the pendulum bob's radius and mass. The highest periods occur for small masses and large pendulum bobs.

3.2. Speed of the Pendulum Bob

Recall that in Section 3.1 we observed that peaks in angular displacement decayed exponentially over time. This suggests that the pendulum bob's speed is inherently slowing down as well. Figure 10 details the pendulum bob's speed vs. the number of swings (half a complete oscillatory cycle). The data shown is for the case of $r = 0.015$ m for a variety of masses. During each swing the pendulum accelerates to a maximal speed before decelerating. The maximum occurs at roughly halfway through each swing, when the pendulum passes the point of 0 displacement from the vertical. Note that physically the pendulum must hit a speed of zero when it swings from one direction to the other; our data does not reflect this due to the time resolution of the sampled time-points.

Furthermore, from Figure 10b, it does not appear that the speed peaks decay exponentially over time from the beginning of the simulation. After a few swings, the peaks in speed appear to satisfy a linear relationship between the logarithm of speed versus time; however, the peak speed significantly decreases from the first to second swing, see Figure 11. Figure 11 illustrates that for most cases from the second swing on, the peak speeds approximately demonstrate exponential decay; however, there is significantly more decay between the first and second swing than successive peaks thereafter. Figure A3 presents the counterpart data of how the speed of pendulum bobs of the same mass decays over time for a variety of radii. Similarly trends are observed in the data.

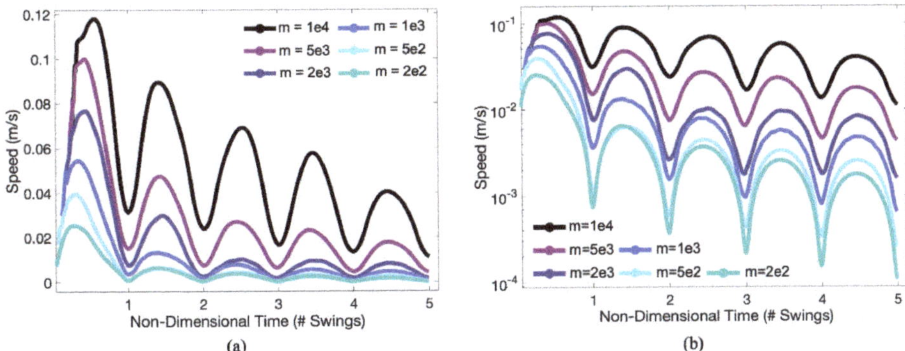

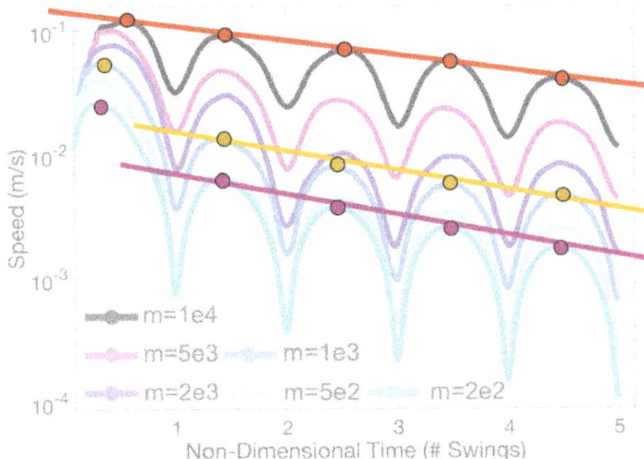

Figure 10. (a) Plot depicting the linear speed of the pendulum bob against non-dimensional time given as the # of swings (half a full displacement cycle) for the case of $r = 0.015$ m and a variety of masses. Speed peaks near the center of each swing. This corresponds to when the pendulum has approximately zero angular displacement from the vertical. The peak speed appears to begin decaying exponentially, starting on the second or third swing in most cases. This can be seen from linear relationships between peak speed and swings in (b).

Figure 11. Plot illustrating that exponential decay appears in peak speed starting with the second swing. There is significantly more decay in peak speed between the first and second swings, than successive swings thereafter.

Next we wished to quantify the amount of damping due to the pendulums immersion in a fluid of kinematic viscosity $\nu = \mu/\rho = 10^{-5}$ m^2/s. To do this we computed the theoretical speed of a pendulum bob void of a fluid environment by energy conservation. We set the original potential energy when the pendulum bob was at time zero and computed the kinematic energy when the pendulum was passing 0 displacement, i.e.,

$$\frac{1}{2}mv_{NF}^2 = mgh_0 \;\Rightarrow\; v_{NF} = \sqrt{2gh_0},$$

where v_{NF} is the velocity of the pendulum bob outside of a fluid environment and h_0 is the initial height of the pendulum bob before it begins swinging. For our initial setup, $h_0 = L(1 - \cos(\pi/2 - \pi/5))$, since the pendulum is released from an initial angle of $\pi/5$ radians from the horizontal.

We then defined the speed ratio to be: $SR = v/v_{NF}$ and the speed damping ratio to be: $SDR = 1 - SR = 1 - v/v_{NF}$. If $v = v_{NF}$, $SDR \approx 0$ suggesting only small damping effects. Figure 12 gives the speed damping ratio as a percentage. For this particular fluid environment, even cases of high mass and small radius result in SDR's of $\sim$88%, i.e., the fluid immersed pendulum bob is moving $\sim$88% slower than its fluid void counterpart. As the radius increases, the SDR increases. Note that smaller masses have less significant increases in SR. As mass increases, SDR decreases for a given radius.

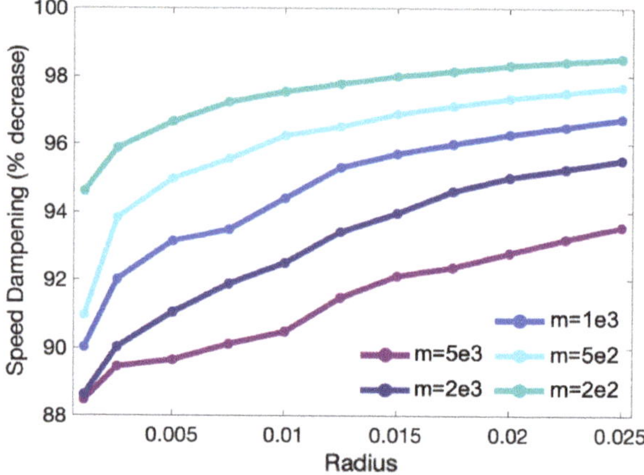

Figure 12. The percentage decrease in speed when comparing pendulum bob speed once it reaches 0 degree angular displacement on the first swing compared between simulated cases in fluid and theoretical value outside of a viscous fluid environment.

Lastly we explored the phase space between pendulum bob speed and its angular displacement. Figure 13a presents the data for the case of $r = 0.001$ m for a spectrum of masses of over 3 orders of magnitude. As suggested by all data previously, the trajectories eventually converge to zero displacement and speed. Interesting, all the trajectories collapse onto an approximate parabolically-capped cone. The last cycle is given for all cases in Figure 13b. Similar topologies are seen in cases of other radii (see Figure A5) over a variety of masses. Furthermore, similar topologies emerge when fixing the mass and exploring trajectories for a variety of radii (see Figure A4).

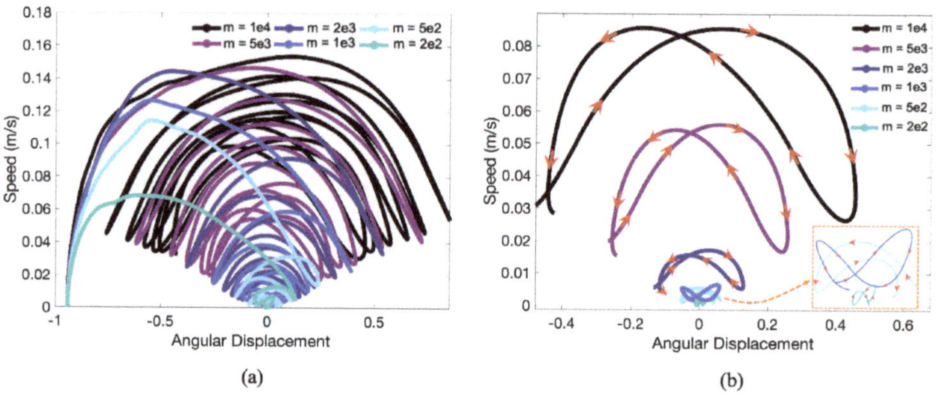

(a)

(b)

Figure 13. (a) Phase space of linear speed of the pendulum bob vs. angular displacement (radians) for a variety of masses in the case of $r = 0.001$ m. (b) A closer look at the last simulated cycle's phase space in each case.

3.3. Forces on the Pendulum Bob

We have observed that a fluid-immersed pendulum experiences exponential decay in angular displacement and speed. As discussed in Section 1, this is due to fluid drag on the pendulum bob. In this section we will explore this drag force. We wish to emphasize that our numerical experiments did not prescribe any specific form of the drag forces *a priori*, or any forces for that matter, beyond gravity acting on the pendulum bob.

First we selected three masses, $m = 5 \times 10^2, 1 \times 10^3$, and 2×10^3 kg, and investigated the drag force acting on the pendulum versus time for a variety of radii. This data is shown in Figure 14. The drag force was calculated by computing the forces perpendicular to the direction of the pendulum arm as the bob swung. A normal unit vector was computed at each sampled time-point and the drag force was computed using a vector projection in the direction opposite to the swinging motion of the pendulum bob.

Figure 14a–c suggest that the drag force exponentially decays as time progressed. This was further confirmed by the linear relationship between the logarithm of drag force vs time in Figure 14d–f. As the radius increases, the surface (circumference) of the circle increases, and thus the amount of drag on the pendulum bob's body increased for a particular mass as well. Hence the drag on the bob is dependent on its geometry (shape and size), as discussed in Section 1. Furthermore, as mass increases, so does the overall drag on the bob. This can also be seen in the counterpart case, where 3 radii are selected ($r = 0.005, 0.015$, and 0.025 m) and mass was varied, see Figure A6.

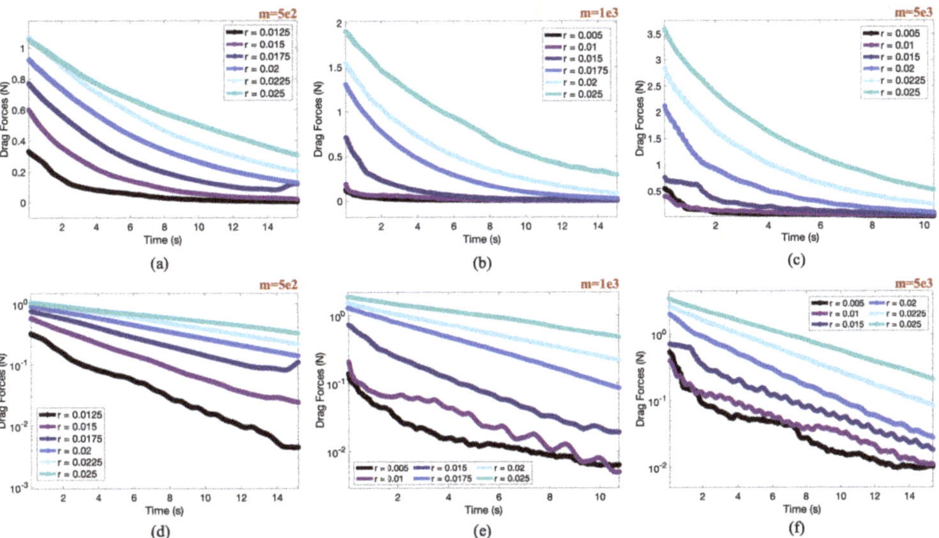

Figure 14. Drag forces (N) over time in seconds for multiple radii for cases with (**a,d**) $m = 5 \times 10^2$ kg, (**b,e**) $m = 1 \times 10^3$ kg, and (**c,f**) $m = 5 \times 10^3$ kg. The semi-log data is provided in (**d–f**) to highlight a linear relationship between the logarithm of the drag force and time. This linear relationship suggests an exponential decay in drag force over time.

Next, we explored the phase space of drag force on the pendulum bob versus angular displacement of the bob. This data is given in Figure 15. Similar to the phase plots of speed versus displacement, the force-displacement trajectories all collapse onto a unique exponentially decaying envelope. Smaller radii (Figure 15a, $r = 0.005$ m) correspond to larger angular displacements and larger spectrum of initial drag forces corresponding to a variety of masses over 3 orders of magnitude. In the case of a larger radii (Figure 15c, $r = 0.025$ m), there are smaller angular displacements, comparatively, and the range of initial drag forces is smaller for the same variety of masses.

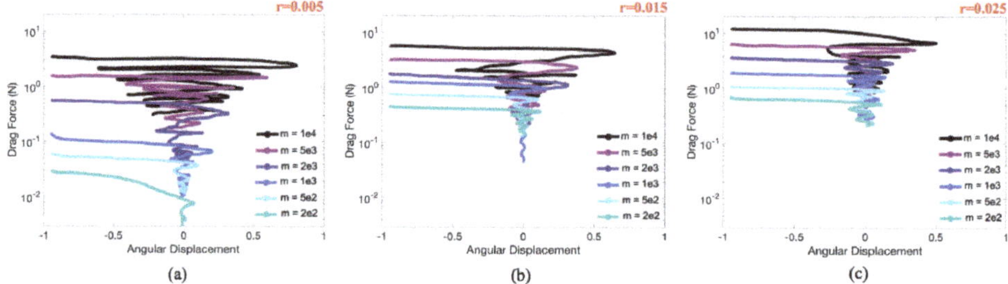

Figure 15. Phase space of drag force (N) versus angular displacement (radians) for a variety of masses in cases of (**a**) $r = 0.005$ m, (**b**) $r = 0.015$ m, and (**c**) $r = 0.025$ m. The data for each case of a specific radius appears to overlap as well as suggesting that as the peaks in angular displacement decay exponentially (see Figure 7), the drag forces also decay exponentially as well.

Finally, we wish to compare force information across all cases of radii and masses considered. The metric we chose to compare for each is the drag coefficient; recall the C_D term from Equation (7). To justify our use of Equation (7), we verified that the pendulum simulations fell into the appropriate Reynolds number range, i.e., $Re > 1$. Recall the Reynolds number, Re, is defined to be

$$Re = \frac{\rho L V}{\mu},\tag{12}$$

where ρ and μ are the fluid's density and dynamic viscosity, and L and V are a characteristic length and velocity scale, respectively. We chose L to be the length the pendulum's arm, but rather than select V to be a constant, we used the time-dependent speed of the pendulum bob for each case. Note that this speed is inherently a function of the radius of the pendulum bob itself, see Figure A3 in Appendix C. Figure 16a illustrates that for $m = 2 \times 10^3$ kg that the peak in Reynolds number is greater than one. Moreover, where Re drops down, i.e., at the beginning and end of each swing, is where speed is near zero. Thus we assume the Drag Force Law derived by Lord Rayleigh will suffice for our purposes here ($F_D \sim V^2$, see Equation (7)). Note that as the pendulums continue to swing, $Re(t)$ will decrease in every case. Eventually there will be a shift into the regime where Lord Rayleigh's Drag Force Law begins to fail and one must consider Stokes Drag Law ($F_D \sim V$, see Equation (4)) when $Re < 1$. Furthermore, we computed the time-averaged Reynolds number over the first swing of the pendulum for all masses and radii considered, see Figure 16b. Generally, as the mass of the bob increases, the average Re increases. On the other hand, as the radius increases, average Re decreases.

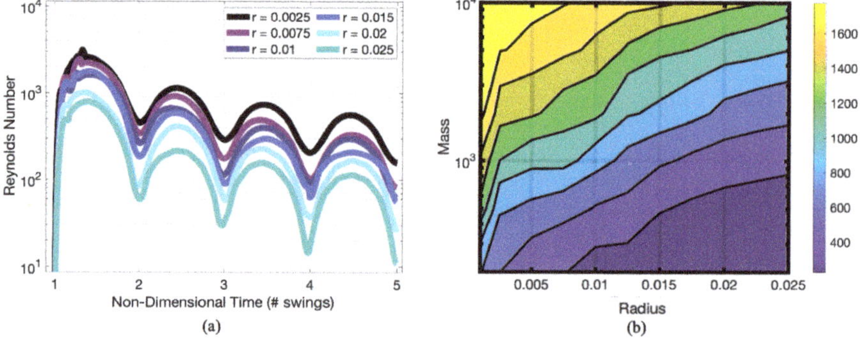

Figure 16. (**a**) Re vs. Time for $m = 2 \times 10^3$ kg and (**b**) a colormap depicting the temporally-averaged Reynolds number during the first swing for different masses and radii. Note that over time as the pendulum slows down, the average Reynolds number will decrease.

Upon calculating Equation (7), we also need to describe A, the cross-sectional area of the circular pendulum, and F_D, the drag force on the pendulum. We define A to be the circumference of the circle, i.e., $A = 2\pi r$, for each given radius, r. Having computed the speed of the pendulum bob and drag force on the body previously, we could solve for the time-dependent drag coefficient C_D at each sample time-point using Equation (7). Figure 17a,b gives $C_D(t)$ over the first swing (half an oscillation cycle) and 4 swings (2 full oscillation cycles), respectively, for the case of $m = 1 \times 10^3$ and a variety of radii. Note that the C_D peaks correspond to when the pendulum bob changes direction and thus reach speeds near zero. Moreover, the time-dependent drag coefficients will increase over time due to the pendulum continually slowing down.

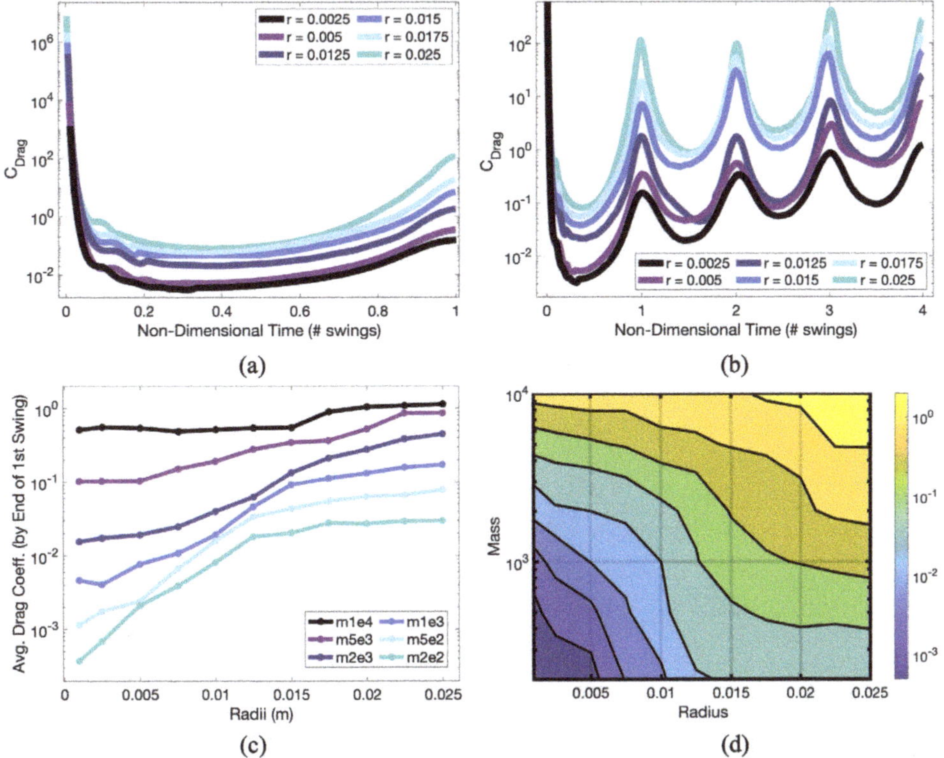

Figure 17. The drag coefficient, C_D, during the first pendulum's first swing (**a**) and first 4-swings (**b**) for a variety of radius in the case of $m = 1e3$ kg. Note that the drag coefficient maximizes when the pendulum reaches near zero speed at the end of a swing. (**c**) The temporally-averaged drag coefficients across the first swing for all mass and radius cases considered. (**d**) A contour map of the temporally-averaged drag coefficients over the first swing from (**c**) as a function of both the pendulum bob's mass and radius. Generally higher drag coefficients are seen for larger mass and size pendulum bobs.

In order to compare all simulations of differing mass and radii, we averaged the time-dependent drag coefficient, $C_D(t)$, over the first swing, as shown in Figure 17c,d. Figure 17d provides a colormap with contour lines of the time-averaged drag coefficient using the data from Figure 17c. Larger radii pendulums tend to have larger drag coefficients and higher mass pendulum bobs also have larger drag coefficients. Notice that a pendulum bob with same shape (circle) and size (radius) could elicit different drag coefficients based on variations in mass. From Section 3.2 we have already observed that variations in mass give rise to variations in speed, which is also required to compute C_D in the first place. The system is highly coupled in its many dynamical features!

Finally, we highlight the relationship between drag coefficient, C_D, and Reynolds number, Re, in Figure 18. For a given mass (mass is denoted by a particular shape in the figure), as average Re increases, C_D decreases. As the system is highly coupled, for a given mass, the average Re only increases as the pendulum bob's radius decreases (radius is denoted by the colormap). On the other hand, for a given radius, as the mass increases, the average C_D and Re also generally increase. The overall trend of decreasing C_D with increasing Re is common in many fluid dynamics phenomena, not only physical experiments, such as flow past rigid objects [57–59], but also in biology, such as tiny insect flight [50,51], or even sports such as baseball [60], American football [61], or football (soccer) [62].

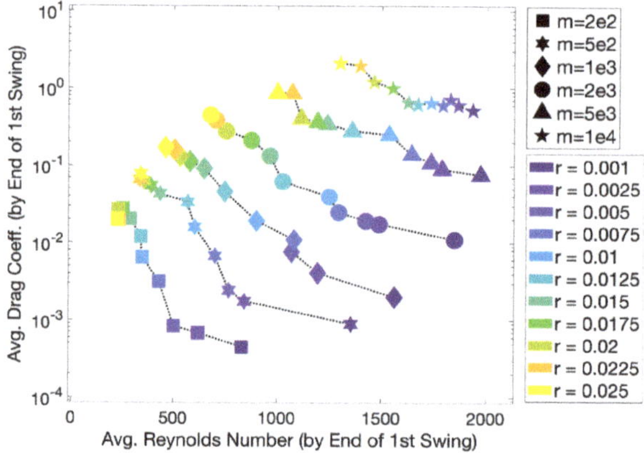

Figure 18. The average drag coefficient, C_D, vs. average Reynolds number for a variety of masses and radii. The averages were computing over the first swing of the pendulum bob.

3.4. Effect the Pendulum Bob Has onto the Fluid

In addition to the data analysis performed in Sections 3.1–3.3, which focused primarily on the Lagrangian structure itself—the pendulum, CFD (FSI) simulations grant us the opportunity to analyze how the underlying fluid reacts to a pendulum swinging through it. Also, we are able to visualize the fluid dynamics and observe the evolution of the fluid's velocity field, $\mathbf{u}(\mathbf{x},t)$, magnitude of velocity, $|\mathbf{u}(\mathbf{x},t)|$, and vorticity ($\nabla \times \mathbf{u}(\mathbf{x},t)$), see Figure 19, for qualitative analysis.

Figure 19 shows the resulting fluid dynamics due to the swinging motion of the pendulum bob of mass, $m = 5 \times 10^2$ kg, and radius, $r = 0.0175$ m, during its first swing. As the pendulum swings, there is a pocket of fast moving fluid directly behind the bob until it passes zero angular displacement, as shown by the Velocity Field and Magnitude of Velocity plots. Note that streamlines are presented on the velocity field's plots and contours are given on the magnitude of velocity plot, as well. Streamlines illustrate the path of massless tracer particles in the flow at an instantaneous point in time, while the contours give a line in which the quantity (here, magnitude of velocity) has constant value. Note that the direction of the pocket of fast moving fluid is towards the pendulum bob; hence objects directly behind the moving bob receive an fluid dynamic benefit. This phenomenon is commonly called *drafting* and has been studied in the context of many sports, such as ice skating [63], running [64,65], swimming [66], or cycling [67], as well as biological locomotion [68–72].

Within the region of fast moving fluid there are two interacting, oppositely spinning vortices behind the bob, as illustrated by the Vorticity plots. When the vortices are shed off the bob entirely, i.e., once the bob swings past zero displacement, the vortex pair continues to move vertically downwards, rather than upwards and to the right with the bob. These visualizations were produced using the raw *.vtk*-data produced during the FSI simulations using the open-source software VisIt [56]. We wish to emphasize that one cannot gain knowledge of fluid dynamics from the reduced-order ODE model, Equation (3) alone.

Moreover, because of the FSI simulations we are able to analyze the fluid data further and determine regions that have varying levels of fluid mixing. During data post-processing we could compute the finite-time Lyapunov exponents (FTLE), which can be used characterize the rate of separation in the trajectories of two infinitesimally close fluid blobs. Maxima in the FTLE (called *ridges*) have been used to determine Lagrangian Coherent Structures (LCSs), which are used to determine distinct flow structures in the fluid [73–76]. LCSs are a tool to divide the fluid's complex dynamics into distinct regions to better understand transport properties of flow [77–79]. In this paper, we computed forward-time FTLE field, whose maximal ridges give LCSs corresponding to regions of repelling fluid trajectories and low values give rise to regions of attraction [76]. This data is presented in Figure 19 as well. Our desire here is not to emphasize fluid mixing metrics, but merely point out that through CFD one is able to investigate deeper dynamics of a system, even one as well studied as a damped pendulum.

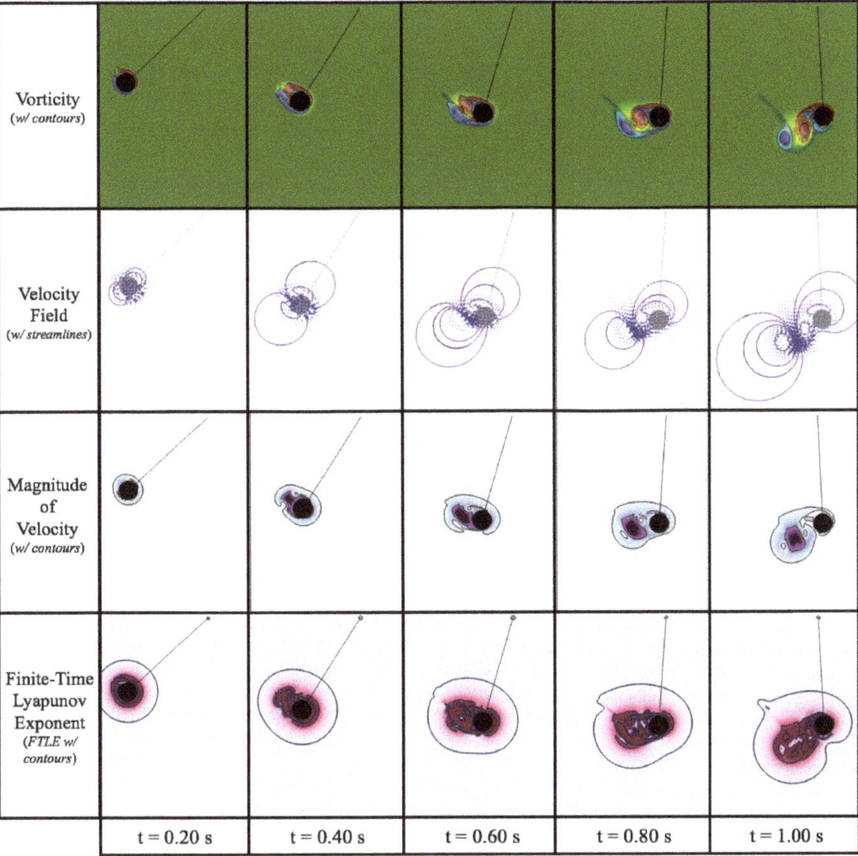

Figure 19. Colormaps (and its contours) illustrating the time evolution of the fluid's vorticity, magnitude of velocity, and finite-time Lyapunov Exponent (FTLE), as well as the velocity field (and its streamlines) resulting from the pendulum bob's first swing in the case of $m = 5 \times 10^2$ kg and $r = 0.0175$ m.

Furthermore, we wish to point out that the resulting fluid dynamics are diverse. Not every pendulum bob sheds vortices in the same way as the case of $(m, r) = (5 \times 10^2 \text{ kg}, 0.0175 \text{ m})$ (as illustrated in Figure 19). Figure 20 illustrates differences in vortex formation and shedding for the case of $m = 5 \times 10^2$ kg for a variety of radii during the first swing. In particular the overall size and magnitude of vortices formed is less in the smaller radius cases; however, once shed, the vortex dynamics are different. This would give rise to different dynamics in drafting behind the bob. In the larger radius cases ($r > 0.015$ m), the vortices move vertically downwards upon being shed, while they are significantly different in the smaller cases: in the $r = 0.010$ m case, the vortex-pair travels along with the pendulum bob, and for $r = 0.005$ m two sets of vortex-pairs move on either side of the pendulum bob. Thus, modifying the size of the pendulum results in different dynamics of the underlying fluid, even though all pendulums swing along the same circular arc; however, they do so at different speeds.

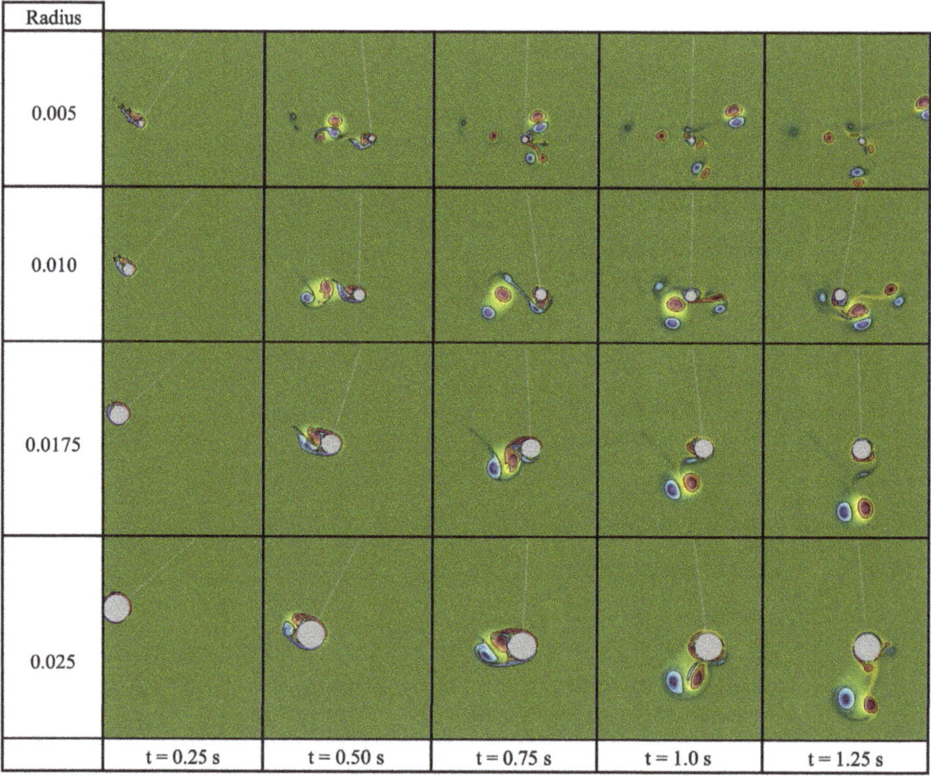

Figure 20. Comparing vortex dynamics among pendulum bob of different radii for a mass of $m = 5 \times 10^2$ kg.

Lastly, taking a further look at the $r = 0.010$ m case (for $m = 5 \times 10^2$ kg) reveals that as the pendulum swings, it swings back through its vortex wake, see Figure 21. The act of swinging through its vortex wake has been suggested as a possible mechanism for increased fluid drag on the pendulum bob [25]. However, a vortex could also enhance the speed of the bob if an appropriately spinning vortex interacts with the bob at the right moment in time, see $t = 1.0$ s in Figure 21. The vortex in red is spinning counter-clockwise and may give the bob a boost in speed, as it is moving in that same direction. There are complex interwoven dynamics within the system. Figure 22 provides an additional sequence of snapshots depicting these complex interactions. It shows a pendulum bob

$(m = 1 \times 10^4$ kg, $r = 0.005$ m$)$ swinging through its own vortex wake during the return swing of the first oscillatory cycle. Such complex interaction mechanisms have not been fully explored and warrant further attention from the scientific community.

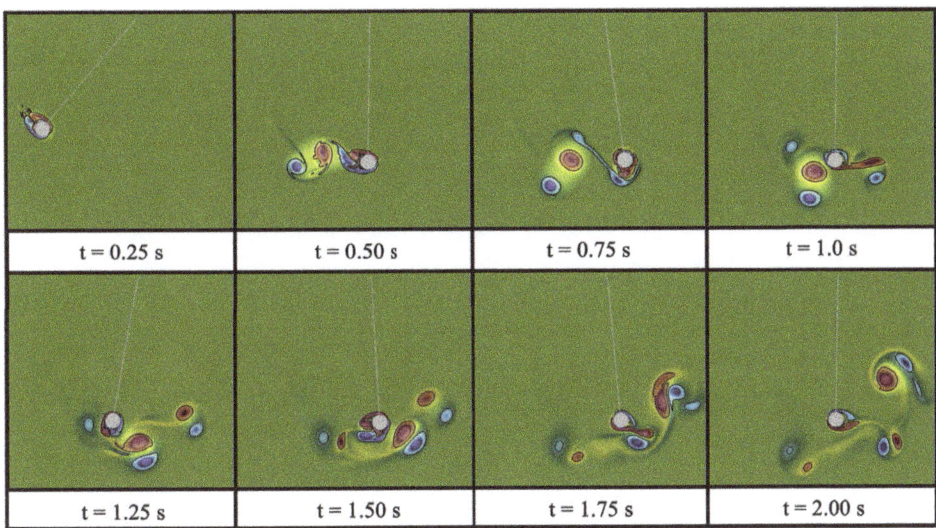

Figure 21. The vortex dynamics of the case $(m, r) = (5 \times 10^2$ kg, 0.0175 m$)$ within the first 2 s of oscillation.

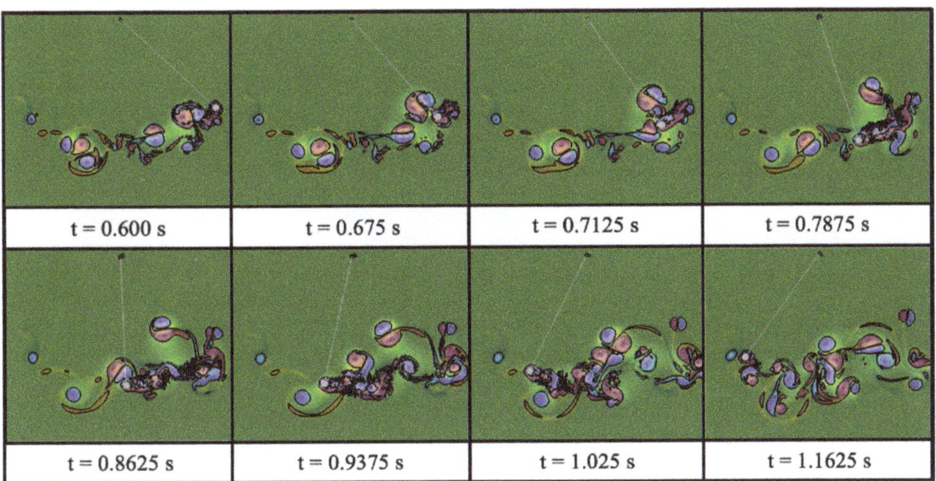

Figure 22. The vortex dynamics of the case $(m, r) = (1 \times 10^4$ kg, 0.005 m$)$ on the return swing during its first oscillatory cycle.

3.5. Numerical Comparison & Validation

Lastly, in this section we will compare and validate the canonical damped physical pendulum equation against our fluid-structure interaction (FSI) model. Recall the damped physical pendulum (Equation (3)) is given by

$$\frac{d^2\theta}{dt^2} + \frac{b}{I}\frac{d\theta}{dt} + \frac{mgL}{I}\sin\theta = 0. \tag{13}$$

Our FSI model did not assume any knowledge of the existence of this reduced ordinary differential equations (ODE) model. Instead it placed a spherical (circular) pendulum bob into a fluid environment, tethered it to a fixed location, and under the influence of gravity alone, it swung. This was to mimic a physical experiment, but performed *in silica*, rather than in a laboratory setting.

To compare Equation (13) and our FSI model, we first matched the parameter values. For each radius and mass considered in our FSI model, we were able to compute the exponential decay of the peaks in angular displacement amplitude using a linear least squares framework to fit a line through the logarithm of the peak values in angular displacement over time (linear regression), see Figure 23a as an illustrative example. The slope of each line was $\gamma = -\frac{b}{2I}$ for that particular mass and radius. Hence for the parameter b/I in Equation (13), we multiplied each slope γ by -2, i.e.,

$$b/I = -2\gamma. \tag{14}$$

Note that the term $\frac{mgL}{I}$ is the approximate natural (undamped) angular frequency squared of the pendulum bob, i.e.,

$$\omega_N^2 = \frac{mgL}{I}. \tag{15}$$

Note that this is not the true natural, undamped angular frequency, as we did not invoke the small angle approximation, i.e., $\sin\theta \approx \theta$ for small θ [9]. Moreover, due to the presence of the fluid, the pendulum bob was not in an undamped setting, so we could not directly calculate ω_N from our numerical experiments. However, we used the pendulum's period, as previously computed in Section 3.1, and γ to compute ω_N^2. For clarity with the undamped case, we define T_D and ω_D to the damped period and angular frequency of our pendulum bob from the FSI experiments. Recall the relationship between ω_D and T_D,

$$\omega_D = \frac{2\pi}{T_D}, \tag{16}$$

and the relationship between ω_N, ω_D, and γ,

$$\omega_D^2 = \omega_N^2 - \gamma^2. \tag{17}$$

Hence using Equations (15)–(17), we can compute the natural (undamped) angular frequency,

$$\omega_N^2 = \frac{mgL}{I} = \frac{4\pi^2}{T_D^2} + \gamma^2. \tag{18}$$

Using Equations (14) and (18) we found the appropriate parameter values for the reduced ODE model, as computed from the FSI simulations. Figure 23b–f provides comparison of the FSI and ODE models' angular displacement over time for a variety of pendulum bob radii and masses. For comparative purposes, the ODE model was initialized at the 5th peak of the FSI model and its solution was computed by propagating both forwards and backwards in time. Moreover we also plotted the exponential decay, using the 5th peak amplitude as the coefficient, and plotted it in a similar manner both forwards and backwards in time from the 5th peak.

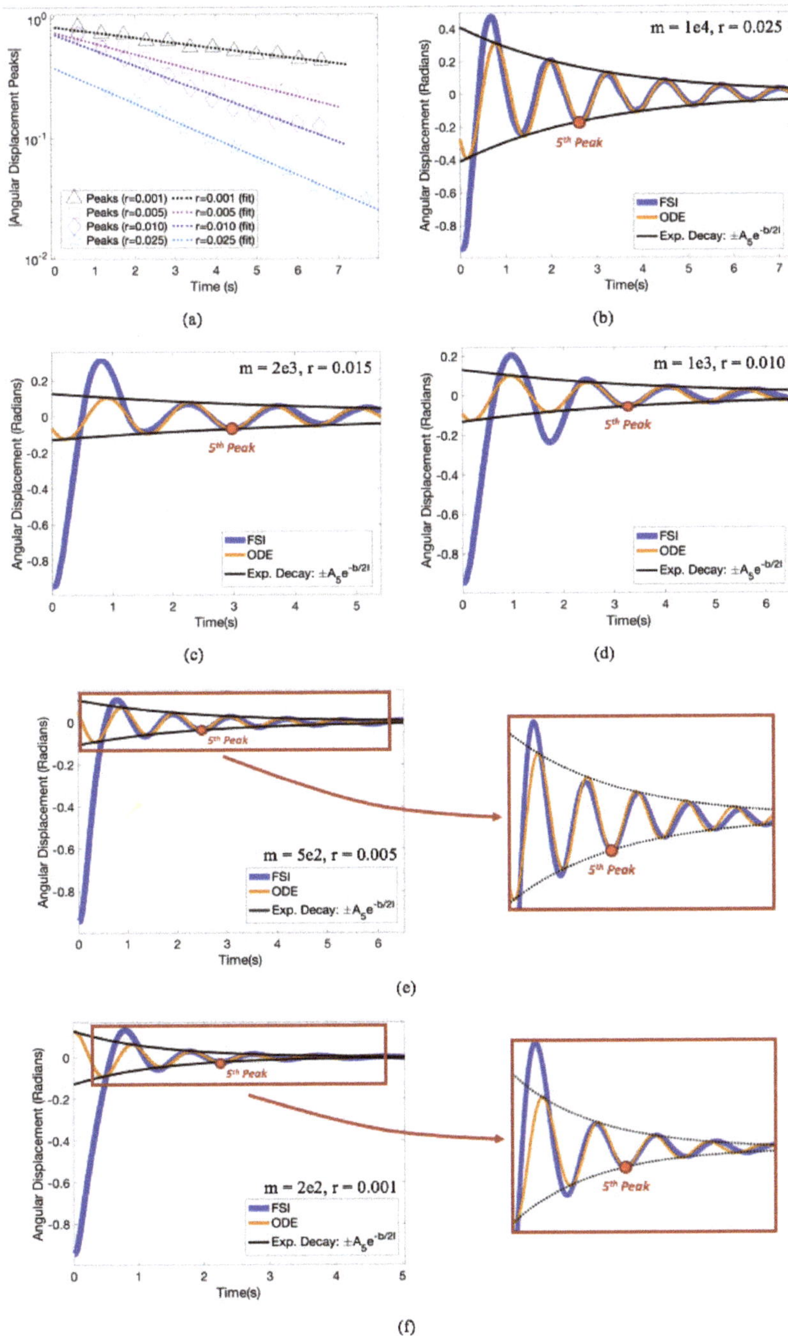

Figure 23. (**a**) Slopes of the least squares (linear regression) fits through the peaks of angular displacement over time to compute the exponential decay, $\gamma = -\frac{b}{2I}$, for a variety of radii in the $m = 5 \times 10^3$ kg case. (**b–f**) Comparison of the FSI and ODE models' angular displacement over time for a variety of masses and radii.

We note that the ODE model only agrees with our FSI model after a few oscillations. If we propagated the ODE model forward in time from the original position of the FSI pendulum, angular displacements were not consistent, see Figure 24. Furthermore, if we used the first peak amplitude (initial angular displacement) as the coefficient on the exponential decay, the FSI model appeared to not agree with its own decay rate. However, the decay rates are consistent, as seen in Figure 23, but the FSI model does not start obeying such decay until after a few oscillations. Thus, the decay rates, $\gamma = -\frac{b}{2I}$, were calculated starting with the third peak rather than the initial displacement. We chose the third peak rather than the second as the linear relationship was more prevalent from that point on. This is the same phenomenon from Figure 11 in Section 3.2, where the exponential decay in peak speeds did not start until what appeared to be the second swing.

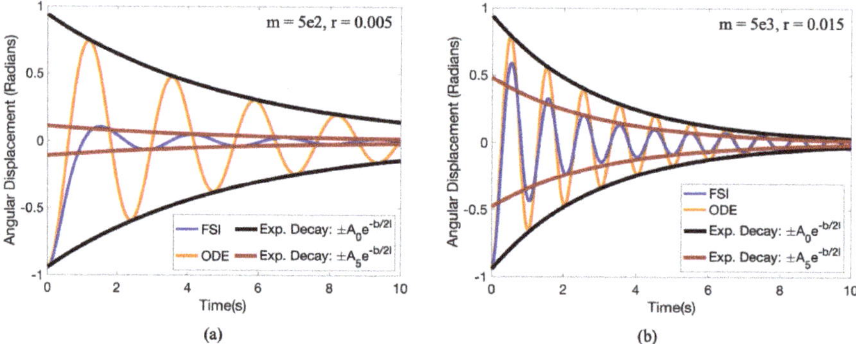

Figure 24. Depicting the dynamics if the ODE model started from the original angular displacement of the FSI pendulum rather than the the 5th peak for the case $(m, r) = (5 \times 10^2 \text{ kg}, 0.005 \text{ m})$ and $(m, r) = (5 \times 10^3 \text{ kg}, 0.015 \text{ m})$ for (**a**,**b**), respectively. A visualization of the exponential decay is also provided with the coefficient either being A_0, the original angular angular displacement, or A_5, the displacement of the 5th peak.

Overall the reduced ODE model agrees with the FSI model after a few oscillations for different size pendulum bob radii as well as over a spectrum of masses; however, the dynamics during the first swing are substantially different (see Figure 24). This is possibly due to a different fluid drag law on the pendulum, before it settles into the regime where the drag can be modeled as linearly proportional to its velocity.

Lastly, we computed the damping parameter, b, by itself as a function of the mass and radius of the pendulum bob. To compute this, we first found the effective moment of inertia of each case using Equation (18), e.g.,

$$I = \frac{mgL}{\frac{4\pi^2}{T_D^2} + \gamma^2}, \tag{19}$$

and then calculated

$$b = -2\gamma I. \tag{20}$$

Note that we chose to calculate the effective moment of inertia, I, for each simulation here. Our FSI pendulum bob geometry was not a solid structure, but rather, had a singular mass source at its center, a shell composed of neutrally-buoyant points tethered together, and from the IB formulation, its shell enclosed fluid within. This fluid had the same properties as the backward fluid environment in which the pendulum is immersed. Moreover, as the *principle of added mass* states that inertia is added to a fluid system when an object is accelerating (or decelerating) through it [21]. Thus, the appropriate moment of inertia, I, to use in the ODE model to match the FSI model is non-trivial.

The damping parameter, b, increased as mass increased for a particular radius, see Figure 25. Moreover, as the radius increased for a given mass, b increased as well. While the system appears to be more sensitive to changes in mass, recall that the mass was varied over two orders of magnitudes in value, while the radius varied roughly over one.

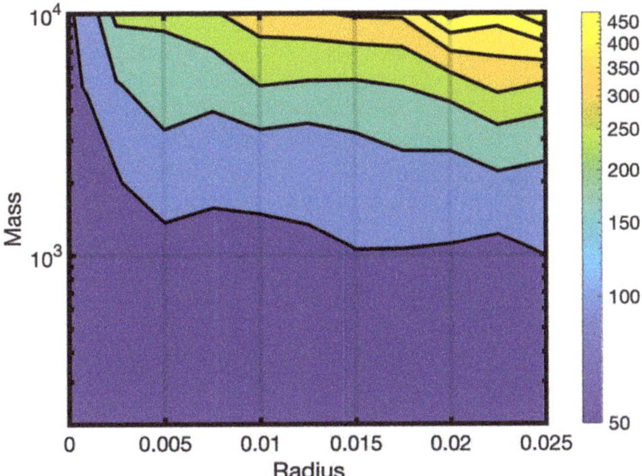

Figure 25. Values of the damping parameter, b, as a function of the mass and radius of the pendulum bob.

4. Discussion and Conclusions

Two-dimensional immersed boundary simulations were used to model the swinging motion of a circular pendulum bob under the influence of gravity that was contained within a viscous, incompressible fluid environment. In addition, to the authors knowledge, this is the first fluid-structure interaction (FSI) simulation that explores the motion of an ordinary pendulum system which also offers an open-source complement. The angular displacement data collected from the motion of the pendulum bob was directly compared against the reduced-order damping ODE model that is familiar to most STEM students. In general, the oscillatory dynamics agree between the ODE model and the FSI model (see Section 3.5). However, there were discrepancies in the decay rates between the first few swing's maximal angular displacements and speeds with those following thereafter. The mechanisms underlying these observations are not fully understood. There appear to be interesting dynamics to probe further involving the pendulum bob's mass and radius, the vortex wake it creates, the interactions of those vortices with each other and the bob itself, and the resulting drag on the bob during large amplitude oscillations.

Moreover, the ODE model's linearly-proportional-velocity damping term, i.e., Stokes Drag Law, was appropriate once the pendulum has swung a few times after a large initial angular displacement. During the first initial swing there was an enhancement of drag on the pendulum bob, potentially obeying Rayleigh's Drag Law and/or from the principle of added mass [24] and/or other complex drag mechanisms involving vortex shedding [23].

Furthermore, we were able to determine the approximate damping parameter, b, that fit the ODE model from our FSI model. The damping parameter, b, was found to be dependent on both mass and radius of the pendulum (see Section 3.5). Also, through the FSI simulations, we were able to quantify how the drag coefficient, C_D, increased with both increasing mass and radius of the pendulum bob (see Section 3.3). On that note, we also illustrated how the pendulum's period of oscillation was a function of both the mass and radius of the pendulum bob (see Section 3.1).

Beyond the dynamics of the pendulum structure itself, using a FSI model allowed us to peek into the resulting dynamics of the underlying fluid (see Section 3.4), which we would not have otherwise been able to do using the reduced-ODE model alone. While the purpose of this study was to explore the dynamics of a FSI pendulum model, which inherently did not assume any particular drag laws *a priori*, and compare the results to the ODE model, this framework can be used to further probe into new scientific frontiers that could unravel the enhanced contributions to fluid drag, whether from traveling through your own vortex wake, as suggested by Mathai et al., 2019 [25], vortex shedding as suggested by Bolster et al., 2010 [23], or added mass, as suggested by Bandi et al., 2013 [24]. There are still complex relationships to decrypt between oscillatory amplitude, geometry and mass of the pendulum bob, and fluid scale. The aforementioned studies performed physical experiments with pendulums and used sophisticated visualization techniques, either the Baker electrolytic technique [80] or *particle image velocimetry* (PIV) [81,82] to visualize the underlying fluid dynamics.

Furthermore, using this FSI framework it is possible to couple multiple pendulum together and study the resulting complex dynamics and/or investigate the motion of a variety of geometric objects being swung as pendulum bobs. We have seen that the size of the pendulum bob affects the underlying fluid dynamics, as observed through different vortex dynamics in Section 3.4; however, one has yet to explore how shape affects the fluid dynamics or motion of the bob itself.

While a student's first brief foray into fluids may have been through the concept of damped simple harmonic motion involving a pendulum, we hope that this manuscript provides the following context for students in an introductory fluid mechanics course:

- A connection to where students may have seen fluid drag laws previously, i.e., the Stokes Drag Law and Pendulum Motion. Furthermore, it illustrates for students that famous laws of physics were discovered with systems that seem as "basic" as that of a pendulum.

- The differences that may arise between modeling a system using a reduced-order ODE model and attempting to computationally model all aspects of the system to a higher degree. We hope this shows students that reduced models are valuable in that they are usually easier to solve while (hopefully) capturing a bulk of a system's dynamics. However, there are clear disadvantages as illustrated by the discrepancies that arise between the reduced order model and computational model—many dynamics are not captured in the reduced-model, e.g., the vortex wake or drafting, that maybe particularly interesting or important to understanding the system as a whole.

- Similarly, the full dynamical richness of a system may only be explored by investigating its explicit fluid mechanics, even in a system as seemingly "simple" as a single pendulum immersed in a fluid. Moreover, to even study systems involving fluids and objects immersed therein, it requires either sophisticated experimental techniques or computational expertise. This work shows that a computer can be an immensely powerful tool for performing science. More than that, programming knowledge is highly sought after in this day and age [83,84].

- The observation that even systems that are routinely studied in some introductory courses, like a pendulum, may still have open, exciting research questions that scientists and engineers actively pursue.

To conclude, the pendulum may be an old, historic device that has been studied for millennia; however, under the hood, there are a lot of hidden, complex dynamics left to discover.

Supplementary Materials: The following are available online at http://www.mdpi.com/2311-5521/5/2/48/s1.

Author Contributions: Conceptualization, M.M. and N.A.B.; Methodology and Software, N.A.B.; Validation, N.A.B.; Formal Analysis, Investigation, and Data Curation, M.M. and N.A.B.; Writing—Original Draft Preparation, N.A.B.; Writing—Review & Editing, M.M. and N.A.B.; Visualization, M.M. and N.A.B.; Funding Acquisition, N.A.B. All authors have read and agreed to the published version of the manuscript.

Funding: Computational resources were provided by the NSF OAC #1826915 and the NSF OAC #1828163. Support for N.A.B. was provided by the TCNJ Support of Scholarly Activity (SOSA) Grant, the TCNJ Department of Mathematics and Statistics, and the TCNJ School of Science.

Acknowledgments: The authors would like to thank Christina Battista, Robert Booth, Karen Clark, Jana Gevertz, Christina Hamlet, Alexander Hoover, Laura Miller, Matthew Mizuhara, Arvind Santhanakrishnan, Emily Slesinger, Edward Voskanian, and Lindsay Waldrop for comments and discussion.

Conflicts of Interest: The authors declare no conflict of interest.

Abbreviations

The following abbreviations are used in this manuscript:

CFD	Computational Fluid Dynamics
FSI	Fluid-Structure Interaction
Re	Reynolds Number
IB	Immersed Boundary Method
ODE	Ordinary Differential Equation

Appendix A. Instructor Resources

Teaching Resources:

Associated supplemental files contain slides, movies, and open-source codes pertaining to the paper. It encompasses the following:

1. **Pendulum_Classroom_Supplement.pptx/.pdf**: presentations which may be used in class; slides that tell the story of the paper. Note that the *.pptx* file has embedded movies in *.mp4* format.
2. **Movies**: directory containing movies (*.mp4* format) pertaining to each simulation shown in the manuscript.
3. Note that an open-source fluid-structure interaction model of a point-mass pendulum can be found at: https://github.com/nickabattista/IB2d in the sub-directory:

$$\texttt{IB2d} \rightarrow \texttt{matIB2d} \rightarrow \texttt{Examples} \rightarrow \texttt{Examples\_Education} \rightarrow \texttt{Pendulum}.$$

4. Visualization software used: VisIt (https://visit.llnl.gov/) (v. 2.12.3)

Appendix B. Immersed Boundary Method

The immersed boundary method [30] was used to model the motion of a pendulum under gravitational acceleration, see Section 2. Although, IB is capable of solving fully coupled fluid-structure interaction systems involving flexible or squishy structures, here we use it to model the stiff boundaries of a pendulum bob immersed within an incompressible, viscous fluid. The fluid motion is governed by the Navier-Stokes equations, given as

$$\rho \left(\frac{\partial \mathbf{u}(\mathbf{x},t)}{\partial t} + \mathbf{u}(\mathbf{x},t) \cdot \nabla \mathbf{u}(\mathbf{x},t) \right) = -\nabla p(\mathbf{x},t) + \mu \Delta \mathbf{u}(\mathbf{x},t) + \mathbf{f}(\mathbf{x},t) \tag{A1}$$

$$\nabla \cdot \mathbf{u}(\mathbf{x},t) = 0, \tag{A2}$$

where $\mathbf{u}(\mathbf{x},t) = (u(\mathbf{x},t), v(\mathbf{x},t))$ is the fluid velocity, $p(\mathbf{x},t)$ is the pressure, $\mathbf{f}(\mathbf{x},t)$ is the force per unit volume (area in 2*D*) applied to the fluid by the immersed boundary, i.e., the pendulum. The independent variables are the position, $\mathbf{x} = (x,y)$, and time, t. Equations (A1) and (A2) are conservation laws for the fluid, i.e., the conservation of momentum and mass, respectively. Note that Equation (A2) is known as the *incompressibility* condition.

The interaction equations between the fluid and the immersed structure are given by

$$\mathbf{f}(\mathbf{x},t) = \int \mathbf{F}(r,t)\delta(\mathbf{x} - \mathbf{X}(r,t))dr \tag{A3}$$

$$\mathbf{U}(\mathbf{X}(r,t),t) = \frac{\partial \mathbf{X}(r,t)}{\partial t} = \int \mathbf{u}(\mathbf{x},t)\delta(\mathbf{x} - \mathbf{X}(r,t))d\mathbf{x}, \tag{A4}$$

where $\mathbf{X}(r, t)$ gives the Cartesian coordinates at time t of the material point labeled by Lagrangian parameter r, $\mathbf{F}(r, t)$ is the force per unit area imposed onto the fluid by elastic deformations in the boundary, as a function of the Lagrangian position, r, and time, t. Equation (A3) applies a force from the immersed boundary to the fluid grid through a delta-kernel integral transformation. Equation (A4) sets the velocity of the boundary equal to the local fluid velocity.

As suggested in Section 2.2, the deformation force equation, $\mathbf{F}(r,t)$, is specific to the system being explored. For this pendulum model, it takes the following form

$$\mathbf{F}(r, t) = \mathbf{F}_{spr} + \mathbf{F}_{target} + \mathbf{F}_{Mass}, \tag{A5}$$

that is the summation of forces arising from spring, target point, and massive point deformations pertaining over each Lagrangian point being modeled with one or more of this model features.

IB Algorithm

In our pendulum model, we imposed periodic boundary conditions on a square domain. To solve Equations (A1)–(A4) we need to update the velocity, pressure, and both the position of the boundary and forces acting on it from the previous time-step data, time n. IB traditionally does this in the following steps [30,39]:

Step 1: Calculate the force density, $\mathbf{F}^n$ on the immersed boundary, from its current boundary configuration at time n, $\mathbf{X}^n$.

Step 2: Use Equation (A3) to spread the force from the Lagrangian boundary to the Eulerian (fluid) mesh to compute $\mathbf{f}^n$

Step 3: Solve the Navier-Stokes equations, A1 and A2, on the Eulerian grid, thus updating $\mathbf{u}^{n+1}$ and p^{n+1} from $\mathbf{u}^n$, p^n, and $\mathbf{f}^n$.

Step 4: Update the Lagrangian point positions, $\mathbf{X}^{n+1}$, using the local fluid velocities, $\mathbf{U}^{n+1}$, computed from $\mathbf{u}^{n+1}$ and (A4).

We quickly note that to approximate the integrals in Equations (A3) and (A4), discretized (and regularized) delta functions were used. We chose to use the delta functions described in [30], i.e., $\delta_h(\mathbf{x})$,

$$\delta_h(\mathbf{x}) = \frac{1}{h^3} \phi\left(\frac{x}{h}\right) \phi\left(\frac{y}{h}\right) \phi\left(\frac{z}{h}\right), \tag{A6}$$

where $\phi(r)$ is defined as

$$\phi(r) = \begin{cases} \frac{1}{8}(3 - 2|r| + \sqrt{1 + 4|r| - 4r^2}), & 0 \leq |r| < 1 \\ \frac{1}{8}(5 - 2|r| + \sqrt{-7 + 12|r| - 4r^2}), & 1 \leq |r| < 2 \\ 0 & 2 \leq |r|. \end{cases} \tag{A7}$$

Appendix C. Additional Pendulum Data

In this appendix we provide complementary data to the data presented in Section 3, e.g., if we provided a figure that contained a variety of masses for a particular radius (as in Figure 6), here we will provide the opposite—a variety of radii for particular cases of mass (as in Figure A1 below). These figures are provided for additional clarity in regards to the comparisons being discussed and analyzed.

First we provide the angular displacement (in radians) over time (in seconds) for cases of pendulums with the same mass, but different radii in Figure A1. This data is to illustrate clearly that pendulum bobs with the same mass can experience different oscillatory patterns for different radii. Moreover, it appears that by increasing mass orders of magnitude, from 2×10^2 kg to 2×10^3 kg to 1×10^4 kg could result in the pendulum bob undergoing different regimes of oscillation—either underdamped to overdamped.

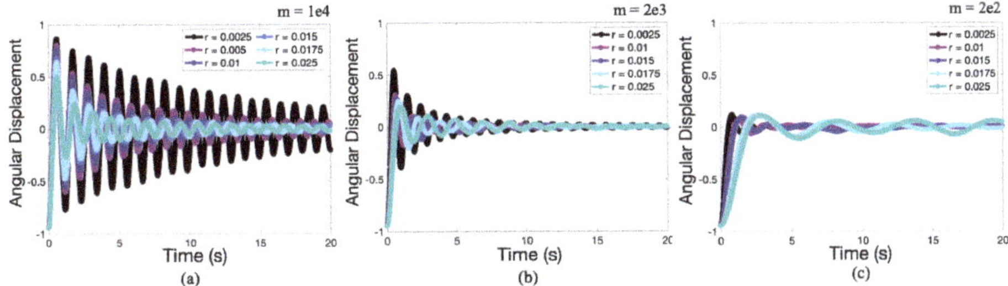

Figure A1. Depicting the angular displacement (radians) vs. time (s) for pendulums with the same mass but different radii. (**a**–**c**) give data for a specific mass, either $m = 1 \times 10^4$ kg, 2×10^3 kg, or 2×10^2 kg, respectively, and a variety of radii in each.

Next we provide a plot of the height the pendulum reaches (in meters) as a function of the peak number in angular displacement for $m = 1 \times 10^4$ kg and a variety of radii in Figure A2. This illustrates that as the radius increases, the height decreases. Not only does the height decreases as the radius increases, the linear speed of the pendulum also decreases as well, as given in Figure A3. Our simulations suggest that smaller pendulum bobs generally move faster than larger ones for a given mass. In both of these figures, among all cases, both the speed and height decay exponentially as illustrated by the semi-logarithmic plots in Figures A2b and A3b. These data are provided to suggest that as the size of the pendulum increases, there must be more drag force acting on the bob to decelerate their speed and thus not allow them to reach as great of heights (angular displacements) as other smaller bobs.

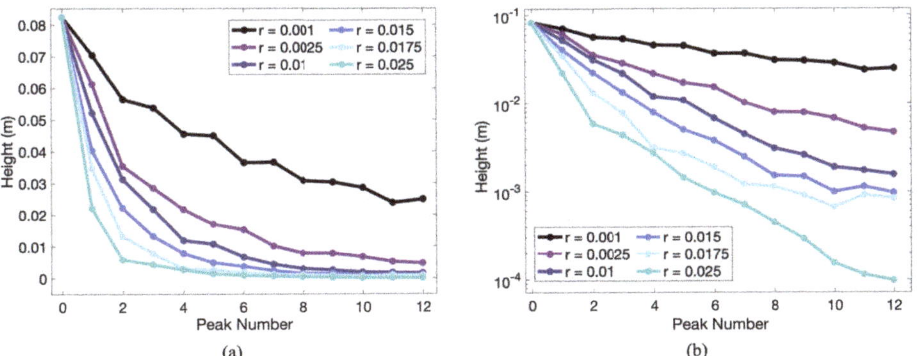

Figure A2. (**a**) Plot illustrating the decay of the height (m) that the pendulum bob reaches as the pendulum continues to swing for the case of $m = 1 \times 10^4$ kg for a variety of radii. The peak amplitude decays exponentially as illustrated by the linear relationship between the logarithm of the amplitude against peak number, as shown in (**b**).

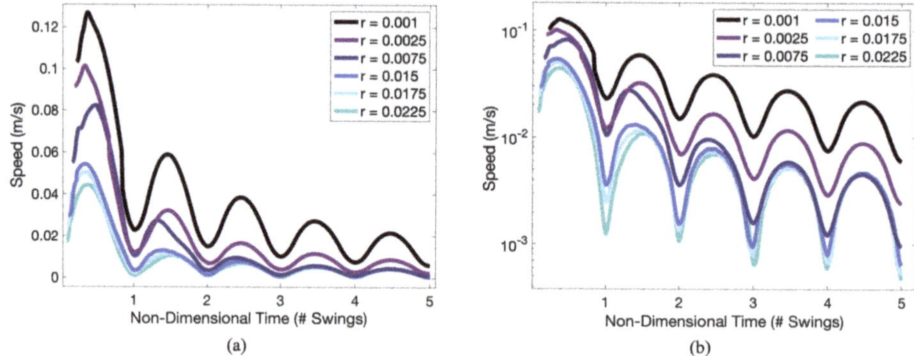

Figure A3. (**a**) Plot depicting the linear speed of the pendulum bob against non-dimensional time given as the # of swings (half a full displacement cycle) for the case of $m = 1 \times 10^3$ kg for a variety of radii. Speed peaks in the middle of a swing corresponding to when the pendulum has zero angular displacement from the vertical and the peak speed appears to decay exponentially, given by the linear relationship in (**b**).

Furthermore we also provide more detailed phase space explorations of linear speed (m/s) versus angular displacement (radians) in Figures A4 and A5. Figure A4 provides the phase space of linear speed versus angular displacement for the case of $m = 5 \times 10^3$ kg and a variety of radii, while Figure A5 selects four radii ($r = 0.001$ m, $r = 0.005$ m, $r = 0.0015$ m, and $r = 0.025$ m) and varies mass. Similar topological structures are observed, where the data collapses onto a parabolically-capped cone. This is intuitive as both the peaks in angular displacement and speed decrease over time; however, what is particularly interesting is that the cone angle looks to be approximately conserved among all cases.

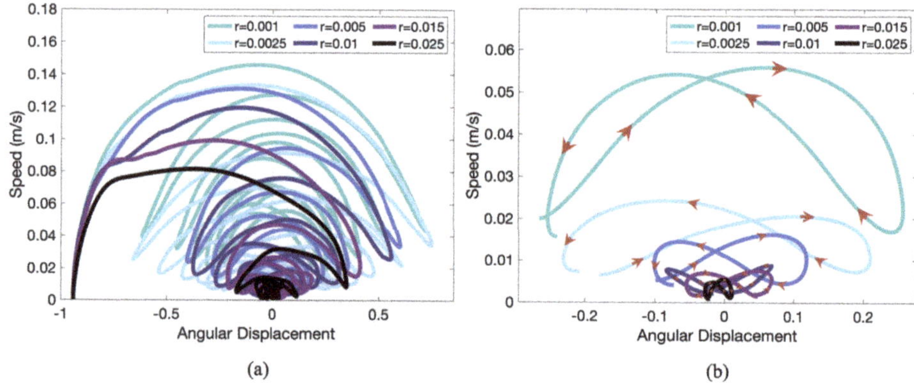

Figure A4. (**a**) Phase space of linear speed of the pendulum bob vs. angular displacement (radians) for a variety of radii in the case of $m = 5 \times 10^3$ kg. (**b**) A closer look at the last simulated cycle's phase space for each case.

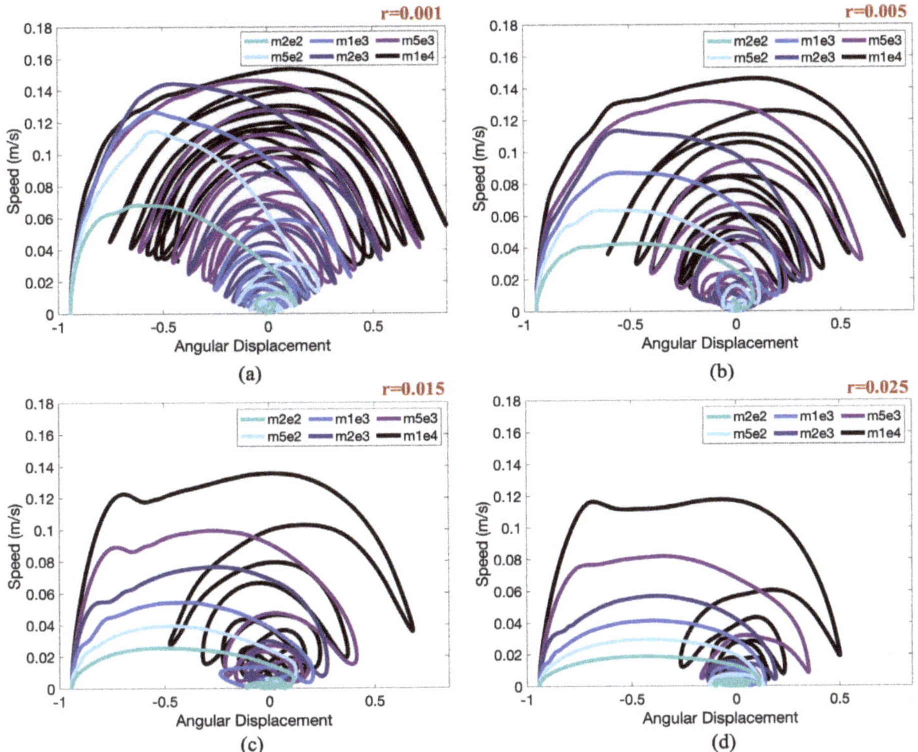

Figure A5. Phase space of linear speed of the pendulum bob vs. angular displacement (radians) for a variety of masses for cases: (**a**) $r = 0.001$ m, (**b**) $r = 0.005$ m, (**c**) $r = 0.015$ m, and (**d**) $r = 0.025$ m.

Finally we provide data depicting the drag force (N) over time for 3 different radii ($r = 0.015$ m, $r = 0.020$ m, and $r = 0.025$ m) over a variety of masses. Compared to Figure 14, we notice that for the same radius but different masses, the drag forces begin to overlap with time. Specifically, the drag forces corresponding to larger masses decay more rapidly. This could also be surmised from Figure 17c,d, which show an increased average drag coefficient during the higher mass cases for a specific radius.

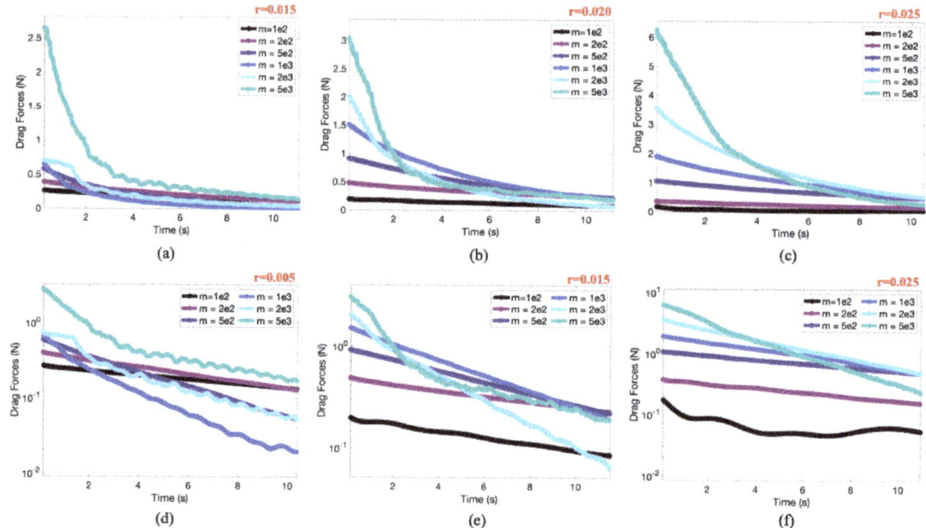

Figure A6. Drag forces (N) over time in seconds for a variety of masses for cases with (**a**,**d**) $r = 0.015$ m, (**b**,**e**) $r = 0.020$ m, and (**c**,**f**) $r = 0.025$ m. The semi-log data is provided in (**d**–**f**) to highlight a linear relationship between the logarithm of the drag force and time. This linear relationship suggests an exponential decay in drag force over time.

References

1. Milne, J. Pendulum Seismometers. *Nature* **1888**, *37*, 570–571. [CrossRef]
2. Morton, W.S.; Lewis, C.M. *China: Its History and Culture*; McGraw-Hill, Inc.: New York, NY, USA, 2005.
3. Matthews, M.R. *Time for Science Education: How Teaching the History and Philosophy of Pendulum Motion Can Contribute to Science Literacy*; Springer: New York, NY, USA, 2000.
4. Blackwell, N. Experimental stone-cutting with the Mycenaean pendulum saw. *Antiquity* **2018**, *92*, 217–232. [CrossRef]
5. Bennett, M.; Schatz, M.F.; Rockwood, H.; Wiesenfeld, K. Huygens' Clocks. *Proc. R. Soc. Lond. A* **2002**, *458*, 563–579. [CrossRef]
6. Boettcher, W.; Merkle, F.; Weitkemper, H.H. History of extracorporeal circulation: The conceptional and developmental period. *J. Extra Corpor. Technol.* **2003**, *3*, 172–183.
7. Scott, G.R. *The History of Torture throughout the Ages*; Kessinger Publishing, LLC: Whitefish, MT, USA, 2009.
8. Poe, E.A. The Pit and the Pendulum. In *The Gift: A Christmas and New Year's Present for 1843*; Leslie, E., Ed.; Carey & Hart: Philadelphia, PA, USA, 1843; Chapter 12, pp. 133–152.
9. Halliday, D.; Resnick, R.; Walker, J. *Fundamentals of Physics*, 7th ed.; John Wiley & Sons: New York, NY, USA, 2004.
10. Stokes, G.G. On the Effect of the Internal Friction of Fluids on the Motion of Pendulums. *Trans. Camb. Philos. Soc.* **1851**, *9*, 8–106.
11. Batchelor, G.K. *Introduction to Fluid Mechanics*; Cambridge University Press: Cambridge, UK, 2000.
12. Happel, J.; Brenner, H. *Low Reynolds Number Hydrodynamics*; Springer: New York, NY, USA, 1981.
13. Buckingham, E. On physically similar systems; illustrations of the use of dimensional equations. *Phys. Rev.* **2014**, *4*, 245–376. [CrossRef]
14. Landau, L.D.; Lifshitz, E.M. *Fluid Mechanics*, 1st ed.; Pergamon: London, UK, 1959.
15. Mahajan, S. Chapter 4: Fluid Drag. Notes from MIT IAP Course in 2006: Lies and Damn Lies: The Art of Approximation in Science. 2006. Available online: http://www.inference.org.uk/sanjoy/mit/book:04.pdf (accessed on 3 January 2020).
16. Nelson, R.A.; Olsson, M.G. The Pendulum-Rich physics from a simple system. *Am. J. Phys.* **1986**, *2*, 54. [CrossRef]

17. Peters, R.D. Nonlinear Damping of the 'Linear' Pendulum. 2003. Available online: https://arxiv.org/abs/physics/0306081 (accessed on 15 December 2019).
18. Peters, R.D. The Pendulum in the 21st Century-Relic or Trendsetter. *Sci. Educ.* **2004**, *13*, 279–295. [CrossRef]
19. Quiroga, G.D.; Ospina-Henao, P.A. Dynamics of damped oscillations: Physical pendulum. *Eur. J. Phys.* **2017**, *38*, 065005. [CrossRef]
20. Hsu, H.; Capart, H. Enhanced upswing in immersed collisions of tethered spheres. *Phys. Fluids* **2007**, *19*, 101701. [CrossRef]
21. Neill, D.; Livelybrooks, D.; Donnelly, R.J. A pendulum experiment on added mass and the principle of equivalence. *Am. J. Phys.* **2007**, *75*, 226. [CrossRef]
22. Sullivan, I.; Niemela, J.; Hershberger, R.; Bolster, D.; Donnelly, R. Dynamics of thin vortex rings. *J. Fluid Mech.* **2008**, *609*, 319–347. [CrossRef]
23. Bolster, D.; Hershberger, R.E.; Donnelly, R.J. Oscillating pendulum decay by emission of vortex rings. *Phys. Rev. E* **2010**, *13*, 046317. [CrossRef] [PubMed]
24. Bandi, M.M.; Concha, A.; Wood, R.; Mahadevan, L. A pendulum in a flowing soap film. *Phys. Fluids* **2013**, *25*, 041702. [CrossRef]
25. Mathai, V.; Loeffen, L.; Chan, T.; Wildeman, S. Dynamics of heavy and buoyant underwater pendulums. *J. Fluid Mech.* **2019**, *862*, 348–363. [CrossRef]
26. Farnell, D.J.; David, R.; Barton, D.C. Numerical simulations of a filament in a flowing soap film. *Int. J. Numer. Methods Fluids* **2004**, *44*, 313–330. [CrossRef]
27. Orchini, A.; Kellay, H.; Mazzino, A. Galloping instability and control of a rigid pendulum in a flowing soap film. *J. Fluids Struct.* **2015**, *56*, 124–133. [CrossRef]
28. Peskin, C. Flow patterns around heart valves: A numerical method. *J. Comput. Phys.* **1972**, *10*, 252–271. [CrossRef]
29. Peskin, C. Numerical analysis of blood flow in the heart. *J. Comput. Phys.* **1977**, *25*, 220–252. [CrossRef]
30. Peskin, C.S. The immersed boundary method. *Acta Numer.* **2002**, *11*, 479–517. [CrossRef]
31. Fauci, L.; Fogelson, A. Truncated Newton methods and the modeling of complex immersed elastic structures. *Commun. Pure Appl. Math* **1993**, *46*, 787–818. [CrossRef]
32. Lai, M.C.; Peskin, C.S. An Immersed Boundary Method with Formal Second-Order Accuracy and Reduced Numerical Viscosity. *J. Comp. Phys.* **2000**, *160*, 705–719. [CrossRef]
33. Cortez, R.; Minion, M. The Blob Projection Method for Immersed Boundary Problems. *J. Comp. Phys.* **2000**, *161*, 428–453. [CrossRef]
34. Griffith, B.E.; Peskin, C.S. On the order of accuracy of the immersed boundary method: Higher order convergence rates for sufficiently smooth problems. *J. Comput. Phys.* **2005**, *208*, 75–105. [CrossRef]
35. Mittal, R.; Iaccarino, C. Immersed boundary methods. *Annu. Rev. Fluid Mech.* **2005**, *37*, 239–261. [CrossRef]
36. Griffith, B.E.; Hornung, R.; McQueen, D.; Peskin, C.S. An adaptive, formally second order accurate version of the immersed boundary method. *J. Comput. Phys.* **2007**, *223*, 10–49. [CrossRef]
37. Griffith, B.E. An Adaptive and Distributed-Memory Parallel Implementation of the Immersed Boundary (IB) Method. 2014. Available online: https://github.com/IBAMR/IBAMR (accessed on 21 October 2014).
38. Griffith, B.E.; Luo, X. Hybrid finite difference/finite element version of the immersed boundary method. *Int. J. Numer. Meth. Eng.* **2017**, *33*, e2888. [CrossRef]
39. Battista, N.A.; Strickland, W.C.; Miller, L.A. IB2d: A Python and MATLAB implementation of the immersed boundary method. *Bioinspir. Biomim.* **2017**, *12*, 036003. [CrossRef]
40. Battista, N.A.; Strickland, W.C.; Barrett, A.; Miller, L.A. IB2d Reloaded: A more powerful Python and MATLAB implementation of the immersed boundary method. *Math. Method. Appl. Sci.* **2018**, *41*, 8455–8480. [CrossRef]
41. Miller, L.A. Fluid Dynamics of Ventricular Filling in the Embryonic Heart. *Cell Biochem. Biophys.* **2011**, *61*, 33–45. [CrossRef]
42. Griffith, B.E. Immersed boundary model of aortic heart valve dynamics with physiological driving and loading conditions. *Int. J. Numer. Methods Biomed. Eng.* **2012**, *28*, 317–345. [CrossRef] [PubMed]
43. Battista, N.A.; Lane, A.N.; Liu, J.; Miller, L.A. Fluid Dynamics of Heart Development: Effects of Trabeculae and Hematocrit. *Math. Med. Biol.* **2017**, *35*, 493–516. [CrossRef] [PubMed]
44. Battista, N.A.; Douglas, D.R.; Lane, A.N.; Samsa, L.A.; Liu, J.; Miller, L.A. Vortex Dynamics in Trabeculated Embryonic Ventricles. *J. Cardiovasc. Dev. Dis.* **2019**, *6*, 6. [CrossRef] [PubMed]

45. Bhalla, A.; Griffith, B.E.; Patankar, N. A forced damped oscillation framework for undulatory swimming provides new insights into how propulsion arises in active and passive swimming. *PLoS Comput. Biol.* **2013**, *9*, e1003097. [CrossRef]

46. Bhalla, A.; Griffith, B.E.; Patankar, N. A unified mathematical frame- work and an adaptive numerical method for fluid-structure interaction with rigid, deforming, and elastic bodies. *J. Comput. Phys.* **2013**, *250*, 446–476. [CrossRef]

47. Hamlet, C.; Fauci, L.J.; Tytell, E.D. The effect of intrinsic muscular nonlinearities on the energetics of locomotion in a computational model of an anguilliform swimmer. *J. Theor. Biol.* **2015**, *385*, 119–129. [CrossRef]

48. Hoover, A.P.; Griffith, B.E.; Miller, L.A. Quantifying performance in the medusan mechanospace with an actively swimming three-dimensional jellyfish model. *J. Fluid. Mech.* **2017**, *813*, 1112–1155. [CrossRef]

49. Miles, J.G.; Battista, N.A. Naut your everyday jellyfish model: Exploring how tentacles and oral arms impact locomotion. *Fluids* **2019**, *4*, 169. [CrossRef]

50. Miller, L.A.; Peskin, C.S. When vortices stick: An aerodynamic transition in tiny insect flight. *J. Exp. Biol.* **2004**, *207*, 3073–3088. [CrossRef]

51. Miller, L.A.; Peskin, C.S. A computational fluid dynamics of clap and fling in the smallest insects. *J. Exp. Biol.* **2005**, *208*, 3076–3090. [CrossRef]

52. Jones, S.K.; Laurenza, R.; Hedrick, T.L.; Griffith, B.E.; Miller, L.A. Lift- vs. drag-based for vertical force production in the smallest flying insects. *J. Theor. Biol.* **2015**, *384*, 105–120. [CrossRef] [PubMed]

53. Engineers Edge, LLC. Kinematic Viscosity Table Chart of Liquids, 2000–2020. Available online: https://www.engineersedge.com/fluid_flow/kinematic-viscosity-table.htm (accessed on 23 October 2019).

54. Kim, Y.; Peskin, C.S. 2D parachute simulation by the immersed boundary method. *SIAM J. Sci. Comput.* **2006**, *28*, 2294–2312. [CrossRef]

55. Kallemov, B.; Bhalla, A.; Griffith, B.E.; Donev, A. An immersed boundary method for rigid bodies. *Comm. Appl. Math. Comp. Sci.* **2016**, *11*, 79–141. [CrossRef]

56. Childs, H.; Brugger, E.; Whitlock, B.; Meredith, J.; Ahern, S.; Pugmire, D.; Biagas, K.; Miller, M.; Harrison, C.; Weber, G.H.; et al. VisIt: An End-User Tool For Visualizing and Analyzing Very Large Data. In *High Performance Visualization–Enabling Extreme-Scale Scientific Insight*; Bethel, E.W., Childs, H., Hansen, C., Eds.; Chapman and Hall/CRC: Boca Raton, FL USA, 2012; pp. 357–372.

57. Jones, A.M.; Knudsen, J.G. Drag coefficients at low Reynolds numbers for flow past immersed bodies. *AIChE J.* **1961**, *7*, 20–25. [CrossRef]

58. Hall, N. Drag of a Sphere. National Aeronautics and Space Administration. 2015. Available online: https://www.grc.nasa.gov/WWW/k-12/airplane/dragsphere.html (accessed on 23 March 2020).

59. Barry, D.A.; Parlange, J.Y. Universal expression for the drag on a fluid sphere. *PLoS ONE* **2018**, *13*, e0194907. [CrossRef]

60. Sawicki, G.; Hubbard, M.; Stronge, W.J. How to hit home runs: Optimum baseball bat swing parameters for maximum range trajectories. *Am. J. Phys.* **2003**, *71*, 1152–1162. [CrossRef]

61. Watts, R.G.; Moore, G. The drag force on an American football. *Am. J. Phys.* **2003**, *71*, 791–793. [CrossRef]

62. Alam, F.; Chowdhury, H.; George, S.; Mustary, I.; Zimmer, G. Aerodynamic Drag Measurements of FIFA-approved Footballs. *Procedia Eng.* **2014**, *72*, 703–708. [CrossRef]

63. Rundell, K.W. Effects of drafting during short-track speed skating. *Med. Sci. Sports Exerc.* **1996**, *28*, 765–771. [CrossRef]

64. Zouhal, H.; Abderrahman, A.; Prioux, J.; Knechtle, B.; Bouguerra, L.; Kebsi, W.; Noakes, T.D. Drafting's Improvement of 3000-m Running Performance in Elite Athletes: Is It a Placebo Effect? *Int. J. Sports Phys. Perform.* **2015**, *10*, 147–152. [CrossRef] [PubMed]

65. Beaumont, F.; Bogard, F.; Murer, S.; Polidori, G.; Madaci, F.; Taiar, R. How does aerodynamics influence physiological responses in middle-distance running drafting? *Math. Mod. Eng. Probl.* **2019**, *6*, 129–135. [CrossRef]

66. Silva, A.J.; Rouboa, A.I.; Moreira, A.; Reis, V.M.; Alves, F.; Vilas-Boas, J.P.; Marinho, D.A. Analysis of drafting effects in swimming using computational fluid dynamics. *J. Sport Sci. Med.* **2008**, *7*, 60–66.

67. Blocken, B.; Defraeye, T.; Koninckx, E.; Carmeliet, J.; Hespel, P. CFD simulations of the aerodynamic drag of two drafting cyclists. *Comput. Fluids* **2013**, *71*, 435–445. [CrossRef]

68. Fish, F.E. Energy conservation by formation swimming: Metabolic evidence from ducklings. In *Mechanics and Physiology of Animal Swimming*; Mattock, L., Bone, Q., Rayner, J.M., Eds.; Cambridge University Press: Cambridge, UK, 1994; Chapter 13, pp. 193–204.

69. Fish, F.E. Kinematics of ducklings swimming in formation: Consequences of position. *J. Exp. Zool.* **1995**, *273*, 1–11. [CrossRef]

70. Weimerskirch, H.; Martin, J.; Clerquin, Y.; Alexandre, P.; Jiraskova, S. Energy saving in flight formation. *Nature* **2001**, *413*, 697–698. [CrossRef] [PubMed]

71. Hemelrijk, C.K.; Redi, D.A.; Hildenbrandt, H.; Padding, J.T. The increased efficiency of fish swimming in a school. *Fish Fish.* **2015**, *16*, 511–521. [CrossRef]

72. Daghooghi, M.; Borazjani, I. The hydrodynamic advantages of synchronized swimming in a rectangular pattern. *Bioinspir. Biomim.* **2015**, *10*, 056018. [CrossRef]

73. Shadden, S.C.; Lekien, F.; Marsden, J.E. Definition and properties of Lagrangian coherent structures from finite-time Lyapunov exponents in two-dimensional aperiodic flows. *Physica D* **2005**, *212*, 271–304. [CrossRef]

74. Shadden, S.C. Lagrangian Coherent Structures: Analysis of Time Dependent Dynamical Systems Using Finite-Time Lyapunov Exponent. 2005. Available online: https://shaddenlab.berkeley.edu/uploads/LCS-tutorial/LCSdef.html (accessed on 19 September 2019).

75. Shadden, S.C.; Katija, K.; Rosenfeld, M.; Marsden, J.E.; Dabiri, J.O. Transport and stirring induced by vortex formation. *J. Fluid Mech.* **2007**, *593*, 315–331. [CrossRef]

76. Haller, G.; Sapsis, T. Lagrangian coherent structures and the smallest finite-time Lyapunov exponent. *Chaos* **2011**, *21*, 023115. doi:10.1063/1.3579597. [CrossRef] [PubMed]

77. Shadden, S.C.; Dabiri, J.O.; Marsden, J.E. Lagrangian analysis of fluid transport in empirical vortex ring flows. *Phys. Fluids* **2006**, *18*, 047105. [CrossRef]

78. Lukens, S.; Yang, X.; Fauci, L. Using Lagrangian coherent structures to analyze fluid mixing by cilia. *Chaos* **2010**, *20*, 017511. doi:10.1063/1.3271340. [CrossRef] [PubMed]

79. Cheryl, S.; Glatzmaier, G.A. Lagrangian coherent structures in the California Current System—Sensitivities and limitations. *Geophys. Astrophys. Fluid Dyn.* **2012**, *106*, 22–44.

80. Mazo, R.M.; Hershberger, R.; Donnelly, R.J. Observations of flow patterns by electrochemical means. *Exp. Fluids* **2008**, *44*, 49–57. [CrossRef]

81. Kiger, K.; Westerweel, J.; Poelma, C. Introduction to Particle Image Velocimetry. 2016. Available online: http://www2.cscamm.umd.edu/programs/trb10/presentations/PIV.pdf (accessed on 21 October 2016).

82. Dantec. Measurement Principles of PIV. 2016. Available online: http://www.dantecdynamics.com/measurement-principles-of-piv (accessed on 21 October 2016).

83. Heron, P.; McNeill, L. Phys21: Preparing Physics Students for 21st-Century Careers (A Report by the Joint Task Force on Undergraduate Physics Programs). American Physical Society and the American Association of Physics Teachers. 2016. Available online: https://www.compadre.org/JTUPP/docs/J-Tupp_Report.pdf (accessed on 7 January 2020).

84. Heron, P.; McNeill, L. Preparing Physics Students for 21st-Century Careers. *Phys. Today* **2017**, *70*, 38.

 *fluids*

Article

Numerical Computations of Vortex Formation Length in Flow Past an Elliptical Cylinder

Matthew Karlson [1], Bogdan G. Nita [2,*] and Ashwin Vaidya [2]

[1] Department of Mathematics, University of Pittsburgh, Pittsburgh, PA 15261, USA; mdk80@pitt.edu
[2] Department of Mathematics, Montclair State University, Montclair, NJ 07043, USA; vaidyaa@montclair.edu
* Correspondence: nitab@montclair.edu; Tel.: +1-973-655-7261

Received: 29 June 2020; Accepted: 8 September 2020; Published: 10 September 2020

Abstract: We examine two dimensional properties of vortex shedding past elliptical cylinders through numerical simulations. Specifically, we investigate the vortex formation length in the Reynolds number regime 10 to 100 for elliptical bodies of aspect ratio in the range 0.4 to 1.4. Our computations reveal that in the steady flow regime, the change in the vortex length follows a linear profile with respect to the Reynolds number, while in the unsteady regime, the time averaged vortex length decreases in an exponential manner with increasing Reynolds number. The transition in profile is used to identify the critical Reynolds number which marks the bifurcation of the Karman vortex from steady symmetric to the unsteady, asymmetric configuration. Additionally, relationships between the vortex length and aspect ratio are also explored. The work presented here is an example of a module that can be used in a project based learning course on computational fluid dynamics.

Keywords: vortex formation length; wake; vortex shedding

---

1. Introduction

Vortex development in a fluid's flow is a highly nonlinear phenomenon which goes through multiple bifurcations. This topic is rarely introduced in a serious manner in an undergraduate course on fluids. However, there are simple ways to talk about vortex development by combining qualitative and quantitative approaches that go beyond simply examining classical images or the use of sophisticated particle image velocimetry (PIV) techniques. We introduce one such method of talking about the physics of vortex development in this paper, which can be used in the form of a lesson plan for a lecture or to motivate computational projects in more advanced classes in fluids. Student-centered practices such as problem-based and project-based learning (PBL) are more commonly practiced in the arts. Instructional methods related to PBL promote a more inductive approach to learning whereby generalizations and abstractions follow from first understanding specific cases [1]. This approach is in contrast to the deductive strategy taken in the sciences which is a more top-down approach and a possible cause of alienation in several students. The concept of problem-based learning began more than 30 years ago in the context of medical education. PBL has been defined as the "posing of a complex problem to students to initiate the learning process" [2] and as "experiential learning organized around the investigation and resolution of messy, real-world problems" [3]. PBL can be implemented at various scales in a course with a focus from a "teacher to student-centered education with process-oriented methods of learning" [4,5]. The recent popularity of project based learning approach in physics and engineering education is based on research indicating the effectiveness of PBL in enhancing student engagement [4,6,7]. Several recent educational papers have specifically discussed the effectiveness of computational problems in fluid dynamics and the use of software such as Comsol, among others, in improving classroom engagement [8–13]. The current paper is an attempt to present a similar example of a complex problem in fluid dynamics which is apt as a unit which can be introduced as a project.

Flow past a circular cylinder is a very well studied problem in classical fluid mechanics. While the creeping flow regime is completely understood and most related problems are approachable analytically, the inertial regime contains several unanswered questions. The evolution of flow past a cylinder is a particularly interesting and well studied problem [14]. The Reynolds number ($Re = UL/\nu$, where U and L represent the characteristic velocity and length respectivel, and ν refers to the dynamic viscosity) has been shown to capture several critical changes in flow structure. The first of these happens around $Re = 5$ [15], where the flow transitions from the creeping flow to one with a symmetric vortex profile. A second flow bifurcation from the steady symmetric profile to a asymmetric unsteady vortex occurs at around $Re \approx 47$ [16] which lasts until about $Re \approx 150$. Following the notation employed in the recent literature [17], we identify these critical Reynolds numbers as Re_{c1} and Re_{c2}, respectively. We refer the readers to the paper by Faruquee et al. [17] who provide a thorough discussion of the various historical experimental and numerical studies on this topic. The critical values referred to above are sensitive to shape of the obstacle, aspect ratio (AR), blockage ratio of channel diameter to channel width, roughness of the cylinder etc. but the qualitative aspects of the various transitions are still maintained.

In this paper we are concerned with highlighting the dynamics of vortex formation, specifically through an examination of the vortex length in a flow past an elliptical obstacle. Experimental measurements of the length and width of wake vortices past cylinders have been discussed in the literature [14,16,18–24]. The vortex length, usually denoted L_w, is defined as "the streamwise distance between the confluence point (wake stagnation point) and the rear stagnation point of the cylinder" [17]. It is also, more commonly, defined as the distance between the rear stagnation point and the "point downstream where the velocity fluctuation level has grown to a maximum" [14] (Figure 1 provides a schematic explanation of this metric). Experiments and numerics indicate the relation between L_w and maximum velocity fluctuation and the base pressure (at the rear of the obstacle) to be inversely proportional [23,25,26]. Coutanceau and Bouard [27] put forth the linear equation $L_w/d = 0.05\,Re$ for $4.4 < Re < 40$, which, in the steady, symmetric vortex regime, related the growth of the near wake vortex as a function of Re.

The effect of AR, which is the ratio of the semi-minor (a) to semi-major (b) axis of the obstacle, is also of interest in this study. Note that $AR = \frac{a}{b} < 1$ indicates an elongated body while, $AR > 1$ suggests a squat, flat object (such as disk in 3D). In two dimensions the shape itself is identical and therefore the AR is reflective of the orientation of the body with respect to the oncoming flow. Hence in 2D, we can describe $AR < 1$ as indicative of a high drag configuration where the entire length of the body is perpendicular to the flow, while in the case of $AR > 1$, the body assumes a pefectly streamlined position with the length of the body parallel to the flow. Recent numerical simulations [17] reveal that L_w has a tendency to increase with increasing AR for $Re_{c1} < Re < Re_{c2}$. However, for $AR < 0.4$, the wake disappears. For $Re = 40$, Faruquee et al. [17] provide the following equation relating the normalized vortex formation length to AR:

$$\frac{L_w}{d} = 2.2(AR)^2 + 0.5(AR) - 0.43 \tag{1}$$

where d is the hydraulic diameter. This formula indicates a critical minimum AR of 0.34 below which no symmetric vortex forms. This critical AR is sensitive to the Re and has been shown to increase with decreasing Re Table 3 [17]. The calculations are in agreement with the experimental studies [27,28].

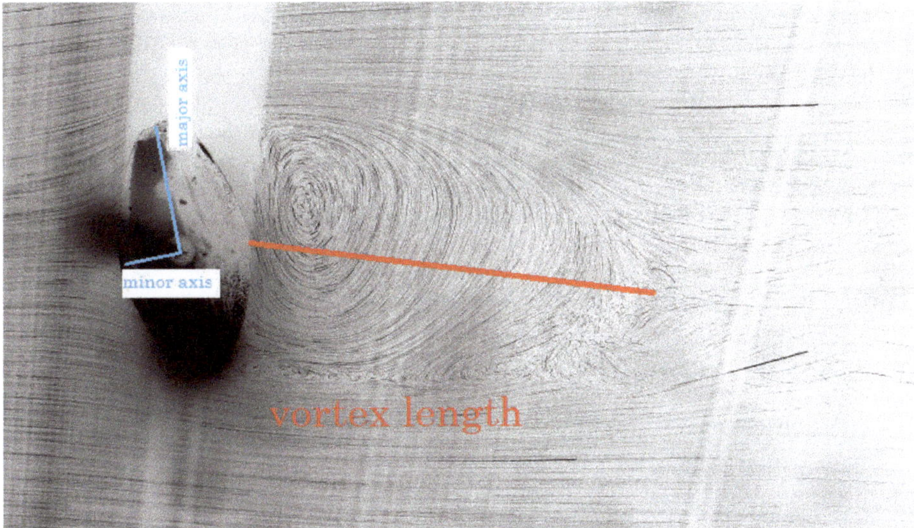

Figure 1. This figure shows a visualization of the primary vortex region in a flow past an ellipsoidal body. Based on past experimental work [24], the vortex length (denoted by the length of the bold red line) is measured as the distance between the rear surface of the body and the "pinch-off" point where the flow velocity vanishes.

In the rest of the paper we numerically investigate the vortex formation length for a class of ellipse shaped obstacles in a flow in the range $10 < Re < 100$ for various ARs. Our investigation goes past Re_{c2} into the asymmetric, periodic vortex shedding regime. Section 2, discusses the numerical method used and Section 3 elaborates the results of our computations and compares them with those in the literature.

Vortex formation in flow past cylinders has been studied extensively for many practical applications. Vortices form when fluids flow past obstacles at sufficiently high speeds and are therefore of particular interest in various branches of engineering as they can pose a threat by inducing harmful vibrations in aircrafts, buildings, and other structures. In the ensuing analysis, we determine the length of the primary vortex region in the wake of an immersed body, since this region is most influential in the dynamics of the body. More specifically, we examine 2D elliptical bodies of different aspect ratio (henceforth denoted AR), which is a ratio of semi-major to semi-minor axis (oblate) and semi-minor to semi-major (prolate). A total of thirteen different values of AR from 0.2 to 2.6, increasing in increments of 0.2, have been considered. This simple numerical approach is a continuation of a previous experiment conducted on a flow past a fixed cylinder in a flow tank, in which the vortex length was determined by means of visualization and serves to elucidate a complex problem by simple means, which we believe makes for an effective class project.

2. Methodology

We used COMSOL Multiphysics to model a 2D flow in a channel past a fixed cylinder. The software uses a finite element method to solve the Navier–Stokes and incompressibility equations given by

$$\rho \left(\frac{\partial \mathbf{u}}{\partial t} + \mathbf{u} \cdot \nabla \mathbf{u} \right) - \mu \nabla \cdot (\nabla \mathbf{u} + \nabla^T \mathbf{u}) + \nabla p = 0 \tag{2}$$

$$\nabla \cdot \mathbf{u} = 0. \tag{3}$$

Here, ρ is the density of the fluid, $\mathbf{u} = (u_x, u_y)$ is the divergence free flow field, t is time, μ is the kinematic viscosity, and p is the pressure. The Reynolds number is defined as $Re = \frac{Ud}{\nu}$, where U is the far-field of free-stream velocity, ranging between 0.1–1 m/s, d is the characteristic length, which in this case was 10^{-1} m and ν is the dynamic viscosity, taken to be 10^{-3} kg/m^3. The solution to the above flow equation yielded the velocity and pressure fields which were then utilized to identify the vortex length for $10 < Re < 100$ and ellipses of varying aspect ratios (AR is the ratio of minor axis to major axis). The major and minor axes of the ellipses were chosen to conform to the desired values of AR but with keeping the area of each ellipse the same as that of the cylinder. Table 1 summarizes the important parameters used in the numerical computations.

Table 1. The table provides some important parameters for the numerical study. Note that the area of the elliptical cylinders is always maintained at 0.0785 m^2 in all calculations.

Channel Height	Channel Length	Cylinder Diameter	Area of Cylinder	AR	Re
0.4 m	2.2 m	0.1 m	0.00785 m^2	0.4–1.4	10–100

The problem of flow past a 2D cylinder is a well studied and benchmarked problem in COMSOL [29]. A description of the code and the methodology of this standard problem can also be found on the COMSOL website: comsol.com/model/download/449401/models.mph.cylinder_flow.pdf. For the purposes of this work, this code was suitably adapted for the study of the elliptical cylinder. The problem was solved using the FSI module in COMSOL which uses a PARADISO solver; it was run for 5 s in increments of 0.01 s using a "fine" mesh consisting of 5662 elements. Convergence tests were performed in a previous study [30,31] for a more complex problem involving 2D and 3D elliptical cylinders with attached wings (or flexible fiber).

The dimensions and geometry of the problem studied here were similar to those used in our previous studies [30]. We also verified our code for the case of flow past a circular cylinder with perfect slip conditions along the top and bottom walls and no slip on the surface of the cylinder, to mimic unbounded flow. Through this calculation we were able to obtain the critical bifurcation of around 47 where the flow transitions from steady to unsteady. This helped us to confirm that the metric used to identify the critical Reynolds number was correct. In a classroom setting the relationship between the Reynolds number and the Strouhal number would be appropriate as an additional validation of the numerical scheme. The remaining computations were conducted in a bounded channel with no-slip conditions on the channel walls and also on the surface of the ellipse. The various panels in Figures 2 and 3 show some sample flow and pressure profiles based on our numerical simulations for various Reynolds numbers.

Two different criteria were used to measure the length of the vortex depending on whether the flow is steady or unsteady. For the steady case, since the flow does not vary in time (see the top panels in Figure 4), the vortex region is symmetric about $y = 0$ and its length remains constant in time. In this case, we examined the horizontal component of the velocity, u_x, and of the pressure gradient, namely, $\frac{dp}{dx}$ as a function of x along $y = 0$. The border or separatrix of the primary vortex pair separates the inner wake flow from the outer uniform flow. At the junction of the separatrix and $y = 0$, the two flows are in opposing directions, the outer flow moving in the positive x while the wake flow moves in the negative x direction. The point at which $u_x = 0$ can therefore be defined as a critical point which marks the end of the primary vortex region. The horizontal distance of this critical point from the rear end of the cylinder is defined as the vortex length. To confirm this hypothesis, a second criterion was also tested. We also observe that the pressure along $y = 0$ increases the fastest as we cross the critical point. Therefore the vortex length can also be defined with respect to the maximum of the magnitude of the pressure gradient. Since $\frac{dp}{dy}$ is negligible, this collapses to the maximum of $\frac{dp}{dx}$ along $y = 0$.

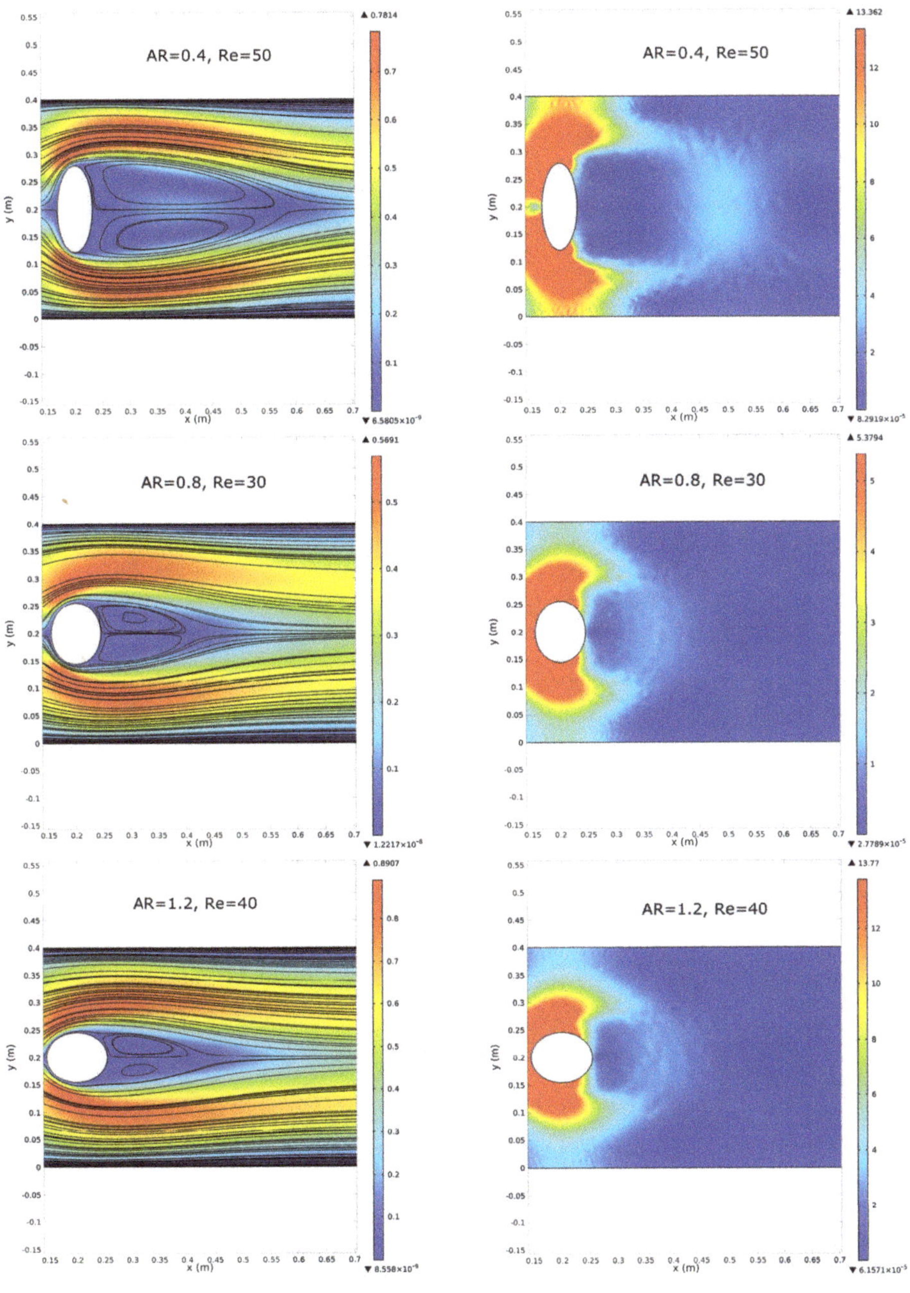

(**a**) Streamlines of the flow in the steady regime

(**b**) Pressure profile in the wake of the ellipse

Figure 2. Numerical simulations of the flow past elliptical objects of various *AR*s: steady flow.

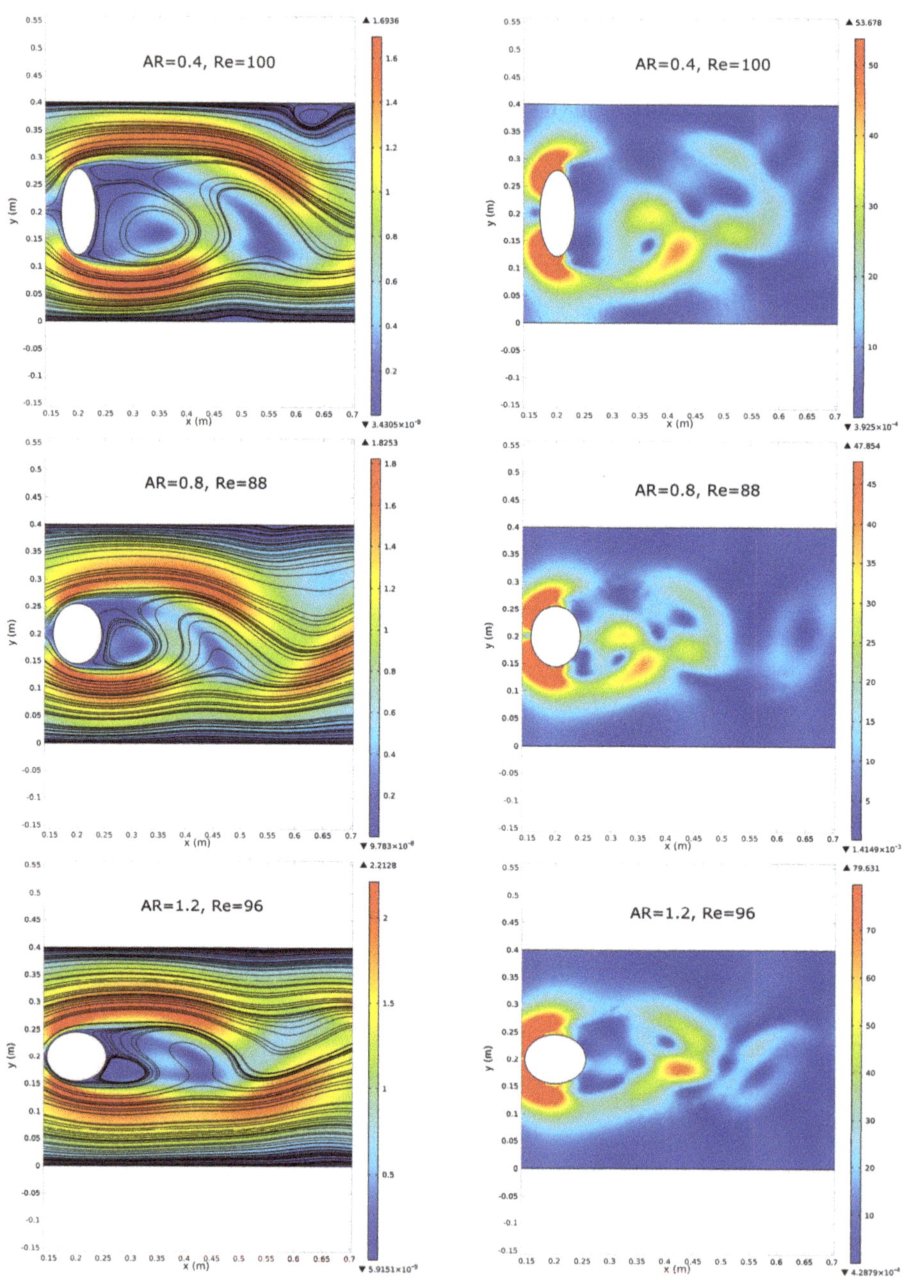

(**a**) Streamlines of the flow in the unsteady regime   (**b**) Pressure profile in the wake of the ellipse

Figure 3. Numerical simulations of the flow past elliptical objects of various *AR*s: unsteady flow.

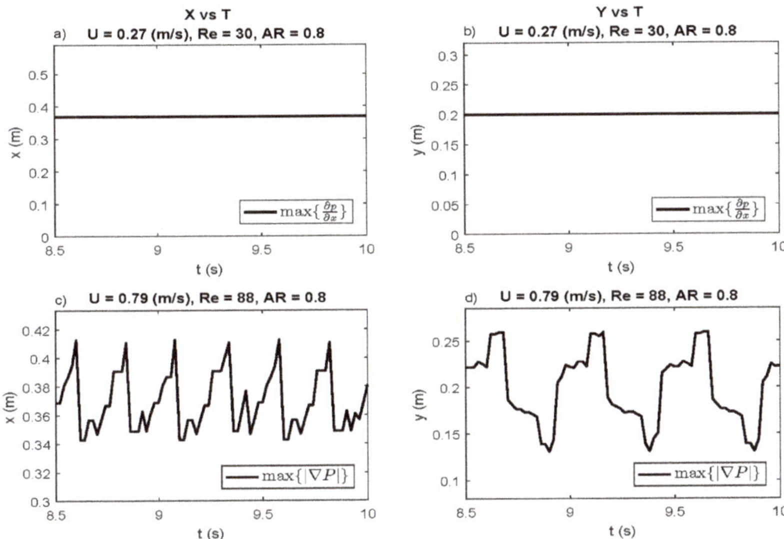

Figure 4. The panels (**a**,**b**) show the position of the wake stagnation point in the case of steady flow, while panels (**c**,**d**) show the periodically changing position of the stagnation point for the oscillating flow regime.

In the case when flow becomes unsteady, the wake region oscillates in space and time (see the bottom panels in Figure 4). The vortex region is asymmetric about $y = 0$ and moves in both spatial directions x and y. In this case the critical point at a given time was defined by the maximum of the magnitude of the pressure gradient, ∇p, which is a generalization of the criterion used in the steady state case and the appropriate length is the time averaged vortex length over a period. We use numerical interpolation to smooth the data and find our zeros and extremum.

3. Results and Discussion

The computations for the steady case show a monotonic increase in vortex length with increasing Re, below a critical threshold, for all ARs (see for example, Figure 5). The rising trend is similar for both criteria employed to measure the vortex length. When the flow becomes unsteady, the mean vortex length is seen to decrease with increasing Re. The maximum vortex length can be associated with the onset of a transition in stability as the flow bifurcates from steady to unsteady Karman vortex. The Figure 5a–d show the mean change in vortex length versus Re for some sample cases of flow past ellipses of AR 0.4, 0.8, 1.0, and 1.2. All graphs, show the same overall profile and the turning point can be used to identify the critical Reynolds number, Re_c which is sensitive to the AR of the body. The Table 2 depicts the L_w/d versus Re profile for $10 < Re < 60$, i.e., for flows that generate a steady vortex, which is in keeping with the linear relationship proposed by Coutanceau and Bouard [27]. The table shows the slopes of best fit lines for various ARs which appear to monotonically decrease with increasing AR and are overall close to the value suggested in the literature. In the regime $60 < Re < 100$, we use an exponential decaying function, $L_w/d = Ae^{\alpha Re}$ to fit the data. Table 2 also shows the value of the exponent, α, along with the goodness of fit for various AR. As in the case of the steady vortex, the dependence of the vortex formation length on Re is sensitive to AR. The missing data for $AR = 1.4$ is due to the termination of our study at $Re = 100$; at this point the $AR = 1.4$ case does not reveal an unsteady vortex. Table 3 also shows the average L_w/d for different AR and Re.

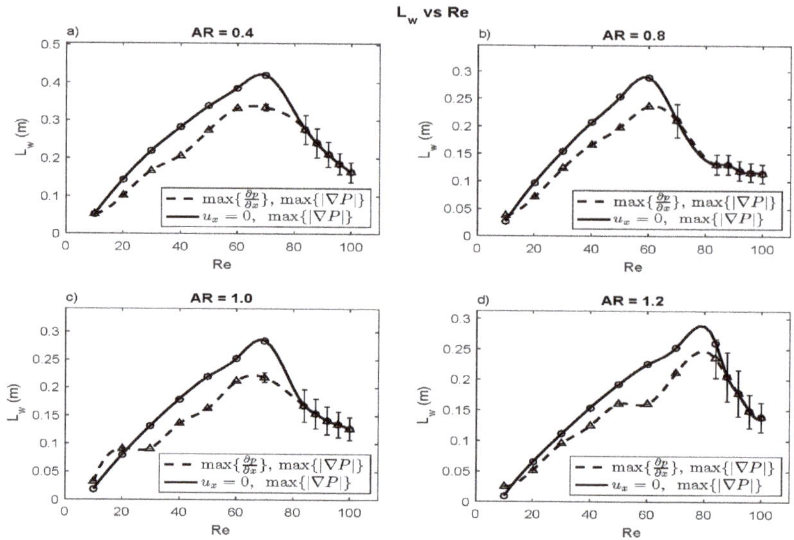

Figure 5. The figures show the change in normalized vortex length, L_w/d, versus Re for $AR = 0.4$ in (**a**), $AR = 0.8$ in (**b**), $AR = 1.0$ in (**c**) and $AR = 1.2$ in (**d**). The maximum point in each graph characterizes the transition Re at which the flow changes from steady symmetric to unsteady asymmetric.

Table 2. This table shows the nature of the L_w vs. Re relationship in the two different flow regimes. The first two rows provide the slope of the best line representing the changing value of L_w in the steady wake. Rows three and four tabulate the exponentially decaying rate of L_w as a function of Re in the unsteady, asymmetric wake.

	$AR = 0.4$	$AR = 0.6$	$AR = 0.8$	$AR = 1.0$	$AR = 1.2$	$AR = 1.4$
Slope	0.108	0.095	0.082	0.068	0.059	0.046
R^2	0.996	0.948	0.922	0.961	0.976	0.959
Exponent	−0.024	−0.023	−0.020	−0.018	−0.034	-
R^2	0.953	0.953	0.951	0.995	0.986	-

A particularly interesting relationship to investigate in this problem is that between the critical Reynolds numbers, Re_c and AR. The transitions from steady to unsteady wake are very sensitive to the specific geometric characteristics of the cylinder. Figure 6 shows the critical Re for ARs ranging between 0.4 and 1.4 estimated from using the velocity and pressure gradient criteria. The Re_c vs. AR curve shows a minimum at $AR \approx 0.6$–0.7. Overall, for bodies with AR in the range 0.4–1.4, Re_c appears to lie in the range 55–95; for a cylinder our calculations show this critical value to lie at the accepted value of $Re_c = 70$. The more streamlined bodies show greater propensity to develop elongated primary vortices. The equation:

$$Re_{crit} = 79.46(AR)^2 - 116.18(AR) + 100.26 \tag{4}$$

captures the quadratic nature of the profile (with R^2 value of 0.96) seen in Figure 6, obtained by fitting the average of the two curves shown in the figure. This equation gives us valuable information about the optimal geometry and its relationship with vortex formation. Engineers and designers wanting to suppress disruptive osculations may find this equation particularly useful.

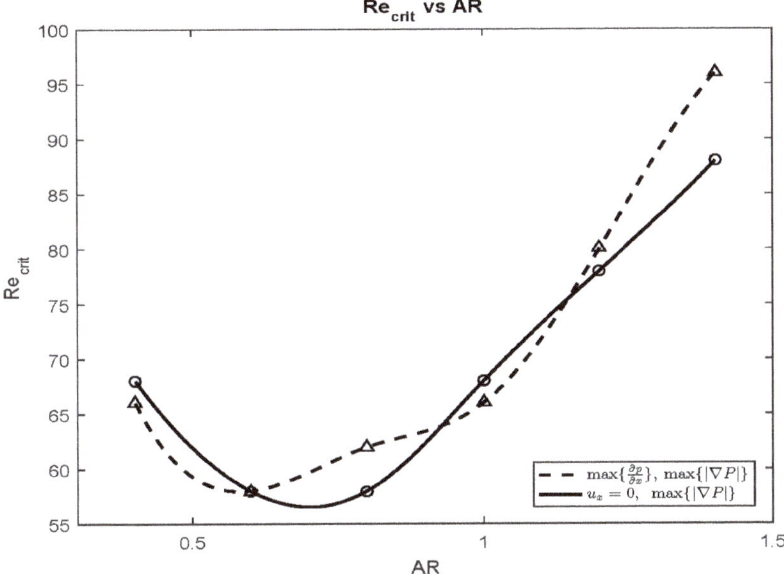

Figure 6. The graph shows the variation of Re_{crit} versus AR.

Table 3. The table shows the average values of the normalized vortex length L_w/d for various values of AR and Re.

Re	10	20	30	40	50	60	70	84	88	92	96	100
AR	L_w/d	L_w/d	L_w/d	L_w/d	L_w/d	L_w/d	L_w/d	L_w/d	L_w/d	L_w/d	L_w/d	L_w/d
0.4	0.513	1	1.67	2.064	2.737	3.306	3.332	2.752	2.399	2.098	1.857	1.628
0.6	0.622	0.872	1.647	1.741	2.365	2.617	1.86	1.239	1.18	1.128	1.087	1.043
0.8	0.369	1.236	1.235	1.663	1.982	2.362	2.097	1.31	1.31	1.196	1.167	1.141
1	0.319	0.908	0.908	1.353	1.621	2.109	2.18	1.678	1.533	1.413	1.343	1.267
1.2	0.251	0.515	0.953	1.244	1.604	1.604	2.115	2.351	2.051	1.783	1.485	1.389
1.4	0.363	0.532	0.961	0.961	1.181	1.419	1.677	1.879	1.894	2.068	2.171	1.923

4. Conclusions

In summary, our overall computations and qualitative profiles are in agreement with previous experimental results [24], where measurements of variations in vortex length were made through flow visualization and imaging techniques of flow past various cylinders in a flow tank. These experiments indicate a similar extremum in vortex length as a function of flow speed (*Re*) and *AR*. The central contribution of this work lies in two consistent ways of defining the vortex formation length and the observation of the vortex length dependence on *Re*, especially in the unsteady flow regime. Questions for future investigation include extending the *Re* regime of the study and experimental verification of the unsteady vortex formation length. Pedagogical treatments in the literature appear to focus primarily on lower level courses in science and engineering. There is little discussion about the teaching and learning of more advanced topics like fluid dynamics [32]. We believe that examples like the ones presented here add value to the science of fluid dynamics in some capacity, but also to the ways in which students can engage with difficult topics in fluid dynamics. Such a treatment lends itself very well to a first serious course in fluids but also in more advanced courses where students have greater proficiency in dealing with computational software like Comsol.

Author Contributions: The computations in this paper were performed while M.K. was a graduate student at Montclair State University. Authors B.G.N. and A.V. designed the study, helped analyze the results and wrote the paper. All authors have read and agreed to the published version of the manuscript.

Funding: This research received no external funding.

Conflicts of Interest: The authors declare no conflict of interest.

References

1. Connor, J. Project Based Learning and Authenticity: An Instructor's View. In Proceedings of the 7th First Year Engineering, Roanoke, VA, USA, 3–4 August 2015.
2. Sahin, M.; Yorek, N. A comparison of problem-based learning and traditional lecture students expectations and course grades in an introductory physics classroom. *Sci. Res. Essays* **2009**, *4*, 753–762.
3. Torp, L.; Sage, S. *Problem as Possibilities, Problem Based Learning for K-16*; Association for Supervision and Curriculum Development: Alexandria, VA, USA, 2002.
4. Ahlfeldt, S.; Mehta, S.; Sellnow, T. Measurement and analysis of student engagement in university classes where varying levels of PBL methods of instruction are in use. *High. Educ. Res. Dev.* **2005**, *24*, 5–20. [CrossRef]
5. Dahlgren, M.A. Portraits of PBL: Course objectives and students' study strategies in computer engineering, psychology and physiotherapy. *Instr. Sci.* **2000**, *28*, 309–329. [CrossRef]
6. Hake, R.R. Interactive-engagement versus traditional methods: A six-thousand-student survey of mechanics test data for introductory physics courses. *Am. J. Phys.* **1998**, *66*, 64–74. [CrossRef]
7. Zyngier, D. Listening to teachers—Listening to students: Substantive conversations about resistance, empowerment and engagement. *Teach. Teach. Theory Pract.* **2007**, *13*, 327–347. [CrossRef]
8. Bot, L.; Gossiaux, P.B.; Rauch, C.P.; Tabiou, S. Learning by doing: A teaching method for active learning in scientific graduate education. *Eur. J. Eng. Educ.* **2005**, *30*, 105–119. [CrossRef]
9. Kwon, H.J. Use of comsol simulation for undergraduate fluid dynamics course. *ASEE Comput. Educ. (CoED) J.* **2013**, *4*, 63.
10. Mokhtar, W. Project-Based Learning (PBL): An Effective Tool to Teach an Undergraduate CFD Course. In Proceedings of the 2011 ASEE Annual Conference & Exposition, Vancouver, BC, Canada, 26–29 June 2011.
11. Mokhtar, W. Using Computational Fluid Dynamics to Introduce Critical Thinking and Creativity in an Undergraduate Engineering Course. *Int. J. Learn.* **2010**, *17*, 441–457.
12. Pawar, S.; San, O. CFD Julia: A learning module structuring an introductory course on computational fluid dynamics. *Fluids* **2019**, *4*, 159. [CrossRef]
13. Vianna, R.S.; Cunha, A.M.; Azeredo, R.B.; Leiderman, R.; Pereira, A. Computing Effective Permeability of Porous Media with FEM and Micro-CT: An Educational Approach. *Fluids* **2020**, *5*, 16. [CrossRef]
14. Williamson, C.H. Vortex dynamics in the cylinder wake. *Annu. Rev. Fluid Mech.* **1996**, *28*, 477–539. [CrossRef]
15. Marris, A.W. A review on vortex streets, periodic wakes, and induced vibration phenomena. *J. Basic Eng.* **1964**, *86*, 185–193. [CrossRef]
16. Williamson, C.H. Defining a universal and continuous Strouhal Reynolds number relationship for the laminar vortex shedding of a circular cylinder. *Phys. Fluids (1958–1988)* **1988**, *31*, 2742–2744. [CrossRef]
17. Faruquee, Z.; Ting, D.S.; Fartaj, A.; Barron, R.M.; Carriveau, R. The effects of axis ratio on laminar fluid flow around an elliptical cylinder. *Int. J. Heat Fluid Flow* **2007**, *28*, 1178–1189. [CrossRef]
18. Linke, W. New measurements on aerodynamics of cylinders particularly their friction resistance. *Phys. Z.* **1931**, *32*, 900.
19. Rosenhead, L. Vortex systems in wakes. *Adv. Appl. Mech.* **1953**, *3*, 185–195.
20. Berger, E.; Wille, R. Periodic flow phenomena. *Annu. Rev. Fluid Mech.* **1972**, *4*, 313–340. [CrossRef]
21. Morkovin, M.V. Flow around circular cylinder a kaleidoscope of challenging fluid phenomena. In Proceedings of the ASME Symposium on Fully Separated Flows, Philadelphia, PA, USA, 18–20 May 1964; pp. 102–118.
22. Oertel, H., Jr. Wakes behind blunt bodies. *Annu. Rev. Fluid Mech.* **1990**, *22*, 539–562. [CrossRef]
23. Zdravkovich, M.M. Flow around circular cylinders. *Fundamentals* **1997**, *1*, 566–571.
24. Chung, B.; Cohrs, M.; Ernst, W.; Galdi, G.P.; Vaidya, A. Wake cylinder interactions of a hinged cylinder at low and intermediate Reynolds numbers. *Arch. Appl. Mech.* **2016**, *86*, 627–641. [CrossRef]
25. Roshko, A. On the wake and drag of bluff bodies. *J. Aeronaut. Sci.* **2012**, *22*, 124–132. [CrossRef]

26. Williamson, C.H.K.; Roshko, A. Measurements of base pressure in the wake of a cylinder at low Reynolds numbers. *Z. Flugwiss. Weltraumforsch.* **1990**, *14*, 38–46.
27. Coutanceau, M.; Bouard, R. Experimental determination of the main features of the viscous flow in the wake of a circular cylinder in uniform translation. Part 1. Steady flow. *J. Fluid Mech.* **1977**, *79*, 231–256. [CrossRef]
28. Taneda, S. Experimental investigation of the wakes behind cylinders and plates at low Reynolds numbers. *J. Phys. Soc. Jpn.* **1956**, *11*, 302–307. [CrossRef]
29. Schafer, M.; Turek, S.; Durst, F.; Krause, E.; Rannacher, R. Benchmark computations of laminar flow around a cylinder. In *Flow Simulation with High-Performance Computers II*; Vieweg+Teubner Verlag: Wiesbaden, Germany, 1996; pp. 547–566.
30. Allaire, R.; Guerron, P.; Nita, B.; Nolan, P.; Vaidya, A. On the equilibrium configurations of flexible fibers in a flow. *Int. J. Non-Linear Mech.* **2015**, *69*, 157–165. [CrossRef]
31. Nolan, P. Numerical Study of Body Shape and Wing Flexibility in Fluid Structure Interaction. Master's Thesis, Montclair State University, Montclair, NJ, USA, 2015.
32. Vaidya, A. Teaching and Learning of Fluid Mechanics. *Fluids* **2020**, *5*, 49. [CrossRef]

Article

On the Existence of Leray-Hopf Weak Solutions to the Navier-Stokes Equations

Luigi C. Berselli [1,*] **and Stefano Spirito** [2]

[1] Dipartimento di Matematica, Università di Pisa, Via F. Buonarroti 1/c, I-56127 Pisa, Italy
[2] DISIM—Dipartimento di Ingegneria e Scienze dell'Informazione e Matematica, Università degli Studi dell'Aquila, Via Vetoio, I-67100 L'Aquila, Italy; stefano.spirito@univaq.it
[*] Correspondence: luigi.carlo.berselli@unipi.it

Abstract: We give a rather short and self-contained presentation of the global existence for Leray-Hopf weak solutions to the three dimensional incompressible Navier-Stokes equations, with constant density. We give a unified treatment in terms of the domains and the relative boundary conditions and in terms of the approximation methods. More precisely, we consider the case of the whole space, the flat torus, and the case of a general bounded domain with a smooth boundary (the latter supplemented with homogeneous Dirichlet conditions). We consider as approximation schemes the Leray approximation method, the Faedo-Galerkin method, the semi-discretization in time and the approximation by adding a Smagorinsky-Ladyžhenskaya term. We mainly focus on developing a unified treatment especially in the compactness argument needed to show that approximations converge to the weak solutions.

Keywords: Navier-Stokes equations; Leray-Hopf weak solutions; existence

Citation: Berselli, L.C.; Spirito, S. On the Existence of Leray-Hopf Weak Solutions to the Navier-Stokes Equations. *Fluids* **2021**, *6*, 42. https://doi.org/10.3390/fluids 6010042

Received: 15 December 2020
Accepted: 5 January 2021
Published: 13 January 2021

Publisher's Note: MDPI stays neutral with regard to jurisdictional claims in published maps and institutional affiliations.

1. Introduction

Let $T > 0$ be an arbitrary finite number representing the time, $\Omega \subset \mathbb{R}^3$ be a domain to be specified later, and $\nu > 0$ be a positive number representing the kinematic viscosity. The incompressible Navier-Stokes equations model the dynamic of a viscous and incompressible fluid at constant temperature and with constant density. They are given by the following system of PDE's posed in $(0, T) \times \Omega$:

$$\begin{cases} \partial_t u + (u \cdot \nabla) u + \nabla p - \nu \Delta u = f & \text{in } (0, T) \times \Omega, \\ \operatorname{div} u = 0 & \text{in } (0, T) \times \Omega. \end{cases} \tag{1}$$

The vector field $u \in \mathbb{R}^3$ is the velocity, $p \in \mathbb{R}$ is the scalar pressure, and to avoid inessential complications, we set the external force $f = 0$ (but all results presented here can be easily extended to the case of a non vanishing external force, see Remark 4). The first equation is the conservation of linear momentum and the second equation, also called the incompressibility constraint, can be considered as the conservation of the mass, since the density is assumed to be constant. The system (1) has to be supplemented with initial and boundary conditions. Regarding the initial condition we impose that

$$u|_{t=0} = u_0, \quad \text{in } \Omega,$$

with u_0 satisfying the compatibility condition $\operatorname{div} u_0 = 0$ in Ω. For the boundary conditions we need to specify the assumptions on the domain. We consider three cases, $\Omega = \mathbb{R}^3$, $\Omega = \mathbb{T}^3$ with $\mathbb{T}^3$ being the three-dimensional flat torus, and $\Omega \subset \mathbb{R}^3$ being a bounded domain, whose boundary will be denoted by $\partial\Omega$; we refer to Assumption 1 for the precise hypotheses on Ω.

For each of the three different cases we impose the different and natural boundary conditions:

$$(i) \quad u \to 0 \text{ for } |x| \to \infty, \quad \text{if } \Omega = \mathbb{R}^3;$$
$$(ii) \quad u \text{ is periodic, if } \Omega = \mathbb{T}^3; \tag{2}$$
$$(iii) \quad u = 0 \text{ on } (0, T) \times \partial\Omega, \quad \text{if } \Omega \text{ is a bounded domain.}$$

Note that the initial datum will be requested to be tangential to the boundary in the case (i), and to satisfy the condition (ii) and (iii) in the other cases. Contrary to the system of compressible Navier-Stokes equations, the pressure p, instead of being obtained through a state equation, is an unknown of the system. This is a consequence of the incompressibility conditions and indeed the pressure can be interpreted as Lagrange multiplier associated with the incompressibility constraint. Note that there are no initial/boundary conditions imposed on the pressure, which (since it appears only as a gradient in the momentum equation) is always determined up to an arbitrary function of time.

Generally speaking, it is very difficult to prove existence and uniqueness of smooth solutions to nonlinear PDE's. Here, with existence we always mean global in time existence, namely existence on any given time interval $(0, T)$, for arbitrary $T > 0$. The available theories for weak solutions provide a framework to give a proper meaning to PDE's, without requiring too much regularity on the solutions and they rely on the theory of generalized functions and distributions. In particular, the landmark idea in the theory of weak solutions is to give up on solving the equations point-wise but trying to solve them in an averaged sense, which is meaningful also from a physical point of view. In the case of fluid mechanics, we expect a very complex behavior by (turbulent) flows appearing in real life, hence we expect to be able to capture only averages of the velocity and pressure, see Reference [1].

The problem of global existence generally becomes easier since the class of available solution is enlarged and several functional analysis tools can be now used. However, the price to pay to have such a relatively simple existence theory is that the uniqueness problem becomes a very difficult one and many calculation which are obvious when dealing with smooth solutions are not possible or hard to be justified. The three-dimensional incompressible Navier-Stokes is a paradigmatic example of a such situation and the introduction of weak solutions dates back about 100 years ago. In fact, in a series of celebrated papers (the time evolution is treated in Reference [2]) Jean Leray introduced the notion of weak solution as a mathematical tool, but also with a strong understanding of the physics behind the equations. The theory of weak solutions is also strictly linked with the name of Eberhard Hopf [3] who gave the first contribution to the problem of existence of weak solutions in a bounded domain, by means of the Faedo-Galerkin method.

It is interesting to observe that many methods and techniques of functional analysis (which are now a common background of graduate students in mathematics) originated from the study of PDE's and especially from those arising in fluid mechanics. In this note we are trying to explain an extremely limited part of the theory: the existence of (Leray-Hopf) weak solutions. This is a topic at the level of most undergraduate students, with a minimal knowledge of Sobolev spaces and functional analysis (mainly weak convergence and weak compactness), as for instance in the widely used (text)books by Brezis [4], just to name one. Note also that we try to present a minimal spot in the abstract theory of Navier-Stokes equations, which can be an "appetizer" for students trying to start a serious understanding of (part of) the mathematical fluid mechanics. It is impossible to review what is done on the subject, even only for the mathematical analysis side. Nevertheless many information, at an introductory or more advanced level, can be found in several books, see for instance, just to name a few in alphabetical order [5–11].

We think that we will not discourage any reader unfolding the (many) mathematical difficulties of the topic, but –instead– we hope that highlighting the challenges which are typical of mathematical fluid mechanics further interest could be stimulated; To this end we quote the following coming from an interview reported in Reference [12] in a essay in memory of Jacob Schwartz:

> *When I asked him [Jacob Schwartz] if there was a subject he had trouble learning, he admitted that there was, namely, fluid dynamics. "It is not a subject that can be expressed in terms of theorems and their proofs," he said.*

Following the above point of view, the first step even in the mathematical analysis of the Navier-Stokes equations is that of giving an appropriate definition of weak solutions, which take into account the functional spaces where it is reasonable find weak solutions, the initial and the boundary conditions. Usually, the functional space to be considered are hinted by the a priori estimate available for the system under consideration. The informal notion of an a priori estimate may be a quantitative bound depending only on the data of the problem, which holds for smooth solutions of the system under consideration, regardless their existence. In particular, for system arising from physics, the a priori estimates usually have a deep physical interpretation.

In the context of the three-dimensional incompressible Navier-Stokes equations the main a priori estimate is indeed the conservation of the energy of the system and is given by the following integral equality:

$$\int |u(t,x)|^2 \, dx + 2\nu \int_0^t \int |\nabla u(s,x)|^2 \, dx ds = \int |u_0(x)|^2 \, dx \qquad t \in [0,T], \qquad (3)$$

where the space integral is over the domain under consideration.

The equality (3) has a very simple *formal* proof. Indeed, let (u,p) be a smooth solution of (1) and (2). By multiplying the momentum equation by u and integrating over Ω we get

$$\int \partial_t u \cdot u - \nu \Delta u \cdot u + (u \cdot \nabla)\, u \cdot u + \nabla p \cdot u \, dx = 0.$$

By integrating by parts and using the divergence free condition and (2) we get

$$-\int \Delta u \cdot u \, dx = \int |\nabla u|^2 \, dx, \qquad \int (u \cdot \nabla)\, u \cdot u \, dx = 0, \qquad \text{and} \qquad \int \nabla p \cdot u \, dx = 0.$$

Then, after integration in time on $(0,t)$ with $t \in (0,T)$ we get (3). Note that (3) gives a quantitative bound depending only on T, and u_0 of square integrals of the velocity field u and its gradient ∇u. The energy equality (3) will serve as motivation for the definition of Leray-Hopf weak solution we will give in Section 3.

Once a reasonable definition of weak solution is given, to prove global existence one usually exploits what it is know as a *compactness argument*, which consists in (1) proving the existence of a sequence of relatively smooth approximating solutions satisfying appropriate uniform estimates; (2) proving that limits of these approximating solutions are effectively weak solution of the problem under consideration. We remark that usually the uniform bounds obtained on the sequence of approximating solutions are the same inferred by the a priori estimates available for the system under consideration; These bounds are then hopefully inherited by weak solutions obtained with a passage to the limit. To be more precise, in the case of the Navier-Stokes equations, the approximation method should be chosen such that the approximate solutions satisfy the energy (in)equality. Due to the limited regularity which can be generally inferred on weak solutions, the validity of any energy balance on the weak solutions to the 3D Navier-Stokes equations is obtained with a limiting process on the approximate solutions and not using the solution u itself as a test function as done to obtain (3), since this argument is only formal and not justified when dealing with genuine Leray-Hopf weak solutions.

In this short note we provide a rather self-contained account on the global existence of weak solutions for the three-dimensional incompressible Navier-Stokes equations and some of the (several) approximation methods used in the literature. Since the convergence argument is essentially the same for every approximation methods and for every choice of the domains and boundary conditions mentioned above, we introduce (for the purpose of the exposition) a notion of *approximating solution* for which we will prove the convergence to a Leray-Hopf weak solution of the problem (1) and (2). This is not the historical path, but is a way we identify to have a unified treatment, which can describe the existence theory within the notion of *approximating solutions*.

Then, we show how several and well-known approximations fit in the framework introduced and, therefore, we recover the existence of Leray-Hopf weak solution by using those methods. In particular, we will consider the most common techniques available for the construction. Further results based on the energy type methods, concerning uniqueness, regularity and the connection with applied analysis of turbulent flows, can be found in the forthcoming monograph [1], which is also written in the spirit of being an introduction for undergraduate students, interested in applied analysis of the Navier-Stokes equations.

Organization of the Paper

The paper is organized as follows: In Section 2 we introduce the functional spaces that we use. Then, in Section 3 we define of Leray-Hopf weak solutions and study their main properties. In Section 4 we give the definition of *approximating solution* and we prove the convergence to a Leray-Hopf weak solution. Finally, in Section 5 we prove that certain approximating schemes fit in the framework of *approximating solution*.

2. Preliminaries

In this section we fix some notations and we recall some basic preliminaries we will need for the analysis. We start by fixing the assumptions on the domain Ω.

Assumption 1. *The domain $\Omega \subset \mathbb{R}^3$ will be of the following type:*
(A1) the whole space, $\Omega = \mathbb{R}^3$;
(A2) the flat torus, $\Omega = \mathbb{T}^3$;
(A3) a bounded connected open set $\Omega \subset \mathbb{R}^3$, locally situated on one side of the boundary $\partial\Omega$, which is at least locally Lipschitz.

2.1. Notation

We will never distinguish between scalar and vector functions unless it is not clear from the context. We will denote by $C_c^\infty(\Omega)$ the space of compactly supported functions which are infinitely differentiable and $\mathcal{D}'(\Omega)$ its dual, which is the space of distributions over Ω. In the case $\Omega = \mathbb{T}^3$ the subscript "c" is not needed and we set $C_c^\infty(\mathbb{T}^3) = C^\infty(\mathbb{T}^3)$. With an abuse of notation we will use $C_c^\infty(\Omega)$ for all the three choices of the domain Ω satisfying Assumption 1. We recall that for any vector $f \in C_c^\infty(\Omega; \mathbb{R}^3)$ the Helmholtz decomposition holds true: there exists two function $g \in C_c^\infty(\Omega; \mathbb{R}^3)$ and $q \in C_c^\infty(\Omega; \mathbb{R})$ such that $f = g + \nabla q$, and g is divergence-free. Given a Banach space E, we denote with $\| \cdot \|_E$ its norm. However, for the classical Lebesgue spaces $L^p(\Omega)$, with $p \in [1, \infty]$, we shall denote their norms with $\| \cdot \|_p$. Finally, we recall that the space $H_0^1(\Omega)$ is the classical Sobolev space obtained as a closure of $C_c^\infty(\Omega)$ in the norm

$$\|v\|_{H^1} := \left(\int_\Omega |v|^2 + |\nabla v|^2 \, dx \right)^{\frac{1}{2}}, \qquad v : \Omega \mapsto \mathbb{R}^k.$$

The subscript "0" is needed only when Ω is a bounded domain. In the case of $\Omega = \mathbb{R}^3$ or $\Omega = \mathbb{T}^3$ we have $H_0^1(\mathbb{R}^3) = H^1(\mathbb{R}^3)$ and $H_0^1(\mathbb{T}^3) = H^1(\mathbb{T}^3)$, but as before, with an abuse of notation, we will use $H_0^1(\Omega)$ for each one of the three choices of the domain Ω satisfying Assumption 1. Moreover, we recall that $H^1(\mathbb{R}^3)$ and $H^1(\mathbb{T}^3)$ can also be characterized in terms of the Fourier Transform and the Fourier Series, respectively. When dealing with a Banach space $(E, \| \cdot \|_E)$ we denote by $x_n \to x$, $x_n \rightharpoonup x$ and $x_n \overset{*}{\rightharpoonup} x$, the strong, weak and weak* convergence, respectively.

Next, let E be a Banach space, then $L^p(0, T; E)$, with $1 \leq p < \infty$, and $L^\infty(0, T; E)$ denote the classical Bochner spaces of strongly measurable (classes of) functions $u : (0, T) \to E$ such that

$$\|u\|_{L^p(E)} := \left(\int_0^T \|u(s)\|_E^p \, ds \right)^{\frac{1}{p}} < \infty,$$

$$\|u\|_{L^\infty(E)} := \operatorname*{ess\,sup}_{t \in [0,T]} \|u(t)\|_E < \infty.$$

Finally, the space of weakly continuous functions in E, which is denoted by $C_w([0, T]; E)$, consists of functions $u : [0, T] \mapsto E$ such that for any $f \in E^*$ the real function of real variable

$$\langle f, u \rangle_{E^* \times E} : [0, T] \ni t \mapsto \langle f, u(t) \rangle_{E^* \times E},$$

is continuous.

Finally, when we write $A \lesssim B$, this means that there exists a constant $c > 0$ (independent on the relevant parameters of the problem) such that $A \leq cB$.

2.2. The Spaces H and V

In the analysis of solutions of the Navier-Stokes equations is useful to consider spaces of divergence-free functions. We start by defining the space

$$\mathcal{V}(\Omega) := \{ \phi \in C_c^\infty(\Omega) : \operatorname{div} \phi = 0 \},$$

Then, we define the spaces

$$H := \overline{\mathcal{V}(\Omega)}^{\|\cdot\|_2}, \qquad V := \overline{\mathcal{V}(\Omega)}^{\|\cdot\|_{1,2}}.$$

We start by noticing that H and V are closed subspace of $L^2(\Omega)$ and $H_0^1(\Omega)$, respectively. Therefore, they are Hilbert space themselves with the inherited scalar products, which are

$$(u, v) := \int_\Omega u \cdot v \, dx \qquad ((u, v)) := \int_\Omega u \cdot v + \nabla u : \nabla v \, dx.$$

Next, although H and V are Hilbert space, hence reflexive, we will not identify them with their duals. We will instead denote by H' and V' the topological dual of H and V endowed with the classical dual norms

$$\|f\|_{H'} := \sup_{\phi \in \mathcal{V}(\Omega), \|\phi\|_2 \leq 1} |\langle f, \phi \rangle_{H' \times H}|,$$

$$\|f\|_{V'} := \sup_{\phi \in \mathcal{V}(\Omega), \|\phi\|_{1,2} \leq 1} |\langle f, \phi \rangle_{V' \times V}|.$$

We stress that H' and V' are not subset of the space of distributions $\mathcal{D}'(\Omega)$ since $\mathcal{D}(\Omega) \not\subset H$.

Finally, we recall that by Sobolev embedding theorem and the interpolation inequality for the L^p-norm, there exists a constant $C > 0$ such that for any $2 \leq p \leq 6$, $\theta = \frac{6-p}{2p} \in [0, 1]$, and any $u \in H_0^1(\Omega)$ it holds that

$$\|u\|_p \leq C \|u\|_2^\theta \|\nabla u\|_2^{1-\theta}. \tag{4}$$

The inequality (4) is a particular case of the well-known Gagliardo-Nirenberg-Sobolev inequality, see Reference [13].

3. Definition of Leray-Hopf Weak Solutions

In this section we give the definition of Leray-Hopf weak solutions and we prove some related properties. The definition is the following.

Definition 1. *A measurable vector field $u : (0,T) \times \Omega \mapsto \mathbb{R}^3$ is a Leray-Hopf weak solution of the Navier-Stokes Equations* (1) *and* (2) *if the following conditions are satisfied.*

1. *It holds that*

$$u \in C_w([0,T];H) \cap L^2(0,T;V); \tag{5}$$

2. *For any $\phi \in \mathcal{V}(\Omega)$ and any $\chi \in C^\infty([0,T))$, it holds that*

$$\int_0^T \left((u(t),\phi)\dot{\chi}(t) - ((u \cdot \nabla)\,u,\phi)\chi(t) - \nu(\nabla u, \nabla \phi)\chi(t) \right) dt - (u_0,\phi)\chi(0) = 0; \tag{6}$$

3. *For any $t \in [0,T]$*

$$\|u(t)\|_2^2 + 2\nu \int_0^t \|\nabla u(s)\|_2^2 \, ds \leq \|u_0\|_2^2. \tag{7}$$

Remark 1. *It is important to point out that it is an open problem whether or not condition* (7) *can be deduced from the conditions* (5) *and* (6). *Note also that in the definition we have* (7) *which is the so-called global energy inequality and not the equality* (3).

Remark 2. *In literature Leray-Hopf weak solutions are often defined in the space $L^\infty(0,T;H)$ rather than $C_w([0,T];H)$ and satisfying* (7) *for a.e. everywhere $t \in (0,T)$ instead that for any $t \in (0,T)$. This is equivalent to Definition* 1, *because in that case the velocity field can redefined on a set of measure zero in time in order to lie in $C_w([0,T];H)$ and satisfying* (7) *for any $t \in (0,T)$, see Reference* [14]. *We preferred to start with a solution already weakly continuous, to avoid the technical step of redefinition.*

We want to show that once we have proved the existence of a vector field satisfying the conditions in the Definition 1, we are actually solving the initial value boundary problem (1) and (2) in the sense of distributions. First of all we notice that from the condition (1), we can deduce that u is divergence-free and satisfies the boundary conditions (2) in the appropriate weak sense. The following lemma guarantee that u attains the initial datum u_0.

Lemma 1. *Let $u_0 \in H$ and u a Leray-Hopf weak solution. Then,*

$$u(t) \rightarrow u_0 \qquad \text{strongly in } H.$$

Proof. For $k \in \mathbb{N}$ and $\bar{t} \in (0,T)$, we consider the following function

$$\chi_k^{\bar{t}}(t) = \begin{cases} 1, & t \in [0,\bar{t}) \\ k(\bar{t}-t)+1, & t \in [\bar{t},\bar{t}+\frac{1}{k}) \\ 0, & t \in [\bar{t}+\frac{1}{k},T). \end{cases}$$

Then, by using $\chi_k^{\bar{t}}$, after sending $k \rightarrow \infty$ and using that $u \in C_w([0,T];H)$ we arrive to the following estimate:

$$|(u(t),\phi) - (u_0,\phi)| \leq \int_0^t |(\nabla u(s),\phi)| + |(u(s) \cdot \nabla)\,u(s),\phi)| \, ds$$

$$\leq t^{\frac{1}{2}} \left(\int_0^T \|\nabla u(s)\|_2^2 \right)^{\frac{1}{2}} \left(\|\phi\|_2^{\frac{1}{2}} + \|\phi\|_\infty \sup_{s \in [0,T]} \|u(s)\|_2 \right).$$

Then, for any fixed $\phi \in \mathcal{V}(\Omega)$ we can send $t \rightarrow 0^+$ and we can conclude that $(u(0),\phi) = (u_0,\phi)$. By using the Helmholtz decomposition we deduce that this is true for any $\phi \in C_c^\infty(\Omega))$ and therefore $u(0) = u_0$ a.e. on Ω. Moreover, the previous calculations also show that

$$u(t) \to u_0 \qquad \text{in } C_w([0,T];H) \qquad \text{as } t \to 0^+.$$

and, again by using the Helmholtz decomposition, the same result is valid also for $\phi \in L^2(\Omega)$. By weak lower semi-continuity of norms in weak convergence we get

$$\|u_0\|_2 \leq \liminf_{t \to 0^+} \|u(t)\|_2^2.$$

Next, by using the energy inequality (7) we also get, by disregarding the non-negative dissipative term and taking the superior limit that

$$\limsup_{t \to 0^+} \|u(t)\|_2^2 \leq \|u_0\|_2^2.$$

This shows that $\|u(t)\|_2 \to \|u_0\|_2$, which combined with the weak convergence implies the strong convergence, since we are in an Hilbert space. Since the norm induced on H is the same as in $L^2(\Omega)$, this proves the strong convergence also in H. $\square$

Finally, we show that to any Leray-Hopf weak solution u it is possible to associate a pressure p such that (u, p) solves the momentum equation in (1) in the sense of distributions.

Lemma 2. *Let u be a Leray-Hopf weak solution of (1) and (2). Then, there exists $p \in \mathcal{D}'((0,T); \times \Omega))$ such that*

$$\partial_t u - v\Delta u + (u \cdot \nabla) u + \nabla p = 0 \quad \text{in } \mathcal{D}'((0,T) \times \Omega).$$

and, for any $t \in (0,T)$, we have $p(t) \in L^2_{loc}(\Omega)$ and $\int_\Omega p(t)\, dx = 0$.

In the case of a general bounded domain Ω satisfying the Assumption 3, the proof of the Lemma 2 is very technical and requires several preliminaries of operator theory. We refer to References [7,10,11,15,16] for the proof. On the other hand in the case of Ω has no physical boundary the proof is straightforward. We consider here the case $\Omega = \mathbb{T}^3$.

Proof. Let u be a Leray-Hopf weak solution in the sense of Definition 1. For a.e. $t \in (0,T)$ consider the elliptic problem

$$-\Delta p(t) = \text{div}(u(t) \cdot \nabla u(t)) \qquad \text{in } \mathbb{T}^3,$$
$$\int_{\mathbb{T}^3} p(t)\, dx = 0. \tag{8}$$

Note that by (5), Gagliardo-Nirenberg Sobolev inequality (4), and standard elliptic regularity we can infer that there exists a unique solution of (8) satisfying $p \in L^{\frac{5}{3}}((0,T) \times \mathbb{T}^3)$. Next, we show that (u, p) solve the Navier-Stokes equations in the sense of distributions. Let $\psi(t,x) = \chi(t)\phi(x)$ with $\chi \in C_c^\infty(0,T)$ and $\phi \in C^\infty(\mathbb{T}^3)$. Let $\phi = P\phi + Q\phi$ be the Helmholtz decomposition, where we denote by $P\phi$ the divergence-free part of ϕ. Then, since P and Q commute with derivatives because there are no physical boundaries, we have that

$$\int_0^T (u(t),\phi)\dot{\chi}(t) - ((u \cdot \nabla) u, \phi)\chi(t) - v(\nabla u, \nabla \phi)\chi(t) + (p(t), \text{div}\, Q\phi)\chi(t)\, dt$$
$$= \int_0^T (u(t), P\phi)\dot{\chi}(t) - ((u \cdot \nabla) u, P\phi)\chi(t) - v(\nabla u, \nabla P\phi)\chi(t)$$
$$- \int_0^T ((u \cdot \nabla) u, Q\phi)\chi(t) - (p(t), \text{div}\, \phi)\chi(t)\, dt \tag{9}$$
$$= - \int_0^T ((u \cdot \nabla) u, Q\phi)\chi(t) - (p(t), \text{div}\, Q\phi)\chi(t)\, dt = 0,$$

where we have used (6) in the second equality and (8) together with the fact that $Q\phi = \nabla q$ for some $q \in C^\infty(\mathbb{T}^3)$ in the last equality. Finally, by an approximation argument, we have that (9) holds for any $\phi \in C_c^\infty((0,T) \times \mathbb{T}^3)$ and we conclude. $\square$

4. Approximate Solutions of the Incompressible Navier-Stokes Equations

In this section, we define the notion of *approximate sequence of solutions* to the Navier-Stokes equations and we prove the convergence to Leray-Hopf weak solutions. We use an approach which is a little different from the one usual used. Our choice, which does not follows the historical path, is motivated by the pedagogical purpose of having a unified treatment for several different methods.

Definition 2. *Let $n \in \mathbb{N}$. We say that $\{u^n\}_n \subset C(0,T;L^2(\Omega))$ is an approximate sequence of solutions with divergence-free initial datum u_0^n if*

1. *It holds that*

$$\{u^n\}_n \text{ is a bounded sequence in } L^\infty(0,T;H) \cap L^2(0,T;V); \tag{10}$$

2. *For any $n \in \mathbb{N}$ and any $\phi \in \mathcal{V}(\Omega)$ there exists $R_\phi^n \in L^1(0,T)$ such that for any $\chi \in C_c^\infty([0,T))$*

$$\int_0^T \left((u^n(t),\phi)\dot{\chi}(t) + ((u^n \cdot \nabla)\,u^n,\phi)\chi(t) + \nu(\nabla u^n, \nabla\phi)\chi(t) \right) dt$$
$$- (u_0^n,\phi)\chi(0) = \int_0^T R_\phi^n(t)\chi(t)\,dt; \tag{11}$$

3. *It holds*

$$R_\phi^n \rightharpoonup 0 \qquad \text{weakly in } L^1(0,T) \text{ as } n \to \infty; \tag{12}$$

4. *For any $n \in \mathbb{N}$ and $t \in (0,T)$ it holds that*

$$\|u^n(t)\|_2^2 + 2\nu \int_0^t \|\nabla u^n(s)\|_2^2\, ds \leq \|u_0^n\|_2^2. \tag{13}$$

Since generally the existence of (smooth) approximating sequences is *rather easy* to be proved, the advantage of this definition is that one has just to check a condition on the data and condition (12) on the remainder (commutator) to show that the approximate solutions converge to a Leray-Hopf weak solution, as is done in the next theorem.

Theorem 1. *Let $u_0 \in H$ and $\{u^n\}_n$ be a sequence of approximate solutions with initial data $\{u_0^n\}_n$ such that*

$$u_0^n \to u_0 \qquad \text{strongly in } H. \tag{14}$$

Then, up to a sub-sequence not relabelled, there exists u such that if Ω satisfies $(A1)$ and $(A2)$ then

$$u^n \to u \qquad \text{strongly in } L^2(0,T;L^2(\Omega)), \tag{15}$$

and if $\Omega = \mathbb{R}^3$

$$u^n \to u \qquad \text{strongly in } L^2(0,T;L_{loc}^2(\mathbb{R}^3)). \tag{16}$$

Moreover, u is a Leray-Hopf weak solution of (1) and (2).

Remark 3. *We stress that because of Remark 1 requiring condition (13) (that is a good energy balance already on the approximate functions) is fundamental in order to obtain the energy inequality (7). Moreover, by inspecting the proof below it will be clear that given $\{u^n\}_n$ satisfying (1) and (2) in Definition 2 then there exists u satisfying (1) and (2) in Definition 1 such that the convergences (15) and (16) hold. This remark will be important in the analysis of the Implicit Euler Scheme in Section 5.3, because the scheme will not fully fit in the framework of Definition 2.*

Remark 4. *We point out that Theorem 1 is not limited to the case of vanishing external force, but the result holds also in the presence of an external body force $f \in L^2(0, T; (H_0^1(\Omega))')$. This can be obtained with minor changes in the proof. In this case Definition 2 must also be integrated with an approximating sequence f^n, and adding the term*

$$\int_0^T (f^n, u^n)\chi(t)\, dt$$

to the left-hand side of (11). This is due to the fact that some of the approximation methods in Section 5 require the body force to be smooth and it is enough to require $f^n \to f$ in $L^2(0, T; (H_0^1(\Omega))')$.

We also note that requiring the body force to be in $L^2(0, T; (H_0^1(\Omega))')$ (and not only in $L^2(0, T; V')$) is needed in order to remain inside the space of distributions and to have an associated pressure as in Lemma 2; We refer to Reference [17] for more details on this issue.

We start with the following straightforward corollary of the classical Arzelà-Ascoli theorem for real functions of a real variable.

Lemma 3. *Let E be a separable Banach space and let $\mathcal{E} \subset E$ be a dense subset. Let $\{F_n\}_n$ be a sequence of measurable functions such that $F_n : [0, T] \mapsto E^*$. Assume that*

1. *the sequence $\{F_n\}_n$ is equi-bounded in E^*,*
2. *for any fixed $\phi \in \mathcal{E}$ the sequence of real functions $\langle F_n, \phi \rangle : [0, T] \ni t \mapsto \langle F_n(t), \phi \rangle, n \in \mathbb{N}$, is equi-continuous.*

Then, $F_n \in C_w([0, T]; E^*)$ and there exists $F \in C_w([0, T]; E^*)$ such that, up to a sub-sequence,

$$F_n \to F \quad in \ C_w([0, T]; E^*).$$

A fundamental step in the proof of existence for nonlinear partial differential equations is the proof of certain compactness which allows to get strong convergence in suitable norms. Observe that the a-priori bounds are useful to get weak or weak-* convergences, by means of application of the Riesz representation theorem and –more generally– of Banach-Alaoglu-Bourbaki theorem. On the other hand, since $T(x_n) \rightharpoonup T(x)$ for a linear operator T, weak convergence allows to consider linear equations, or more precisely, the linear terms in the equations. On the other hand weak convergence is in general not enough to prove that

$$\int_0^T ((u^n \cdot \nabla)\, u^n, \phi)\chi(t)\, dt \to \int_0^T ((u \cdot \nabla)\, u, \phi)\chi(t)\, dt, \qquad \text{as } n \to \infty.$$

Hence, by the a priori estimates we can construct a limit object u, but we still have to show that u is a weak solution of the limiting problem.

To address this point several results have been used. Leray used Helly's theorem on monotone functions and an ingenious application of Riesz theorem with multiple Cantor diagonal arguments. Hopf used an inequality by Friederichs to handle the Galerkin case. Starting from the work of J.L. Lions [18] it became common to use the approach by the so-called Aubin-Lions lemma, which is borrowed from the general theory of abstract equations and is based on obtaining some estimates on the time derivative (at least in negative space) of the solution. This latter approach is very flexible, but it requires some non-trivial functional analysis preliminaries to estimate the time-derivative, since instead one can use directly some properties coming from the proper definition of the approximation. Note that in the Definition 1 of weak solution there is no mention to the time-derivative. We will show how to obtain compactness in a elementary way, directly from the weak formulation and thus avoiding the use of time derivatives in Bochner spaces. We believe this may be a simpler approach, at least for presentation to students. We also point out that in certain applications to more complex fluid problems as for instance fluids in a moving domain or non-Newtonian fluids with rheology with time-dependent constitutive law, the proper definition of the time derivative is technically complicated and an approach avoiding the use of this notion becomes particularly welcome.

The next lemma provides a general criterion for strong convergence, which has the advantage to avoid assumptions on the time derivative. Of course, the lemma holds only on bounded domains and therefore we exclude the whole space case, since in the latter one has to work locally. We also stress that the hypothesis are not optimal since we do not prove an *if and only if*, but the hypotheses are easily verifiable for general nonlinear evolution problems.

The lemma below is very similar to the one proved by Landes and Mustonen in Reference [19] and for an application to the Navier-Stokes equations see Landes [20]. For an optimal version (at least in general Hilbert spaces) we refer to Rakotoson and Temam [21].

Lemma 4. *Let $U \subset \mathbb{R}^3$ be any bounded domain or $U = \mathbb{T}^3$. Let $1 < p < \infty$ and assume that $g \in L^\infty(0, T; L^1(U)) \cap L^p(0, T; W_0^{1,p}(U))$ and $\{g^n\}_n$ is a sequence such that*

$$\{g^n\}_n \text{ is bounded in } L^\infty(0, T; L^1(U)) \cap L^p(0, T; W_0^{1,p}(U)),$$

and $g^n(t) \rightharpoonup g(t)$ weakly in $L^1(U)$ for a.e. $t \in [0, T]$. Then, it holds that

$$g^n \to g \qquad \text{in } L^p(0, T; L^p(U)).$$

Proof. We prove the lemma only in the case of $U \subset \mathbb{R}^3$ being a bounded domain with smooth boundary. First, since $g^n(t, \cdot)$ and $g(t, \cdot)$ are in $W_0^{1,p}(U)$, their extensions to zero off U are both in $W^{1,p}(\mathbb{R}^3)$. We denote by $\bar{g}^n$ and $\bar{g}$ these extensions. a.e. $t \in (0, T)$.

Let ρ_ε be a standard spatial mollifier and set $g_\varepsilon^n := \rho_\varepsilon * \bar{g}^n$ and $g_\varepsilon := \rho_\varepsilon * \bar{g}$. Next, we have that

$$|g_\varepsilon^n(t, x) - \bar{g}^n(t, x)| \le \varepsilon \int_{B_1} \rho(y) \int_0^1 |\nabla \bar{g}^n(t, x - \varepsilon \tau y)|\, d\tau dy$$

$$\le \varepsilon \left(\int_{B_1} \rho^{p'}(y)\, dy \right)^{\frac{1}{p'}} \left(\int_{B_1} \left(\int_0^1 |\nabla \bar{g}^n(t, x - \varepsilon \tau y)|\, d\tau \right)^p dy \right)^{\frac{1}{p}}$$

$$\le \varepsilon \left(\int_{B_1} \rho^{p'}(y)\, dy \right)^{\frac{1}{p'}} \left(\int_{B_1} \int_0^1 |\nabla \bar{g}^n(t, x - \varepsilon \tau y)|^p\, d\tau dy \right)^{\frac{1}{p}},$$

and the same estimate holds also for $\bar{g}$. Therefore, the following estimates hold

$$\int_0^T \|g_\varepsilon^n - \bar{g}^n\|_p^p\, dt \lesssim \varepsilon^p \int_0^T \|\nabla \bar{g}^n\|_p^p\, dt,$$

$$\int_0^T \|g_\varepsilon - \bar{g}\|_p^p\, dt \lesssim \varepsilon^p \int_0^T \|\nabla \bar{g}\|_p^p\, dt,$$

with bounds depending only on ρ.

Next, by triangular inequality we have

$$\int_0^T \|g^n - g\|_{L^p(U)}^p\, dt \le \int_0^T \|g^n - g_\varepsilon^n\|_{L^p(U)}^p\, dt + \int_0^T \|g_\varepsilon^n - g_\varepsilon\|_{L^p(U)}^p\, dt + \int_0^T \|g_\varepsilon - g\|_{L^p(U)}^p\, dt$$

$$\le \int_0^T \|\bar{g}^n - g_\varepsilon^n\|_{L^p(\mathbb{R}^3)}^p\, dt + \int_0^T \|g_\varepsilon^n - g_\varepsilon\|_{L^p(U)}^p\, dt + \int_0^T \|g_\varepsilon - \bar{g}\|_{L^p(\mathbb{R}^3)}^p\, dt \qquad (17)$$

$$\le \varepsilon^p \int_0^T \left(\|\nabla \bar{g}^n\|_p^p + \|\nabla \bar{g}\|_p^p \right) dt + \int_0^T \int_U |g_\varepsilon^n(t, x) - g_\varepsilon(t, x)|^p\, dxdt.$$

The first term from the right-hand side can be made arbitrarily small by choosing ε small enough.

To conclude, we first note that by definition of convolution, there exists $C = C(\varepsilon, U)$ such that

$$|g_\varepsilon(t, x)| + |g_\varepsilon^n(t, x)| \leq C.$$

Next, since clearly it holds that for the extended functions it holds

$$\bar{g}^n(t) \rightharpoonup \bar{g}(t) \quad \text{weakly in } L^1(\mathbb{R}^3),$$

we also have that

$$g_\varepsilon^n(t, x) \to g_\varepsilon(t, x), \quad \text{a.e. in } t \in (0, T) \text{ and for all } x \in \mathbb{R}^3.$$

This follows by fixing $\epsilon > 0$, a time t such that $\bar{g}^n(t) \rightharpoonup \bar{g}(t)$, $x \in \mathbb{R}^3$, and noticing that

$$g_\varepsilon^n(t, x) - \bar{g}_\varepsilon(t, x) = \int_{\mathbb{R}^3} (\bar{g}^n(t, y) - \bar{g}(t, y)) \rho_\epsilon(x - y) \, dy \to 0,$$

as $n \to \infty$.

This shows that, for any fixed $\varepsilon > 0$, the last term in last inequality in (17) goes to zero as $n \to \infty$, by using Dominated Convergence Theorem. The proof is concluded since we showed that $\|g^n - g\|_{L^p(0,T;L^p(U))}$ can be made arbitrarily small. $\square$

The following theorem is the main result of this section.

Proof of the Theorem 1. Let $\{u^n\}_n$ be a sequence of approximate solutions. By condition (10) of Definition 2 we can infer that up to a sub-sequence (not relabelled) there exists $u \in L^\infty(0, T; H) \cap L^2(0, T; V)$ such that

$$u^n \overset{*}{\rightharpoonup} u \quad \text{weakly-* in } L^\infty(0, T; L^2(\Omega)), \tag{18}$$

$$u^n \rightharpoonup u \quad \text{weakly in } L^2(0, T; L^2(\Omega)), \tag{19}$$

$$\nabla u^n \rightharpoonup \nabla u \quad \text{weakly in } L^2(0, T; L^2(\Omega)). \tag{20}$$

For $k \in \mathbb{N}$ and $\bar{t} \in (0, T)$, we consider the following function

$$\chi_k^{\bar{t}}(t) = \begin{cases} 1, & t \in [0, \bar{t}) \\ k(\bar{t} - t) + 1, & t \in [\bar{t}, \bar{t} + \frac{1}{k}) \\ 0, & t \in [\bar{t} + \frac{1}{k}, T), \end{cases} \tag{21}$$

Let $\phi \in C_c^\infty(\Omega)$ with $\operatorname{div} \phi = 0$ and $s, t \in (0, T)$. By using the function $\chi_k^{\bar{t}}(t)$ with first with $\bar{t} = t$ and then with $\bar{t} = s$, together with the fact that $u^n \in C(0, T; L^2(\Omega))$ we can infer that

$$(u^n(t), \phi) - (u^n(s), \phi) + \int_s^t \left(((u^n(\tau) \cdot \nabla) u^n(\tau), \phi) + v(\nabla u^n(\tau), \nabla \phi) + R_\phi^n(\tau) \right) d\tau = 0.$$

Next, for $\phi \in C_c^\infty(\Omega)$, let $F^n(t) := ((u^n(t) \cdot \nabla) u^n(t), \phi) + v(\nabla u^n(t), \nabla \phi) + R_\phi^n(t)$. Then, condition (10), the Gagliardo-Nirenberg-Sobolev inequality (4), and the hypothesis on R_ϕ^n in (11) imply that the family $\{F^n\}_n$ is equi-integrable and then the function $t \mapsto (u^n(t), \phi)$ is equi-continuous. Since $\mathcal{V}(\Omega)$ is dense in H and H is reflexive, we can conclude by using Lemma 3 that

$$u^n \to u \quad \text{in } C_w([0, T]; H). \tag{22}$$

By using (22) and (20) we can prove that u satisfies the energy inequality (7). Indeed, for any $t \in (0, T)$ we have that

$$\|u(t)\|_2^2 \leq \liminf_{n \to \infty} \|u^n(t)\|_2^2,$$

$$\int_0^t \|\nabla u(s)\|_2^2 \, ds \leq \liminf_{n \to \infty} \int_0^t \|\nabla u^n(s)\|_2^2 \, ds.$$

Then,

$$\|u(t)\|_2^2 + 2\nu \int_0^t \|\nabla u(s)\|_2^2 \, ds \le \liminf_{n\to\infty} \|u^n(t)\|_2^2 + \liminf_{n\to\infty} 2\nu \int_0^t \|\nabla u^n(s)\|_2^2 \, ds$$

$$\le \liminf_{n\to\infty} \left(\|u^n(t)\|_2^2 + 2\nu \int_0^t \|\nabla u^n(s)\|_2^2 \, ds \right)$$

$$\le \lim_{n\to\infty} \|u_0^n\|_2^2 = \|u_0\|_2^2.$$

where we have used (14). In order to conclude it remains only to prove (15) and (16). If Ω is the flat torus or a bounded domain, then (15) follows directly by Lemma 4. If $\Omega = \mathbb{R}^3$ we need a localization argument. We first note that by the Helmholtz decomposition (22) holds also in $C_w([0,T]; L^2(\mathbb{R}^3))$. Next, for $k \in \mathbb{N}$ let $\psi \in C_c^\infty(B_{k+1}(0))$ such that $\psi = 1$ on B_k and define $g^n := u^n \psi$. The sequence $\{g^n\}_n$ satisfies the hypothesis of Lemma 4, and therefore, after a diagonal argument, it follows that there exists a sub-sequence not relabelled such that

$$g^n \to g = u\,\psi \qquad \text{strongly in } L^2(0,T; L^2(B_{k+1})). \tag{23}$$

Then, condition (23) easily implies (16). $\square$

5. Approximation Methods

After the general result of the previous section, we are now going to show that a general class of methods used to construct weak solutions will fit the in the framework of Theorem 1, as described in Section 4.

5.1. Leray Approximation Scheme

We start describing the original scheme introduced by Leray in Reference [2] (even if we use a completely different compactness argument to show the convergence of approximations). In this case we consider $\Omega = \mathbb{R}^3$. We fix a sequence $\{\varepsilon_n\}$ of positive numbers going to zero and let ρ_{ε_n} be a standard mollifier (only) in the space variables. For $v : (0,T) \times \mathbb{R}^3 \mapsto \mathbb{R}^3$ we set $\Psi_n(v) := \rho_{\varepsilon_n} * v$, where the convolution is only in the space variables. Let $u_0 \in H$ and let $n \in \mathbb{N}$. Define $u_0^n = \Psi_n(u^0)$ and u^n as the solution of the following Cauchy problem:

$$\begin{cases} \partial_t u^n - \nu \Delta u^n + \left(\Psi_n(u^n) \cdot \nabla\right) u^n + \nabla p^n = 0 & \text{in } (0,T) \times \mathbb{R}^3, \\ \operatorname{div} u^n = 0 & \text{in } (0,T) \times \mathbb{R}^3, \\ u^n|_{t=0} = u_0^n & \text{on } \{t=0\} \times \mathbb{R}^3. \end{cases} \tag{24}$$

We want to prove that for any $n \in \mathbb{N}$ the function u^n exists, is smooth, and $\{u^n\}_n$ is an approximate sequence of solutions in the sense of Definition 2.

Theorem 2. *Let $u_0 \in H$. Then, it holds that*

1. *for any fixed $n \in \mathbb{N}$ there exists a unique $u^n \in C([0,T]; H^3(\mathbb{R}^3))$ solution of (24);*
2. *there exists a Leray-Hopf weak solution u and a possible sub-sequence of $\{u^n\}_n$ such that*

$$u^n \to u \qquad \text{strongly in } L^2(0,T; L^2_{loc}(\mathbb{R}^3)).$$

Proof. Let us prove (1). The proof is very classical so we only sketch it. By using a fixed point argument we can prove that there exists a time $T_1 = T_1(\|u_0^n\|_{H^3}) > 0$ such that there exists a unique $u^n \in C([0,T^*); H^3(\mathbb{R}^3))$ solution of (24) for $T^* \ge T_1$. Let us suppose that T^* is the maximal time of existence of u^n and, if $T^* < T$ then $\lim_{t\to T^{*-}} \|u^n(s)\|_{H^3} = \infty$.

To obtain a global solution we exploit a standard energy estimates argument. Indeed, we first note that by multiplying (24) by u^n and integrating by parts we get

$$\|u^n(t)\|_2^2 + 2\nu \int_0^t \|\nabla u^n(s)\|_2^2 \, ds = \|u_0^n\|_2^2, \tag{25}$$

with an equality which is valid for all $t < T^*$. Note that this is exactly the same calculation we have done formally to obtain the energy inequality (3) in the introduction. In particular, from (25) we obtain

$$\sup_{t \in (0,T)} \|u^n(t)\|_2 \leq \|u_0^n\|_2. \tag{26}$$

Next, by using that $H^3(\mathbb{R}^3)$ is an algebra and by using the standard properties of mollifiers, it is easy to prove that

$$
\begin{aligned}
\frac{d}{dt}\|u^n(t)\|_{H^3}^2 + \nu\|\nabla u^n(t)\|_{H^3}^2
&\lesssim \left| \left((\Psi_n(u^n(t)) \cdot \nabla) u^n(t), u^n(t) \right)_{H^3} \right| \\
&\lesssim \|\Psi_n(u^n(t))\|_{H^3} \|\nabla u^n(t)\|_{H^3} \|u^n(t)\|_{H^3} \\
&\lesssim \|\Psi_n(u^n(t))\|_{H^3}^2 \|u^n(t)\|_{H^3}^2 + \frac{\nu}{2}\|\nabla u^n(t)\|_{H^3}^2 \\
&\lesssim \frac{1}{\epsilon_n^6}\|u^n(t)\|_2^2 \|u^n(t)\|_{H^3}^2 + \frac{\nu}{2}\|\nabla u^n(t)\|_{H^3}^2 \\
&\lesssim \frac{1}{\epsilon_n^6}\|u_0^n\|_2^2 \|u^n(t)\|_{H^3}^2 + \frac{\nu}{2}\|\nabla u^n(t)\|_{H^3}^2.
\end{aligned}
$$

where in the last inequality we have used (26). Therefore, we have that

$$\frac{d}{dt}\|u^n(t)\|_{H^3}^2 \leq C_n \|u_0^n\|_2^2 \|u^n(t)\|_{H^3}^2$$

and, by using the Gronwall Lemma, we conclude that necessarily $T^* = T$ (this argument shows that in fact u^n is defined for all $t > 0$, for any fixed $n \in \mathbb{N}$).

Next, to show (2) it is enough to prove that $\{u^n\}_n$ satisfies the conditions in Definition 2. Clearly, from (1) we have that $\{u^n\}_n \subset C([0, T]; L^2(\mathbb{R}^3))$. By using the standard property of mollifiers $\|\Psi_n(u^0)\|_2 \leq \|u_0\|_2$, and from the energy estimate (25) we get that $\{u^n\}_n$ is bounded uniformly in $L^\infty(0, T; H) \cap L^2(0, T; H)$. Moreover, (25) is exactly (13) and then it remains only to verify (11). For $\phi \in C_c^\infty(\Omega)$ and $t \in (0, T)$ the function $t \mapsto R_\phi^n(t)$ is defined as follows

$$R_\phi^n(t) := \left(\left([\Psi_n(u^n(t)) - u^n(t)] \cdot \nabla \right) u^n(t), \phi \right).$$

With this choice of R_ϕ^n, the equation (11) is satisfied and it remains only to prove that convergence stated in condition (12) of Definition 2. First, note that by Hölder inequality

$$|R_\phi^n(t)| \leq \|\nabla u^n(t)\|_2 \|\phi\|_\infty \|\Psi_n(u^n(t)) - u^n(t)\|_2. \tag{27}$$

Then, for any fixed $(t, x) \in (0, T) \times \mathbb{R}^3$, by a direct calculation (using again the properties of mollifiers) we have

$$
\begin{aligned}
|\Psi_n(u^n(t, x)) - u^n(t, x)|
&\leq \epsilon_n \int_{B_1} \rho(y) \int_0^1 |\nabla u^n(t, x - \epsilon \tau y)| \, d\tau \, dy \\
&\leq \epsilon_n \left(\int_{B_1} \rho^2(y) \, dy \right)^{\frac{1}{2}} \left(\int_{B_1} \left(\int_0^1 |\nabla u^n(t, x - \epsilon \tau y)| \, d\tau \right)^2 dy \right)^{\frac{1}{2}} \\
&\leq \epsilon_n \left(\int_{B_1} \rho^2(y) \, dy \right)^{\frac{1}{2}} \left(\int_{B_1} \int_0^1 |\nabla u^n(t, x - \epsilon \tau y)|^2 \, d\tau \, dy \right)^{\frac{1}{2}}.
\end{aligned}
$$

Then, by a further integration

$$\int_{\mathbb{R}^3} |\Psi_n(u^n(t,x)) - u^n(t,x)|^2 \, dx \leq \varepsilon_n^2 \int_{\mathbb{R}^2} |\nabla u^n(t,x)|^2 \, dx,$$

and going back to (27) we have that

$$|R_\phi^n(t)| \leq \varepsilon_n \|\nabla u^n(t)\|_2^2 \|\phi\|_\infty.$$

Since the sequence $\{\nabla u^n\}_n$ is bounded uniformly with respect to n in $L^2(0, T; L^2(\mathbb{R}^3))$, we have that $R_\phi^n \to 0$ in $L^1(0, T)$. $\square$

5.2. Faedo-Galerkin Method

The next scheme we consider is the Faedo-Galerkin method. The variant we present is close to the one considered by Hopf and is at the basis of several computational methods, which are used also in fields different from fluid dynamics. In particular, we will see that the unified treatment is possible under the assumption of having a basis which is orthogonal in both L^2 and H^1, as is the case of the spectral basis made by eigenfunctions of the Stokes operator. Observe that in the space-periodic case this basis is explicitly constructed by considering complex exponentials, while in the case of a smooth bounded domain, the existence is obtained via the standard theory of compact operators, showing existence of countable non-decreasing positive $\{\lambda_j\}$ and smooth $\{\psi_j\}$ such that it holds for all $j \in \mathbb{N}$

$$\begin{aligned}
-\Delta \psi_j + \nabla \pi_j &= \lambda_j \psi_j && \text{in } \Omega, \\
\operatorname{div} \psi_j &= 0 && \text{in } \Omega, \\
\psi_j &= 0 && \text{on } \partial\Omega.
\end{aligned}$$

We consider $\Omega \subset \mathbb{R}^3$ with smooth boundary $\partial\Omega$ or the three-dimensional flat torus. Let be given an orthonormal basis $\{\psi_m\}_{m\in\mathbb{N}}$ of H, such that $\psi_m \in V(\Omega)$. The Faedo-Galerkin method is based on the construction of approximate solutions of the type

$$u^n(t,x) = \sum_{j=1}^{n} c_j^n(t)\psi_j(x) \qquad n \in \mathbb{N}, \tag{28}$$

which solve the Navier-Stokes equations projected equations over the finite dimensional space $V_n = \operatorname{Span}(\psi_1, \ldots, \psi_n) \subset V$. This means that for $n \in \mathbb{N}$, the approximate problem to be solved is given by

$$\begin{cases}
\dfrac{d}{dt}(u^n, \psi_m) + \nu(\nabla u^n, \nabla \psi_m) + ((u^n \cdot \nabla)\, u^n, \psi_m) = 0 & t \in (0, T), \\
(u^n(0), \boldsymbol{\psi}_m) = (u_0, \psi_m) & t = 0,
\end{cases} \tag{29}$$

for $m = 1, \ldots, n$, which is a Cauchy problem for a system of n ODE's in the coefficients $\{c_j^n(t)\}_{j=1}^n$. Let P^n be projection operator from H into V_n:

$$P^n : f \in H \mapsto P^n f := \sum_{m=1}^{n} (f, \psi_m)\psi_m.$$

Then, the ODE's (29) reduce to the following system of PDE's:

$$\begin{cases}
\partial_t u^n + P^n\big((u^n \cdot \nabla)\, u^n\big) - \nu \Delta u^n = 0 & \text{in } (0, T) \times \Omega, \\
u^n|_{t=0} = P^n u^0 & \text{in } \Omega.
\end{cases} \tag{30}$$

In the next theorem we prove that u^n is smooth and exists on $(0, T)$, and that $\{u^n\}_n$ is an approximate sequence of solutions.

Theorem 3. *Let $u_0 \in H$. Then, it holds that*

1. *For any fixed $n \in \mathbb{N}$ there exists a unique $u^n \in C^1((0, T); C^\infty(\Omega)) \cap C^0([0, T]; C^\infty(\Omega))$ solution of* (30);
2. *There exists a Leray-Hopf weak solution u and a possible sub-sequence of $\{u^n\}_n$ such that*

$$u^n \to u \qquad \text{strongly in } L^2(0, T; L^2(\Omega)).$$

Proof. We prove (1). By the theory of ordinary differential equations one easily obtains that there exists a unique solution $c_j^n(t) \in C^1(0, T^n)$, for some $0 < T^n \leq T$, being (29) a nonlinear (quadratic) system in the coefficients $c_j^n(t)$. Moreover, u^n is defined through (28) and satisfies (30). Then, by multiplying (30) by u^n and integrating by parts we get

$$\|u^n(t)\|_2^2 + 2\nu \int_0^t \|\nabla u^n(s)\|_2^2 \, ds = \|u_0^n\|_2^2 \leq \|u_0\|_2^2, \tag{31}$$

where we have used that $\|u_0^n\|_2 = \|P^n u_0\|_2 \leq \|u_0\|_2$. Therefore, for any $n \in \mathbb{N}$ we have that

$$\sum_{j=1}^n |c_j^n(t)|^2 = \|u^n(t)\|_2^2 \leq \|u_0\|_2^2,$$

which easily implies that necessarily $T^n = T$.

To prove (2) we show that $\{u^n\}_n$ satisfy the conditions in Definition 2. Clearly, the sequence $\{u^n\}_n$ is in $C([0, T]; L^2(\Omega))$ and by (31) it verifies the condition (1) and the energy inequality (13). To check that (11) is verified, let $\phi \in \mathcal{V}(\Omega)$ and, for any $t \in (0, T)$, define

$$R_\phi^n(t) := \left(P^n((u^n(t) \cdot \nabla) u^n(t)) - (u^n(t) \cdot \nabla) u^n(t), \phi\right).$$

Note that we have that

$$|R_\phi^n(t)| \lesssim \|u^n(t)\|_3 \|\nabla u^n(t)\|_2 \|\phi - P^n \phi\|_6$$
$$\lesssim \|u^n(t)\|_2^{\frac{1}{2}} \|u^n(t)\|_{H^1}^{\frac{3}{2}} \|\phi - P^n \phi\|_{H^1},$$

where we have used the Gagliardo-Nirenberg-Sobolev inequality (4) and that P^n is a projection in both H and V, since in this case it holds

$$\left\|\nabla \left[f - \sum_{m=1}^n (f, \psi_m)\psi_m\right]\right\|^2 = \sum_{m=n+1}^\infty \lambda_m |(f, \psi_m)|^2 \|\psi_m\|^2 \xrightarrow{n \to +\infty} 0,$$

for all $f \in H_0^1(\Omega)$. Then, by using Hölder inequality, and Gagliardo-Nirenberg Sobolev inequality and taking into account that $T < \infty$ we have that $R_\phi^n \to 0$ in $L^1(0, T)$. $\square$

As already specified if $\Omega = \mathbb{T}^3$, then one can take $\{\psi_m\}_{m \in \mathbb{N}}$ to be the Fourier basis. Then, the Faedo-Galerkin method consists in finding the approximated sequence of type (28) solving the Navier-Stokes equations projected over the first n Fourier modes. On the other hand, in the case $\Omega = \mathbb{R}^3$ one possible choice is to use the method of invading domains, that is to consider the problem in the ball $B(0, R)$ with zero boundary conditions on $\partial B(0, R)$ and to construct a solution u_R by the Galerkin method. It turns out that the energy estimate (3) is valid for u_R, providing uniform estimates (on u_R which is considered as a function over the whole space, after extension by zero off of Ω); this allows to pass to the limit as $R \to +\infty$, more or less in the same way as before.

5.3. Implicit Euler Scheme

The scheme we consider in the present subsection deals with the time-discretization and represents a first step also in the numerical analysis of the Navier-Stokes equations.

We consider the case of Ω being a bounded domain satisfying the hypothesis $(A3)$. Let $n \in \mathbb{N}$ and define the time-step $\kappa_n := T/n$ and the net $I^M = \{t_m\}_{m=0}^n$, such that $t_i - t_{i-1} = \kappa_n$, for any $i = 1, ..., n$.

Moreover, given $u_0 \in H$, consider a sequence (In the space periodic setting this can be obtained simply with a mollification with kernel ρ_ϵ, with $\epsilon = \frac{1}{\sqrt{n}}$.) of initial data $\{u_0^n\}_n \subset V$ such that

$$\|\nabla u_0^n\|_2 \lesssim \sqrt{n}\|u_0^n\|_2, \qquad \text{and} \qquad u_0^n \to u_0 \qquad \text{strongly in } H. \tag{32}$$

For $m \in \{1, ..., n\}$, given $\widetilde{u}_n^{m-1}$ the iterate $\widetilde{u}_n^m$ is obtained by solving the boundary value problem

$$\begin{cases} \dfrac{\widetilde{u}_n^m - \widetilde{u}_n^{m-1}}{\kappa_n} - \nu\Delta\widetilde{u}_n^m + (\widetilde{u}_n^m \cdot \nabla)\,\widetilde{u}_n^m + \nabla\widetilde{p}_n^m = 0 & \text{in } \Omega, \\[2mm] \nabla \cdot \widetilde{u}_n^m = 0 & \text{in } \Omega, \\[1mm] \widetilde{u}_n^m = 0 & \text{on } \partial\Omega, \end{cases}$$

with $\widetilde{u}_n^0 = u_0^n$. (In the case of a non-zero force one has to set $\widetilde{f}_n^m = \kappa_n^{-1}\int_{t_{m-1}}^{t_m} f(t)\,dt$ in the right-hand side of the momentum equation which defines $\widetilde{u}_n^m$.)

For any fixed $n \in \mathbb{N}$, we define the following sequences of functions defined on $[0, T]$ with values in V and in $L^2(\Omega)$:

$$u^n(t) = \sum_{m=1}^n \chi_{[t_{m-1}, t_m)}(t)\left(\widetilde{u}_n^{m-1} + \frac{(t - t_{m-1})}{\kappa_n}(\widetilde{u}_n^m - \widetilde{u}_n^{m-1})\right), \quad u^n(t_n) = \widetilde{u}_n^n,$$

$$v^n(t) = \sum_{m=1}^n \chi_{(t_{m-1}, t_m]}(t)\,\widetilde{u}_n^m, \quad v^n(t_0) = u_0^n. \tag{33}$$

$$p^n(t) = \sum_{m=1}^n \chi_{(t_{m-1}, t_m]}(t)\,\widetilde{p}_n^m.$$

We are now ready to prove the following theorem, which is referred in literature as an "alternate proof" by semi-discretization, see Reference [11].

Theorem 4. *Let $u_0 \in H$. Then, it holds that*

1. *For any fixed $n \in \mathbb{N}$, there exist $\{\widetilde{u}_n^m\}_{m=1}^n \subset H_0^1(\Omega)$ such that for any $m = 1, ..., n$ and any $\psi \in V$*

$$(\widetilde{u}_n^m, \psi) - (\widetilde{u}_n^{m-1}, \psi) + \kappa_n\,\nu(\nabla\widetilde{u}_n^m, \nabla\psi) + \kappa_n((\widetilde{u}_n^m \cdot \nabla)\,\widetilde{u}_n^m, \psi) = 0; \tag{34}$$

2. *There exists a Leray-Hopf weak solution u such that the sequence $\{u^n\}_n$ and $\{v^n\}_n$, defined in (33), satisfy*

$$\begin{aligned} u^n \to u & \qquad \text{strongly in } L^2(0, T; L^2(\Omega)), \\ u^n - v^n \to 0 & \qquad \text{strongly in } L^2(0, T; L^2(\Omega)). \end{aligned}$$

Proof. For the proof of *(1)* we refer to Reference [11]. The idea is the following: For any fixed $n \in \mathbb{N}$ and any $m = 1, ..., n$, the existence of $\widetilde{u}_n^m \in V$ solution of (34) is obtained by applying the Brouwer fixed point theorem to the following modified version of the steady Navier-Stokes equations, where the given iterate $\widetilde{u}_n^{m-1}$ is considered an external force:

$$\begin{cases} \dfrac{\widetilde{u}_n^m}{\kappa_n} - \nu\Delta\widetilde{u}_n^m + (\widetilde{u}_n^m \cdot \nabla)\,\widetilde{u}_n^m + \nabla\widetilde{p}_n^m - \dfrac{\widetilde{u}_n^{m-1}}{\kappa_n} = 0 & \text{in } \Omega, \\[2mm] \nabla \cdot \widetilde{u}_n^m = 0 & \text{in } \Omega, \\[1mm] \widetilde{u}_n = 0 & \text{in } \partial\Omega. \end{cases}$$

In particular, by the definitions (33), we have that $\{u^n\}_n \subset C(0,T;L^2(\Omega))$ and $\{v^n\}_n \subset L^2(0,T;L^2(\Omega))$.

Next, we prove part (2).

By taking $\psi = \widetilde{u}_n^m$ in (34) and by using the elementary inequality $(a-b,a) = \frac{a^2-b^2}{2} + \frac{(a-b)^2}{2}$ valid for all $a,b \in \mathbb{R}$, we have that

$$\|\widetilde{u}_n^m\|_2^2 - \|\widetilde{u}_n^{m-1}\|_2^2 + \|\widetilde{u}_n^m - \widetilde{u}_n^{m-1}\|_2^2 + \kappa_n \nu \|\nabla \widetilde{u}_n^m\|_2^2 = 0. \tag{35}$$

Then, for any fixed $m \in \{1,\dots,n\}$ we have that

$$\|\widetilde{u}_n^m\|_2^2 \le \|u_0^n\|_2^2 \le \|u_0\|_2^2, \tag{36}$$

$$\kappa\nu \sum_{i=1}^m \|\nabla \widetilde{u}_n^i\|_2^2 \le \|u_0^n\|_2^2 \le \|u_0\|_2^2, \tag{37}$$

$$\sum_{i=1}^m \|\widetilde{u}_n^i - \widetilde{u}_n^{i-1}\|_2^2 \le \|u_0^n\|_2^2 \le \|u_0\|_2^2. \tag{38}$$

By using (36)–(38) and (33) we easily have that

$$\{u^n\}_n \text{ is bounded in } L^\infty(0,T;H), \tag{39}$$

$$\{v^n\}_n \text{ is bounded in } L^\infty(0,T;H) \cap L^2(0,T;V). \tag{40}$$

We want to prove a uniform bound in $L^2(0,T;V)$ also for $\{u^n\}_n$. By a direct calculation we have that

$$
\begin{aligned}
\int_0^T \|\nabla u^n(t)\|_2^2\, dt &= \sum_{m=1}^n \int_{t_{m-1}}^{t_m} \left(1 - \frac{(t-t_{m-1})}{\kappa_n}\right)^2 \|\nabla \widetilde{u}_n^{m-1}\|_2^2\, dt \\
&\quad + 2\sum_{m=1}^n \int_{t_{m-1}}^{t_m} \left(1 - \frac{(t-t_{m-1})}{\kappa_n}\right)\left(\frac{(t-t_{m-1})}{\kappa_n}\right)(\nabla \widetilde{u}_n^{m-1}, \nabla \widetilde{u}_n^m)\, dt \\
&\quad + \sum_{m=1}^n \int_{t_{m-1}}^{t_m} \left(\frac{(t-t_{m-1})}{\kappa_n}\right)^2 \|\nabla \widetilde{u}_n^m\|_2^2\, dt \\
&\le \frac{\kappa_n}{2} \sum_{m=1}^n \|\nabla \widetilde{u}_n^{m-1}\|_2^2 + \frac{\kappa_n}{2} \sum_{m=1}^n \|\nabla \widetilde{u}_n^m\|_2^2 \\
&\le \frac{\kappa_n}{2} \|\nabla \widetilde{u}_n^0\|_2^2 + \kappa_n \sum_{m=1}^n \|\nabla \widetilde{u}_n^m\|_2^2.
\end{aligned}
$$

By using (32) we obtain

$$\int_0^T \|\nabla u^n(t)\|_2^2\, dt \lesssim \kappa_n \sum_{m=1}^n \|\nabla \widetilde{u}_n^m\|_2^2 + \kappa_n \|\nabla u_0^n\|_2^2$$

$$\lesssim \kappa_n \sum_{m=1}^n \|\nabla \widetilde{u}_n^m\|_2^2 + \|u_0\|_2^2 \lesssim \|u_0\|_2^2.$$

where we have also used that $\kappa_n = T/n$ and (37). Therefore we have that $\{u^n\}_n$ is bounded in $L^2(0,T;V)$ and then, taking into account (39), $\{u^n\}_n$ satisfies the condition (1) in Definition 2. Next, we show that $\{u^n\}_n$ satisfies the condition (2) of Definition 2. First, for all $\phi \in \mathcal{V}(\Omega)$ and $\chi \in C_c^\infty([0,T))$ we have, by using (33) and (34), that

$$\int_0^T \Big((u^n(t),\phi)\dot{\chi}(t) + ((v^n(t)\cdot\nabla)v^n(t),\phi)\chi(t) + \nu(\nabla v^n(t),\nabla\phi)\chi(t) \Big)\, dt - (u_0^n,\phi)\chi(0) = 0.$$

If we define

$$R_\phi^n := (v^n(t) - u^n(t), \nu\Delta\phi) + ((v^n(t) - u^n(t)) \otimes v^n(t) + u^n(t) \otimes (v^n(t) - u^n(t)), \nabla\phi),$$

then $\{u^n\}_n$ satisfies the formulation (11) and we only need to prove that (12). To this end we note that

$$\int_0^T |R_\phi^n(t)|\, dt \leq c\|\nabla\phi\|_\infty \left(\int_0^T \|u^n(s) - v^n(s)\|_2^2\, ds\right)^{\frac{1}{2}} \left(\int_0^T \|u^n(s)\|^2 + \|v^n(s)\|_2^2\, ds\right)^{\frac{1}{2}}$$

$$+ \nu\, T^{\frac{1}{2}}\|\nabla^2\phi\|_2 \left(\int_0^T \|u^n(s) - v^n(s)\|_2^2\, ds\right)^{\frac{1}{2}}.$$

By a direct calculation, we have that

$$\int_0^T \|u^n(t) - v^n(t)\|_2^2\, dt = \frac{\kappa_n}{3} \sum_{m=1}^n \|\widetilde{u}_n^m - \widetilde{u}_n^{m-1}\|_2^2 \leq C\kappa_n, \tag{41}$$

and therefore

$$\int_0^T |R_\phi^n(t)|\, dt \leq C\|\nabla\phi\|_\infty \kappa_n,$$

and (12) follows.

In conclusion, we have proved that $\{u^n\}_n$ satisfies the conditions (1) and (2) of Definition 2 and thanks to Theorem 1 and Remark 3, there exists u satisfying the condition (1) and (2) in Definition 1. Then, in order to conclude, we only need to prove that u satisfies also the energy inequality. First, we note that by using (35) and (33), a direct calculation implies that for any $t \in (0, T)$

$$\|v^n(t)\|_2^2 + 2\nu \int_0^t \|\nabla v^n(s)\|_2^2\, ds \leq \|u_0^n\|_2^2. \tag{42}$$

By using (40) and (41) we can infer that v^n converges to the same limit of u^n, namely that

$$\begin{aligned}
v^n &\to u & &\text{strongly in } L^2(0, T; L^2(\Omega)), \\
\nabla v^n &\rightharpoonup \nabla u & &\text{weakly in } L^2(0, T; L^2(\Omega)).
\end{aligned} \tag{43}$$

For $k \in \mathbb{N}$ and $t \in (0, T)$, let χ_k^t be the same function already defined in (21). Noticing that $-\dot{\chi}_k^t$ is positive, after multiplying (42) and integrating in time we get that, for any $t \in (0, T)$, it holds

$$\frac{1}{k} \int_t^{t+k} \|v^n(s)\|_2^2\, ds + 2\nu \int_0^T \chi_k^t(s)\|\nabla v^n(s)\|_2^2\, ds \leq \|u_0^n\|_2^2 \int_0^T (-\dot{\chi}_k^t(s))\, ds = \|u_0^n\|_2^2.$$

By using (43) we get

$$\frac{1}{k} \int_t^{t+k} \|u(s)\|_2^2\, ds + 2\nu \int_0^T \chi_k^t(s)\|\nabla u(s)\|_2^2\, ds \leq \|u_0\|_2^2,$$

and by Lebesgue differentiation and dominated convergence theorems we obtaine that for a.e. $t \in (0, T)$

$$\|u(t)\|_2^2 + 2\nu \int_0^t \|\nabla u(s)\|_2^2\, ds \leq \|u_0\|_2^2. \tag{44}$$

Let $\mathcal{N} \subset (0, T)$ the set of measure zero where (44) does not hold and fix $t \in \mathcal{N}$. Then, there exists $\{t_k\}_k \subset (0, T) \setminus \mathcal{N}$ such that $t_k \to t$ and

$$\|u(t_k)\|_2^2 + 2\nu \int_0^{t_k} \|\nabla u(s)\|_2^2\, ds \leq \|u_0\|_2^2.$$

Since $u \in C_w(0, T; H)$ and $\|\nabla u(\cdot)\|_2^2 \in L^1(0, T)$ it follows that

$$\|u(t)\|_2^2 + 2\nu \int_0^t \|\nabla u(s)\|_2^2 \, ds \leq \liminf_{k \to \infty} \|u(t_k)\|_2^2 + \lim_{k \to \infty} 2\nu \int_0^{t_k} \|\nabla u(s)\|_2^2 \, ds \leq \|u_0\|_2^2,$$

and therefore (44) holds for any $t \in (0, T)$. $\square$

5.4. Smagorinsky-Ladyžhenskaya Model

In this section we show how the approximation by adding a nonlinear stress tensor produce weak solutions. We consider for $n \in \mathbb{N}$ the following boundary initial value problem

$$\begin{cases} \partial_t u^n + (u^n \cdot \nabla) u^n + \nabla p^n - \nu \Delta u^n - \dfrac{1}{n} \operatorname{div}(|Du^n| Du^n) = 0 & \text{in } (0, T) \times \Omega, \\[2mm] \operatorname{div} u^n = 0 & \text{in } (0, T) \times \Omega, \\[2mm] u^n = 0 & \text{on } (0, T) \times \partial\Omega, \\[2mm] u^n(0) = u_0 & \text{in } \Omega, \end{cases} \tag{45}$$

where $Du^n = \frac{\nabla u^n + (\nabla u^n)^T}{2}$. This system has been introduced for numerical approximation of turbulent flows by Smagorinsky [22] and its analysis as a possible approximation for the Navier-Stokes equations started with the studies by Ladyženskaya [23], cf. also Reference [24] for the role of this method in the analysis of Large Eddy Simulation models. For the analysis also of related models, with general stress tensor given by $S(v) = S(Dv) = |Dv|^{p-2} Dv$, with various values of p, see References [18,25] and also the more recent Reference [26].

Theorem 5. *Let $u_0 \in H$. Then, it holds that*

1. *For any fixed $n \in \mathbb{N}$ there exists a unique $u^n \in C([0, T]; L^2(\Omega))$ solution of (45);*
2. *There exists a Leray-Hopf weak solution u and a possible sub-sequence of $\{u^n\}_n$ such that*

$$u^n \to u \qquad \text{strongly in } L^2(0, T; L^2(\Omega)).$$

Proof. By using the theory of monotone operators (cf. References [8,18]) there exists a unique $u^n \in C(0, T; L^2(\Omega))$ weak solution of (45) with $u^n \in L^\infty(0, T; H) \cap L^2(0, T; V)$ and $Du^n \in L^3(0, T; L^3(\Omega))$ such that it holds

$$\|u^n(t)\|_2^2 + 2\nu \int_0^t \|\nabla u^n(s)\|_2^2 \, ds + \frac{2}{n} \int_0^t \|Du^n(s)\|_3^3 \, ds \leq \|u_0^n\|_2^2.$$

Observe that by Korn inequality $\|Du^n\|_3 \sim \|\nabla u^n\|_3$.

To prove the second part of Theorem 5 we show that $\{u^n\}_n$ satisfy the conditions in Definition 2. Define the remainder $R_\phi^n(t)$ by

$$R_\phi^n(t) := -\frac{1}{n} \int_\Omega |Du^n(t)| Du^n(t) \cdot D\phi \, dx.$$

By means of the Hölder inequality we get

$$|R_\phi^n(t)| \leq \frac{1}{n} \int_\Omega |Du^n(t)|^2 |D\phi| \, dx \leq \frac{1}{n} \|Du^n\|_3^2 \|D\phi\|_3.$$

Consequently, it also holds

$$\int_0^T |R_\phi^n(t)| \, dt \leq \frac{T^{1/3}}{n^{1/3}} \left(\frac{1}{n} \int_0^T \|Du^n\|_3^3 \, dt \right)^{2/3} \|D\phi\|_3,$$

showing that $R_\phi^n \to 0$ in $L^1(0, T)$. Since the other conditions in Definition 2 are trivially satisfied, an application of Theorem 1 finally ends the proof. $\square$

Author Contributions: Both authors contributed equally to Conceptualization, Writing—review & editing. Both authors have read and agreed to the published version of the manuscript.

Funding: This research received no external funding.

Acknowledgments: LCB and SS are member of the group GNAMPA of INdAM.

Conflicts of Interest: The authors declare no conflict of interest.

References

1. Berselli, L.C. *Three-Dimensional Navier-Stokes Equations for Turbulence*; Mathematics in Science and Engineering; Elsevier/Academic Press: London, UK, 2020.
2. Leray, J. Sur le mouvement d'un liquide visqueux emplissant l'espace. *Acta Math.* **1934**, *63*, 193–248. [CrossRef]
3. Hopf, E. Über die Anfangswertaufgabe für die hydrodynamischen Grundgleichungen. *Math. Nachr.* **1951**, *4*, 213–231. [CrossRef]
4. Brezis, H. *Functional Analysis, Sobolev Spaces and Partial Differential Equations*; Universitext; Springer: New York, NY, USA, 2011.
5. Constantin, P.; Foias, C. *Navier-Stokes Equations*; Chicago Lectures in Mathematics; University of Chicago Press: Chicago, IL, USA, 1988.
6. Doering, C.R.; Gibbon, J.D. *Applied Analysis of the Navier-Stokes Equations*; Cambridge Texts in Applied Mathematics; Cambridge University Press: Cambridge, UK, 1995.
7. Galdi, G.P. *An Introduction to the Mathematical Theory of the Navier-Stokes EQUATIONS. Steady-State Problems*; Springer Monographs in Mathematics; Springer: New York, NY, USA, 2011.
8. Ladyžhenskaya, O.A. *The Mathematical Theory of Viscous Incompressible Flow*, 2nd ed.; Revised and Enlarged. Translated from the Russian by Richard A. Silverman and John Chu. Mathematics and its Applications; Gordon and Breach Science Publishers: New York, NY, USA, 1969; Volume 2.
9. Robinson, J.C.; Rodrigo, J.L.; Sadowski, W. The Three-Dimensional Navier-Stokes Equations. In *Cambridge Studies in Advanced Mathematics*; Cambridge University Press: Cambridge, UK, 2016; Volume 157, Classical Theory.
10. Sohr, H. The Navier-Stokes equations. *Birkhäuser Advanced Texts: Basler Lehrbücher. [Birkhäuser Advanced Texts: Basel Textbooks]*; Birkhäuser Verlag: Basel, Switzerland, 2001; An Elementary Functional Analytic Approach.
11. Temam, R. *Navier-Stokes Equations*; Theory and numerical analysis, Reprint of the 1984 edition; AMS Chelsea Publishing: Providence, RI, USA, 2001.
12. Anastasio, S.; Douglas, R.G.; Foias, C.; Ching, W.-M.; Davis, M.; Sharir, M.; Wigler, M.; Fisher, J.A.; Lax, P.D.; Nirenberg, L.; et al. In memory of Jacob Schwartz. *Not. Am. Math. Soc.* **2015**, *62*, 473–490. [CrossRef]
13. Nirenberg, L. On elliptic partial differential equations. *Ann. Scuola Norm. Sup. Pisa Cl. Sci.* **1959**, *13*, 115–162.
14. Galdi, G.P. An introduction to the Navier-Stokes initial-boundary value problem. In *Fundamental Directions in Mathematical Fluid Mechanics*; Birkhäuser: Basel, Switzerland, 2000; pp. 1–70
15. Simon, J. On the existence of the pressure for solutions of the variational Navier-Stokes equations. *J. Math. Fluid Mech.* **1999**, *1*, 225–234. [CrossRef]
16. Tartar, L. *Topics in Nonlinear Analysis*; Publications Mathématiques d'Orsay: Orsay, France, 1978.
17. Simon, J. On the identification $H = H'$ in the Lions theorem and a related inaccuracy. *Ric. Mat.* **2010**, *59*, 245–255. [CrossRef]
18. Lions, J.-L. *Quelques Méthodes de Résolution des Problèmes aux Limites non Linéaires*; Dunod, Gauthier-Villars: Paris, France, 1969.
19. Landes, R.; Mustonen, V. A strongly nonlinear parabolic initial-boundary value problem. *Ark. Mat.* **1987**, *25*, 29–40. [CrossRef]
20. Landes, R. A remark on the existence proof of Hopf's solution of the Navier-Stokes equation. *Arch. Math. (Basel)* **1986**, *47*, 367–371. [CrossRef]
21. Rakotoson, J.M.; Temam, R. An optimal compactness theorem and application to elliptic-parabolic systems. *Appl. Math. Lett.* **2001**, *14*, 303–306. [CrossRef]
22. Smagorinsky, J. General circulation experiments with the primitive equations. *Mon. Weather Rev.* **1963**, *91*, 99–164. [CrossRef]
23. Ladyžhenskaya, O.A. New equations for the description of motion of viscous incompressible fluids and solvability in the large of boundary value problems for them. *Proc. Stek. Inst. Math.* **1967**, *102*, 95–118.
24. Berselli, L.C.; Iliescu, T.; Layton, W.J. *Mathematics of Large Eddy Simulation of Turbulent Flows*; Scientific Computation; Springer-Verlag: Berlin, Germany, 2006.
25. Málek, J.; Nečas, J.; Rokyta, M.; ůžička, M.R. *Weak and Measure-Valued Solutions to Evolutionary PDEs*; Volume 13 of Applied Mathematics and Mathematical Computations; Chapman & Hall: London, UK, 1996.
26. Breit, D. *Existence Theory for Generalized Newtonian Fluids*; Mathematics in Science and Engineering; Elsevier/Academic Press: London, UK, 2017.

MDPI

St. Alban-Anlage 66

4052 Basel

Switzerland

Tel. +41 61 683 77 34

Fax +41 61 302 89 18

www.mdpi.com

Fluids Editorial Office

E-mail: fluids@mdpi.com

www.mdpi.com/journal/fluids